Rafal Ablamowicz
William E. Baylis
Thomas Branson
Pertti Lounesto
Ian Porteous
John Ryan
J.M. Selig
Garret Sobczyk

Lectures on Clifford (Geometric) Algebras and Applications

Rafal Abłamowicz
Garret Sobczyk
Editors

Birkhäuser
Boston • Basel • Berlin

Rafał Abłamowicz
Department of Mathematics
Tennessee Technological University
Cookeville, TN 38505
U.S.A.

Garret Sobczyk
Dept. de Física y Matemáticas
Universidad de las Américas—Puebla
Santa Catarina Mártir
Cholula, Puebla, 72820
México

Library of Congress Cataloging-in-Publication Data

Lectures on Clifford (geometric) algebras and applications / [edited by] Rafał Abłamowicz and Garret Sobczyk.
p. cm.
Includes bibliographical references and index.
ISBN 0-8176-3257-3 (alk. paper) – ISBN 3-7643-3257-3 (alk. paper)
1. Clifford algebras. I. Abłamowicz, Rafał. II. Sobczyk, Garret, 1943-

QA199.L43 2003
512'.57–dc21

2003052205
CIP

AMS Subject Classifications: 11E88, 15-75, 15A66

Printed on acid-free paper

Birkhäuser

ISBN 0-8176-3257-3 SPIN 10935828
ISBN 3-7643-3257-3

Cover design by Joseph Sherman, Camden, CT.
Printed in the United States of America.

9 8 7 6 5 4 3 2 1

Birkhäuser Boston • Basel • Berlin
A member of BertelsmannSpringer Science+Business Media GmbH

Professor Pertti Lounesto presenting Lecture 1 on May 18, 2002, during the "6th Conference on Clifford Algebras and their Applications in Mathematical Physics," May 20–25, 2002, Tennessee Technological University, Cookeville, TN.

Dedicated to Professor Pertti Lounesto
Mathematician, Teacher, Friend, and Colleague

Contents

Preface

Advances in technology over the last 25 years have created a situation in which workers in diverse areas of computer science and engineering have found it necessary to increase their knowledge of related fields in order to make further progress. Clifford (geometric) algebra offers a unified algebraic framework for the direct expression of the geometric ideas underlying the great mathematical theories of linear and multilinear algebra, projective and affine geometries, and differential geometry. Indeed, for many people working in this area, geometric algebra is the natural extension of the real number system to include the concept of direction. The familiar complex numbers of the plane and the quaternions of four dimensions are examples of lower-dimensional geometric algebras.

During "The 6th International Conference on Clifford Algebras and their Applications in Mathematical Physics" held May 20–25, 2002, at Tennessee Technological University in Cookeville, Tennessee, a *Lecture Series on Clifford Geometric Algebras* was presented. Its goal was to to provide beginning graduate students in mathematics and physics and other newcomers to the field with no prior knowledge of Clifford algebras with a bird's eye view of Clifford geometric algebras and their applications. The lectures were given by some of the field's most recognized experts. The enthusiastic response of the more than 80 participants in the Lecture Series, many of whom were graduate students or postdocs, encouraged us to publish the expanded lectures as chapters in book form. The book, which contains many up-to-date references to the rapidly evolving literature, should also serve as a handy reference to professionals who want to keep abreast with the latest developments and applications in the field.

In Chapter 1, Pertti Lounesto states that "Clifford algebra is by definition the minimal construction designed to control the geometry in question." He then gives a concise but coherent introduction to geometric algebras by thoroughly examining the Clifford product of two vectors, the definition of a bivector, and how these concepts are used to represent reflections and rotations in the plane and in higher dimensional spaces. The geometric significance of quaternions is explained in three and four dimensions. He carefully shows us how the more advanced concepts of spinors, exterior algebra and contraction, the Grassmann–Caley algebra and shuffle product, naturally evolve from the more basic concepts. An important feature is that he shows how the lower dimensional Clifford algebras can be represented in terms of the familiar algebra of square matrices. His lecture ends with several categorical definitions of Clifford algebras of a quadratic form, and their deformations to Clifford algebras of an arbitrary bilinear form.

In Chapter 2, Ian Porteous shows us that the "control of the geometry" that each Clifford algebra has follows directly from its close relationship to the corresponding classical group. He begins with a systematic study of Clifford algebras by showing how they can be constructed from matrix algebras over the real numbers, complex numbers, or quaternions with the famous periodicity of eight. He catalogs all this information into useful tables of their most important features, such as tables of the spinor groups, groups of motion, conjugation types, the general linear groups, and the corresponding dimensions of the associated Lie algebras. He gives a brief history and discussion of Vahlen matrices and conformal transformations, which have recently found diverse applications in pattern recognition and image processing.

John Ryan, in Chapter 3, introduces the basic concepts of Clifford analysis which extends the well-known complex analysis of the plane to three and higher dimensions. A generalized Cauchy–Riemann operator, called the Dirac operator, makes it possible to introduce the analogues of holomorphic functions. Cauchy's theorem and Cauchy's integral formula all generalize nicely to the higher dimensions of Clifford analysis. He shows that holomorphic functions are preserved under the action of conformal Möbius transformations. One of the most important topics in classical complex analysis is the study of boundary value problems. These problems generalize nicely to the study of boundary behavior in Hardy spaces, which is closely related to the study of monogenic functions on the n-ball and its boundary, the $(n-1)$-sphere. By introducing the Fourier transform, and complex Clifford analysis, monogenic functions defined on the n-ball can be holomorphically extended to larger "tube" domains.

The last three chapters are built upon the core material set down in the first three chapters.

Much of the recent interest in Clifford (geometric) algebras can be traced back to the work of David Hestenes and coworkers in late 1960's and early 1970's, who viewed Clifford's geometric algebra as a unified language for mathematics and physics. In Chapter 4, William Baylis explores some of the extensive applications that have been made to physics. One of the earliest applications was to electromagnetism and the efficient formulation of Maxwell's equation as a single equation when expressed in the Pauli algebra, which is the geometric algebra of the physical space, or in the even more powerful Dirac algebra, which is the algebra of Minkowski spacetime. Both of these geometric algebras can be applied to directly to the study of special relativity because of the ease of representing not only the rotations of 3-dimensional space, but the more general Lorentz transformations of space and time. The relationships and interplay between these two important algebras are clearly explained, and a set of fully integrated exercises provides many opportunities for the reader to get a hands-on knowledge of the basic ideas. Some of the many important topics explored are charge dynamics in uniform fields, directed plane waves, polarization and phase shifters, standing waves, and the potential of a moving charge. The spinorial formulation of classical relativistic physics leads naturally to the quantum-mechanical form and provides insight into spin $\frac{1}{2}$ systems.

Jon Selig, in Chapter 5, applies Clifford algebra to robotics and computer vision. He begins by arguing that Clifford algebra is particularly well suited to modern microprocessor architectures because the problems of computer graphics can be expressed very efficiently in terms of Clifford algebras. Using quaternions to express rotations, he tackles the problem of satellite navigation. Next he uses biquaternions to write down the kinematic equations of the Stewart platform, which is used in aircraft simulators and in novel machine tools. The problem is to determine the position and orientation of the platform from the lengths of the hydraulic actuators. Both the quaternions and biquaternions are examples of different Clifford algebras. By turning to yet another Clifford algebra, he uses homogeneous elements to represent points, lines and planes in three dimensions. The operations in this Clifford algebra are used to represent incidents relationships, which he then uses to solve the correspondence problem in computer vision, and the inverse kinematic problem of the joint angles of a robot that will place the end-effector in a prescribed position.

Chapter 6, by Tom Branson, explains some of the deepest applications of Clifford algebras that have been made in differential geometry. If Clifford (geometric) algebras are as fundamental to mathematics as has been claimed, one would expect that using these algebras should make possible new insights into all areas of mathematics which have their roots in geometry. Branson introduces the reader to all kinds of new geometrical objects constructed in Clifford algebra. Starting with the concept of the Dirac operator, introduced in Chapter 3, and the structure group of orthogonal transformations on Riemannian geometry, introduced in Chapter 2, he goes on to discuss spin geometry and the even more sophisticated objects which arise, such as Stein–Weiss gradients, the Bochner–Weitzenbock formulas, and the Hijazi inequality. This chapter assumes, on the part of the reader, an understanding of differential geometry, a good mastery of the material contained in the first three chapters and is not for the timid of heart. Professor Branson has remarked, "Spin geometry is a fairly deep subject, and is not a subject where you can completely master one line before going on to the next line. At some point the reader has to believe in these objects and let the understanding build."

A particularly valuable feature of this book is the inclusion of an Appendix where the editors have tried, for the first time, to give an up-to-date summary of the existing Clifford (geometric) algebra software that is accessible on the web. We believe that being able to perform quick and reliable computations with these algebras leads to a greater understanding of the theory and applications. The software provides tools for generating examples, checking conjectures, finding counterexamples, and advancing understanding via experimentation.

The reader should note that in Chapter 1 (Lounesto), $C\ell_{p,q}$ denotes the Clifford algebra of the quadratic space $\mathbb{R}^{p,q}$ endowed with a quadratic form

$$Q(\mathbf{x}) = x_1^2 + x_2^2 + \cdots + x_p^2 - x_{p+1}^2 - \cdots - x_n^2$$

where $n = p + q$. Furthermore, $C\ell_n$ is a shorthand for $C\ell_{n,0}$.

In Chapter 2 (Porteous), $C\ell_{p,q}$ denotes the Clifford algebra of the quadratic space $\mathbb{R}^{p,q}$ endowed with a quadratic form

$$Q(\mathbf{x}) = -x_1^2 - x_2^2 - \cdots - x_p^2 + x_{p+1}^2 + \cdots + x_n^2$$

where $n = p + q$. Furthermore, Porteous introduces $C\ell_{p,q,r}$, the Clifford algebra generated multiplicatively and additively by p basis 1-vectors $\mathbf{e}_i, i = 1, \ldots, p$, that square to -1 in $C\ell_{p,q,r}$; q basis 1-vectors $\mathbf{e}_i, i = p+1, \ldots, p+q$, that square to $+1$; and r basis 1-vectors $\mathbf{e}_i$, $i = p+q+1, \ldots, p+q+r$, that square to 0. Likewise, the quadratic space $\mathbb{R}^{0,n}$ is denoted here by $\mathbb{R}^n$ while the associated Clifford algebra being denoted by $C\ell_{0,n}$ and not by $C\ell_n$ as in Chapter 1.

In Chapter 3 (Ryan), $C\ell_n$ is used to denote the Clifford algebra of the negative-definite quadratic form $Q(\mathbf{x}) = -x_1^2 - \cdots - x_n^2$. For reference, this is referred to as $C\ell_{0,n}$ in Lounesto's Chapter 1.

The meaning of $C\ell_{p,q}$ in Chapter 4 (Baylis), is the same as that of $C\ell_{p,q}$ from Chapter 1. In Chapter 5 (Selig), a slightly different notation is used. Namely, $C\ell(p,q,r)$ denotes the Clifford algebra generated by p basis 1-vectors $\mathbf{e}_i, i = 1, \ldots, p$, that square to $+1$; q basis 1-vectors $\mathbf{e}_i, i = p+1, \ldots, p+q$, that square to -1; and r basis 1-vectors $\mathbf{e}_i, i = p+q+1, \ldots, p+q+r$, that square to 0. In Chapter 6 (Branson), the meaning of $C\ell_{p,q}$ is the same as in Chapter 1.

Acknowledgments

Rafał Abłamowicz would like to acknowledge and express appreciation for the generous support provided by the National Science Foundation, Award 0201303. The Award made it possible to bring an unprecedented number of researchers, postdocs, graduate and undergraduate students to the 6th Conference of whom about half attended the Lecture Series. Additional funding, for which he is grateful, came from the College of Arts and Sciences, the Center for Manufacturing Research, and the Provost Office at Tennessee Technological University; the Graduate School at the University of Arkansas in Fayetteville, and the College of Arts and Sciences at George Mason University. Garret Sobczyk would like to thank INIP of the Universidad de Las Americas - Puebla for support in this project. He is a member of Sistema Nacional de Investigadores, Exp. 14587.

Both Editors are grateful to all six contributors for preparing their lectures for publication and for making an effort to standardize notation and definitions. The Editors, regretfully, were unable to consult with late Professor Pertti Lounesto who unfortunately died shortly after the 6th Conference. Pertti was extremely well known in the Clifford algebra community for many years. He was acutely aware of historical detail, and thoroughly knew the literature, as is evident from his extensive web pages on Clifford algebras that are still available at

http://www.helsinki.fi/~lounesto/.

He greatly enjoyed finding counterexamples and otherwise scrutinizing the literature, and insisted upon a high standard of mathematical rigor. His intentions were always to further deeper understanding of Clifford algebras, to give proper historical perspective and credit, and to help young researchers win recognition for their achievements. He is greatly missed.

Thanks are also due to Professor Jacques Helmstetter, Université de Grenoble I, for providing mathematical insight and clarifications to some of the statements in Lecture 1. Finally, the Editors wish to thank Ann Kostant of Birkhäuser for her support and encouragement in this project.

Rafał Abłamowicz
Cookeville, Tennessee, U.S.A.

Garret Sobczyk
Cholula, Mexico

September 15, 2003

1

Introduction to Clifford Algebras

Pertti Lounesto

ABSTRACT Clifford algebras of lower-dimensional Euclidean spaces and Minkowski spacetime are discussed and identified with their matrix images. The geometric significance of quaternions and bivectors is explored in 3D and 4D. Pauli's introduction of spin is reviewed in terms of Clifford algebras. The exterior algebra and contractions are introduced and related to the Clifford algebra. The exterior and shuffle products of the Grassmann–Cayley algebra are applied to the join and meet; it is shown that the shuffle product is like an exterior product stepping downwards from the volume element. Spin groups and conformal groups are discussed in lower dimensions. Then Clifford algebras are defined over arbitrary fields for arbitrary quadratic forms as well as for nonsymmetric bilinear forms.

1.1 Introduction

Clifford algebras are algebras of geometries, the geometry being determined by a quadratic form on a linear space. Clifford algebra is by definition the minimal construction designed to control the geometry in question.

In geometry, information about oriented subspaces can be encoded in blades, which can be added and multiplied. Everybody is familiar with this tool in the special case of 1-dimensional subspaces, commonly manipulated by vectors and not by projection operators, which lose information about orientations.

In physics, the concept of Clifford algebra as such or in the disguise of matrices is a necessity in the description of electron spin, because spinor spaces cannot be constructed by tensorial methods in terms of exterior powers of the defining vector space.

Thus, Clifford algebra is the natural algebraic tool for controlling subspaces and rotations/spins. Today, Clifford algebras are used in such disciplines as molecular chemistry and electromagnetism, and in engineering applications, such as robotics and computer vision.

This lecture was presented at the "Lecture Series on Clifford Algebras and their Applications", May 18 and 19, 2002, as part of the 6th International Conference on Clifford Algebras and their Applications in Mathematical Physics, Cookeville, TN, May 20–25, 2002.
AMS Subject Classification: 15A66, 17B37, 20C30, 81R25.
Keywords: Clifford algebras, bilinear form, quadratic form.

1.2 Clifford algebra of the Euclidean plane

1.2.1 The Clifford product of two vectors

Consider the plane $\mathbb{R} \times \mathbb{R} = \{(x,y) \mid x,y \in \mathbb{R}\}$. Introduce a linear structure by addition $(x_1,y_1)+(x_2,y_2) = (x_1+x_2, y_1+y_2)$ and by scaling $\lambda(x,y) = (\lambda x, \lambda y)$, where $\lambda \in \mathbb{R}$. The linear structure makes the plane $\mathbb{R} \times \mathbb{R}$ into a *linear space* $\mathbb{R}^2$. Take a basis $(\mathbf{e}_1, \mathbf{e}_2)$ of $\mathbb{R}^2$, say $\mathbf{e}_1 = (1,0)$, $\mathbf{e}_2 = (0,1)$. Introduce the length of $\mathbf{r} = x\mathbf{e}_1 + y\mathbf{e}_2$ as $|\mathbf{r}| = \sqrt{x^2+y^2}$. The introduction of length makes $\mathbb{R}^2$ into a *Euclidean plane*, also denoted by $\mathbb{R}^2$. The basis vectors $\mathbf{e}_1, \mathbf{e}_2$ are *unit vectors*, $|\mathbf{e}_1| = 1$, $|\mathbf{e}_2| = 1$.

Related to length there is the *scalar valued product* $\mathbf{a} \cdot \mathbf{b} = a_1b_1 + a_2b_2$ of two vectors $\mathbf{a} = a_1\mathbf{e}_1 + a_2\mathbf{e}_2$, $\mathbf{b} = b_1\mathbf{e}_1 + b_2\mathbf{e}_2 \in \mathbb{R}^2$. The scalar product is symmetric, $\mathbf{a} \cdot \mathbf{b} = \mathbf{b} \cdot \mathbf{a}$. Two vectors $\mathbf{a}, \mathbf{b}$ are *orthogonal*, $\mathbf{a} \perp \mathbf{b}$, if their scalar product vanishes, $\mathbf{a} \cdot \mathbf{b} = 0$. The unit vectors $\mathbf{e}_1, \mathbf{e}_2$ are orthogonal, $\mathbf{e}_1 \perp \mathbf{e}_2$.

Next, we introduce an associative, but noncommutative (nonsymmetric) product for vectors in $\mathbb{R}^2$. If the vector $\mathbf{r}$ is multiplied by itself or squared, $\mathbf{r}\mathbf{r} = \mathbf{r}^2$, we require that the square of the vector equals the square of the length of the vector,

$$\mathbf{r}^2 = |\mathbf{r}|^2.$$

In coordinate form, we introduce a product for vectors such that

$$(x\mathbf{e}_1 + y\mathbf{e}_2)^2 = x^2 + y^2.$$

Using the distributive law, without assuming commutativity, gives

$$x^2\mathbf{e}_1^2 + y^2\mathbf{e}_2^2 + xy(\mathbf{e}_1\mathbf{e}_2 + \mathbf{e}_2\mathbf{e}_1) = x^2 + y^2.$$

This is satisfied if the orthogonal unit vectors $\mathbf{e}_1$, $\mathbf{e}_2$ obey the multiplication rules

$\mathbf{e}_1^2 = \mathbf{e}_2^2 = 1$ $\mathbf{e}_1\mathbf{e}_2 = -\mathbf{e}_2\mathbf{e}_1$	which corresponds to	$\lvert\mathbf{e}_1\rvert = \lvert\mathbf{e}_2\rvert = 1$ $\mathbf{e}_1 \perp \mathbf{e}_2$

Using associativity and anticommutativity we can calculate the square as $(\mathbf{e}_1\mathbf{e}_2)^2 = -\mathbf{e}_1^2\mathbf{e}_2^2 = -1$. Since the square of the product $\mathbf{e}_1\mathbf{e}_2$ is negative, it follows that $\mathbf{e}_1\mathbf{e}_2$ is neither a scalar nor a vector. The product $\mathbf{e}_1\mathbf{e}_2$ is a new kind of quantity called a **bivector**, and represents the oriented plane area of the square with sides $\mathbf{e}_1$ and $\mathbf{e}_2$. Write for short $\mathbf{e}_{12} = \mathbf{e}_1\mathbf{e}_2$.

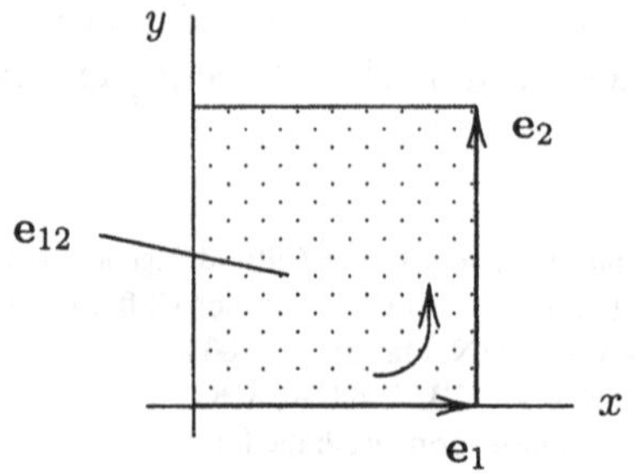

Define for two vectors $\mathbf{a} = a_1\mathbf{e}_1 + a_2\mathbf{e}_2$, $\mathbf{b} = b_1\mathbf{e}_1 + b_2\mathbf{e}_2$ the *Clifford product* $\mathbf{ab} = a_1b_1 + a_2b_2 + (a_1b_2 - a_2b_1)\mathbf{e}_{12}$, a sum of a scalar and a bivector.[1]

1.2.2 The Clifford algebra $C\ell_2$ of $\mathbb{R}^2$

Let $(\mathbf{e}_1, \mathbf{e}_2)$ be an orthonormal basis of the Euclidean plane $\mathbb{R}^2$. The four elements

1	scalar
$\mathbf{e}_1$, $\mathbf{e}_2$	vectors
$\mathbf{e}_{12}$	bivector

form a basis $(1, \mathbf{e}_1, \mathbf{e}_2, \mathbf{e}_{12})$ for the **Clifford algebra** $C\ell_2$ of $\mathbb{R}^2$, that is, an arbitrary element

$$u = u_0 + u_1\mathbf{e}_1 + u_2\mathbf{e}_2 + u_{12}\mathbf{e}_{12} \quad \text{in} \quad C\ell_2$$

is a linear combination of a scalar u_0, a vector $\vec{u} = u_1\mathbf{e}_1 + u_2\mathbf{e}_2$ and a bivector $\overset{\leftharpoonup}{u} = u_{12}\mathbf{e}_{12}$.

Example 1 *Compute* $\mathbf{e}_1\mathbf{e}_{12} = \mathbf{e}_1\mathbf{e}_1\mathbf{e}_2 = \mathbf{e}_2$, $\mathbf{e}_{12}\mathbf{e}_1 = \mathbf{e}_1\mathbf{e}_2\mathbf{e}_1 = -\mathbf{e}_1^2\mathbf{e}_2 = -\mathbf{e}_2$, $\mathbf{e}_2\mathbf{e}_{12} = \mathbf{e}_2\mathbf{e}_1\mathbf{e}_2 = -\mathbf{e}_1\mathbf{e}_2^2 = -\mathbf{e}_1$ *and* $\mathbf{e}_{12}\mathbf{e}_2 = \mathbf{e}_1\mathbf{e}_2^2 = \mathbf{e}_1$. *Note in particular that* $\mathbf{e}_{12}$ *anticommutes with both* $\mathbf{e}_1$ *and* $\mathbf{e}_2$.

The Clifford algebra $C\ell_2$ is a 4-dimensional real linear space with basis elements 1, $\mathbf{e}_1$, $\mathbf{e}_2$, $\mathbf{e}_{12}$ which have the multiplication table

	$\mathbf{e}_1$	$\mathbf{e}_2$	$\mathbf{e}_{12}$
$\mathbf{e}_1$	1	$\mathbf{e}_{12}$	$\mathbf{e}_2$
$\mathbf{e}_2$	$-\mathbf{e}_{12}$	1	$-\mathbf{e}_1$
$\mathbf{e}_{12}$	$-\mathbf{e}_2$	$\mathbf{e}_1$	-1

1.2.3 The exterior product of two vectors

Extracting the scalar and bivector parts of the Clifford product we have as products of two vectors $\mathbf{a} = a_1\mathbf{e}_1 + a_2\mathbf{e}_2$ and $\mathbf{b} = b_1\mathbf{e}_1 + b_2\mathbf{e}_2$:

$\mathbf{a} \cdot \mathbf{b} = a_1b_1 + a_2b_2,$	the scalar product "a dot b",
$\mathbf{a} \wedge \mathbf{b} = (a_1b_2 - a_2b_1)\mathbf{e}_{12},$	the exterior product "a wedge b".

The bivector $\mathbf{a} \wedge \mathbf{b}$ represents the oriented plane segment of the parallelogram with sides $\mathbf{a}$ and $\mathbf{b}$. The area of this parallelogram is $|a_1b_2 - a_2b_1|$, and we will

[1]The Clifford product $\mathbf{ab}$ is not the same as the scalar product $\mathbf{a} \cdot \mathbf{b} = a_1b_1 + a_2b_2$.

take the *magnitude* of the bivector $\mathbf{a} \wedge \mathbf{b}$ to be this area $|\mathbf{a} \wedge \mathbf{b}| = |a_1 b_2 - a_2 b_1|$.

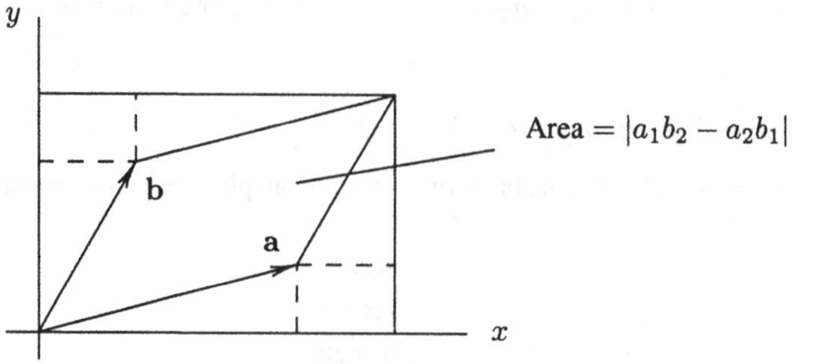

The parallelogram can be regarded as a kind of geometrical product of its sides:

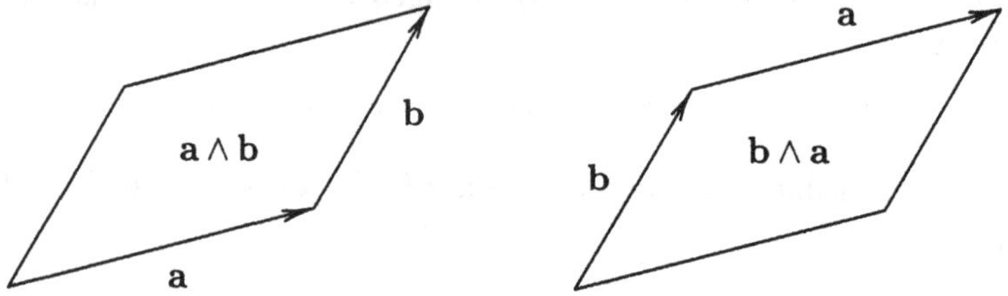

The bivectors $\mathbf{a} \wedge \mathbf{b}$ and $\mathbf{b} \wedge \mathbf{a}$ have the same magnitude but opposite directions of rotation. This can be expressed simply by writing

$$\mathbf{a} \wedge \mathbf{b} = -\mathbf{b} \wedge \mathbf{a}.$$

Using the multiplication table of the Clifford algebra $\mathcal{C}\ell_2$ we notice that the Clifford product

$$(a_1\mathbf{e}_1 + a_2\mathbf{e}_2)(b_1\mathbf{e}_1 + b_2\mathbf{e}_2) = a_1b_1 + a_2b_2 + (a_1b_2 - a_2b_1)\mathbf{e}_{12}$$

of two vectors $\mathbf{a} = a_1\mathbf{e}_1 + a_2\mathbf{e}_2$ and $\mathbf{b} = b_1\mathbf{e}_1 + b_2\mathbf{e}_2$ is the sum of a scalar $\mathbf{a} \cdot \mathbf{b} = a_1b_1 + a_2b_2$ and a bivector $\mathbf{a} \wedge \mathbf{b} = (a_1b_2 - a_2b_1)\mathbf{e}_{12}$.[2] In an equation,

$$\mathbf{ab} = \mathbf{a} \cdot \mathbf{b} + \mathbf{a} \wedge \mathbf{b}. \qquad (a)$$

The commutative rule $\mathbf{a} \cdot \mathbf{b} = \mathbf{b} \cdot \mathbf{a}$, together with the anticommutative rule $\mathbf{a} \wedge \mathbf{b} = -\mathbf{b} \wedge \mathbf{a}$, implies a relation between $\mathbf{ab}$ and $\mathbf{ba}$. Thus,

$$\mathbf{ba} = \mathbf{a} \cdot \mathbf{b} - \mathbf{a} \wedge \mathbf{b}. \qquad (b)$$

Adding and subtracting equations (a) and (b), we find

$$\mathbf{a} \cdot \mathbf{b} = \tfrac{1}{2}(\mathbf{ab} + \mathbf{ba}) \quad \text{and} \quad \mathbf{a} \wedge \mathbf{b} = \tfrac{1}{2}(\mathbf{ab} - \mathbf{ba}).$$

---

[2] The bivector valued exterior product $\mathbf{a} \wedge \mathbf{b} = (a_1b_2 - a_2b_1)\mathbf{e}_{12}$, which represents a plane area, should not be confused with the vector-valued cross product $\mathbf{a} \times \mathbf{b} = (a_1b_2 - a_2b_1)\mathbf{e}_3$, which represents a line segment.

Two vectors $\mathbf{a}$ and $\mathbf{b}$ are parallel, $\mathbf{a} \parallel \mathbf{b}$, when they commute, $\mathbf{ab} = \mathbf{ba}$, that is, $\mathbf{a} \wedge \mathbf{b} = 0$ or $a_1b_2 = a_2b_1$, and orthogonal, $\mathbf{a} \perp \mathbf{b}$, when they anticommute, $\mathbf{ab} = -\mathbf{ba}$, that is, $\mathbf{a} \cdot \mathbf{b} = 0$. Thus,

$$\begin{array}{ccccccc} \mathbf{ab} = \mathbf{ba} & \iff & \mathbf{a} \parallel \mathbf{b} & \iff & \mathbf{a} \wedge \mathbf{b} = 0 & \iff & \mathbf{ab} = \mathbf{a} \cdot \mathbf{b}, \\ \mathbf{ab} = -\mathbf{ba} & \iff & \mathbf{a} \perp \mathbf{b} & \iff & \mathbf{a} \cdot \mathbf{b} = 0 & \iff & \mathbf{ab} = \mathbf{a} \wedge \mathbf{b}. \end{array}$$

1.2.4 Perpendicular projections and reflections

Let us calculate the component of $\mathbf{a}$ in the direction of $\mathbf{b}$ when the two vectors make an angle φ, $0 < \varphi < 180°$. The parallel component $\mathbf{a}_\parallel$ is a scalar multiple of the unit vector $\mathbf{b}/|\mathbf{b}|$:

$$\mathbf{a}_\parallel = |\mathbf{a}| \cos\varphi \frac{\mathbf{b}}{|\mathbf{b}|} = |\mathbf{a}||\mathbf{b}| \cos\varphi \frac{\mathbf{b}}{|\mathbf{b}|^2}.$$

In other words, the parallel component $\mathbf{a}_\parallel$ is the scalar product $\mathbf{a} \cdot \mathbf{b} = |\mathbf{a}||\mathbf{b}| \cos\varphi$ multiplied by the vector $\mathbf{b}^{-1} = \mathbf{b}/|\mathbf{b}|^2$, called the inverse[3] of the vector $\mathbf{b}$. Thus,

$$\mathbf{a}_\parallel = (\mathbf{a} \cdot \mathbf{b}) \frac{\mathbf{b}}{|\mathbf{b}|^2} = (\mathbf{a} \cdot \mathbf{b})\mathbf{b}^{-1}.$$

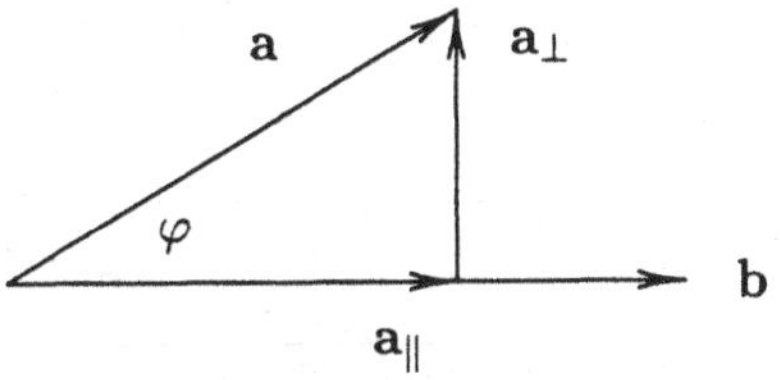

The last formula tells us that the length of $\mathbf{b}$ is irrelevant when projecting into the direction of $\mathbf{b}$.

The perpendicular component $\mathbf{a}_\perp$ is given by the difference

$$\mathbf{a}_\perp = \mathbf{a} - \mathbf{a}_\parallel = \mathbf{a} - (\mathbf{a} \cdot \mathbf{b})\mathbf{b}^{-1} = (\mathbf{ab} - \mathbf{a} \cdot \mathbf{b})\mathbf{b}^{-1} = (\mathbf{a} \wedge \mathbf{b})\mathbf{b}^{-1}.$$

Note that the bivector $\mathbf{e}_{12}$ anticommutes with all the vectors in the $\mathbf{e}_1\mathbf{e}_2$-plane; therefore

$$(\mathbf{a} \wedge \mathbf{b})\mathbf{b}^{-1} = -\mathbf{b}^{-1}(\mathbf{a} \wedge \mathbf{b}) = \mathbf{b}^{-1}(\mathbf{b} \wedge \mathbf{a}) = -(\mathbf{b} \wedge \mathbf{a})\mathbf{b}^{-1}.$$

The area of the parallelogram with sides $\mathbf{a}, \mathbf{b}$ is

$$|\mathbf{a}_\perp \mathbf{b}| = |\mathbf{a} \wedge \mathbf{b}| = |\mathbf{a}||\mathbf{b}| \sin\varphi$$

where $0 < \varphi < 180°$.

The reflection of $\mathbf{r}$ across the line $\mathbf{a}$ is obtained by sending $\mathbf{r} = \mathbf{r}_\parallel + \mathbf{r}_\perp$ to $\mathbf{r}' = \mathbf{r}_\parallel - \mathbf{r}_\perp$, where $\mathbf{r}_\parallel = (\mathbf{r} \cdot \mathbf{a})\mathbf{a}^{-1}$. The mirror image $\mathbf{r}'$ of $\mathbf{r}$ with respect to $\mathbf{a}$ is then

[3] The inverse $\mathbf{b}^{-1}$ of a nonzero vector $\mathbf{b} \in \mathbb{R}^2 \subset C\ell_2$ satisfies $\mathbf{b}^{-1}\mathbf{b} = \mathbf{bb}^{-1} = 1$ in the Clifford algebra $C\ell_2$. A vector and its inverse are parallel vectors.

$$\begin{aligned}\mathbf{r}' &= (\mathbf{r}\cdot\mathbf{a})\mathbf{a}^{-1} - (\mathbf{r}\wedge\mathbf{a})\mathbf{a}^{-1}\\ &= (\mathbf{r}\cdot\mathbf{a} - \mathbf{r}\wedge\mathbf{a})\mathbf{a}^{-1}\\ &= (\mathbf{a}\cdot\mathbf{r} + \mathbf{a}\wedge\mathbf{r})\mathbf{a}^{-1}\\ &= \mathbf{a}\mathbf{r}\mathbf{a}^{-1}\end{aligned}$$

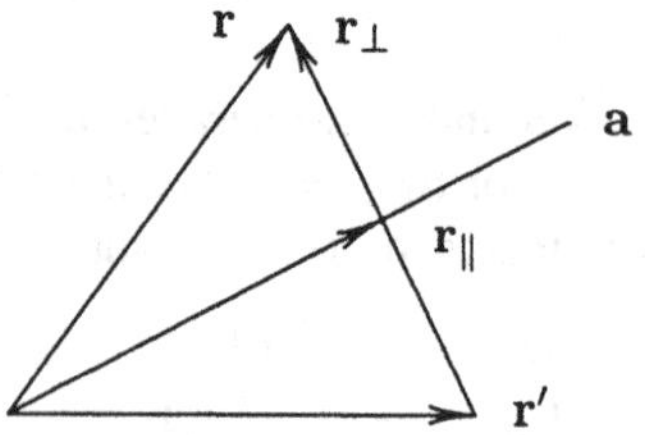

and further

$$\begin{aligned}\mathbf{r}' &= (2\mathbf{a}\cdot\mathbf{r} - \mathbf{r}\mathbf{a})\mathbf{a}^{-1}\\ &= 2\frac{\mathbf{a}\cdot\mathbf{r}}{\mathbf{a}^2}\mathbf{a} - \mathbf{r}.\end{aligned}$$

The formula $\mathbf{r}' = \mathbf{a}\mathbf{r}\mathbf{a}^{-1}$ can be obtained directly using only commutation properties of the Clifford product: decompose $\mathbf{r} = \mathbf{r}_{\|} + \mathbf{r}_{\perp}$, where $\mathbf{a}\mathbf{r}_{\|}\mathbf{a}^{-1} = \mathbf{r}_{\|}\mathbf{a}\mathbf{a}^{-1} = \mathbf{r}_{\|}$, while $\mathbf{a}\mathbf{r}_{\perp}\mathbf{a}^{-1} = -\mathbf{r}_{\perp}\mathbf{a}\mathbf{a}^{-1} = -\mathbf{r}_{\perp}$.

The composition of two reflections, first across $\mathbf{a}$ and then across $\mathbf{b}$, is given by

$$\mathbf{r} \to \mathbf{r}' = \mathbf{a}\mathbf{r}\mathbf{a}^{-1} \to \mathbf{r}'' = \mathbf{b}\mathbf{r}'\mathbf{b}^{-1} = \mathbf{b}(\mathbf{a}\mathbf{r}\mathbf{a}^{-1})\mathbf{b}^{-1} = (\mathbf{b}\mathbf{a})\mathbf{r}(\mathbf{b}\mathbf{a})^{-1}.$$

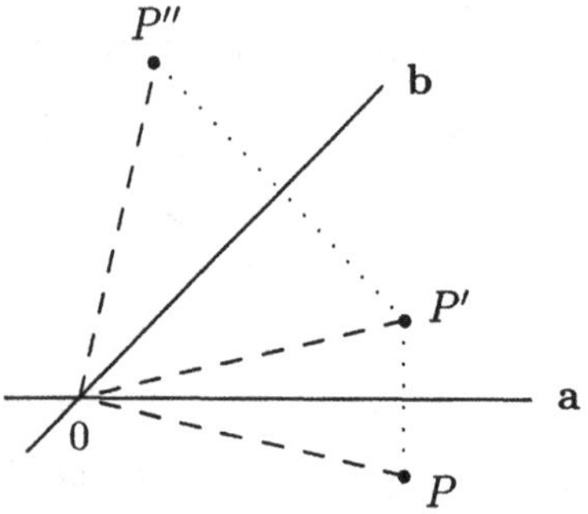

The composite of these two reflections is a rotation by twice the angle between $\mathbf{a}$ and $\mathbf{b}$.

1.2.5 Rotations

The product of a vector $\mathbf{r} = x\mathbf{e}_1 + y\mathbf{e}_2$ and a unit complex number $e^{\mathbf{i}\varphi} = \cos\varphi + \mathbf{i}\sin\varphi$, where for short $\mathbf{i} = \mathbf{e}_{12}$, is another vector in the $\mathbf{e}_1\mathbf{e}_2$-plane:

$$\mathbf{r}\cos\varphi + \mathbf{r}\mathbf{i}\sin\varphi = \mathbf{r}e^{\mathbf{i}\varphi}.$$

The vector $\mathbf{r}\mathbf{i} = x\mathbf{e}_2 - y\mathbf{e}_1$ is perpendicular to $\mathbf{r}$ so that a counterclockwise rotation by $\pi/2$ carries $\mathbf{r}$ to $\mathbf{r}\mathbf{i}$.

Since the unit bivector $\mathbf{i}$ anticommutes with every vector $\mathbf{r}$ in the $\mathbf{e}_1\mathbf{e}_2$-plane, the rotated vector can also be expressed as

$$\mathbf{r}\cos\varphi + \mathbf{r}\mathbf{i}\sin\varphi = \mathbf{r}\cos\varphi - \mathbf{i}\mathbf{r}\sin\varphi = e^{-\mathbf{i}\varphi}\mathbf{r}.$$

Furthermore, we have $\cos\varphi + \mathbf{i}\sin\varphi = (\cos\frac{\varphi}{2} + \mathbf{i}\sin\frac{\varphi}{2})^2$ and thus the rotated vector also has the form $s^{-1}\mathbf{r}s$ where $s = e^{\mathbf{i}\varphi/2}$ and $s^{-1} = e^{-\mathbf{i}\varphi/2}$. The counterclockwise rotation of $\mathbf{r}$ by the angle φ will then result in $\mathbf{r}z = z^{-1}\mathbf{r} = s^{-1}\mathbf{r}s$

where $z = e^{i\varphi}$, $z^{-1} = e^{-i\varphi}$ and $s^2 = z$. There are two complex numbers s and $-s$ which result in the same rotation $s^{-1}\mathbf{r}s = (-s)^{-1}\mathbf{r}(-s)$. In other words, there are two complex numbers which produce the same final result but via different actions.

$$s = e^{i\varphi/2}$$

$$-s = e^{-i(2\pi-\varphi)/2} = e^{i\varphi/2}e^{-i\pi}$$

$$\boxed{e^{i\pi} = -1}$$

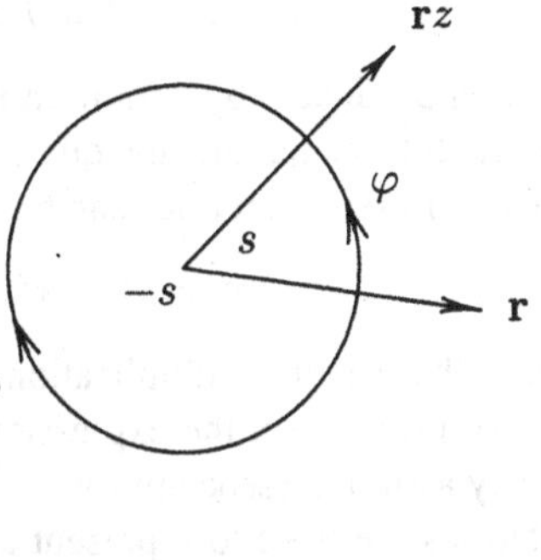

1.2.6 $C\ell_2$ as the matrix algebra $\mathbb{R}(2)$

We can represent the orthogonal unit vectors $\mathbf{e}_1$, $\mathbf{e}_2$ by the matrices

$$\mathbf{e}_1 \simeq \begin{pmatrix} 1 & 0 \\ 0 & -1 \end{pmatrix}, \quad \mathbf{e}_2 \simeq \begin{pmatrix} 0 & 1 \\ 1 & 0 \end{pmatrix}$$

and the bivector $\mathbf{e}_{12} = \mathbf{e}_1\mathbf{e}_2$ by

$$\mathbf{e}_{12} \simeq \begin{pmatrix} 0 & 1 \\ -1 & 0 \end{pmatrix}.$$

These correspondences establish an isomorphism of algebras

$$C\ell_2 \simeq \mathbb{R}(2) = \left\{ \begin{pmatrix} a & b \\ c & d \end{pmatrix} \mid a, b, c, d \in \mathbb{R} \right\}.$$

It should be emphasized that the Clifford algebra $C\ell_2$ has more structure than the matrix algebra $\mathbb{R}(2)$. In the Clifford algebra, there is a distinguished subspace of vectors

$$\mathbb{R}^2 \simeq \left\{ \begin{pmatrix} x & y \\ y & -x \end{pmatrix} \mid x, y \in \mathbb{R} \right\}.$$

1.3 Quaternions

Quaternions are generalized complex numbers of the form

$$w + ix + jy + kz$$

where w, x, y, z are real numbers and the generalized imaginary units[4] i, j, k satisfy the following multiplication rules:

$$i^2 = j^2 = k^2 = -1,$$
$$ij = k = -ji, \quad jk = i = -kj, \quad ki = j = -ik.$$

The quaternion product is by definition **noncommutative**: the order of multiplication matters. It turns out that the quaternion multiplication is associative.

The above multiplication rules can be condensed into the following form:

$$i^2 = j^2 = k^2 = ijk = -1.$$

Alternatively, these latter multiplication rules could be taken as a definition of the quaternion product (in the expression ijk we have omitted parentheses and thereby tacitly assumed associativity).

Two notations are used to represent a quaternion: the hypercomplex number notation $w + ix + jy + kz$ and the 4-tuple notation $(w, x, y, z) \in \mathbb{R}^4$, composed of a scalar $w \in \mathbb{R}$ and a vector $(x, y, z) \in \mathbb{R}^3$. The 4D real linear space $\mathbb{R}^4$, endowed with the quaternion product, is denoted by $\mathbb{H}$, in honor of Hamilton.

1.3.1 Vectors as pure quaternions

A quaternion q is a sum of a scalar $w \in \mathbb{R}$ and a vector $\mathbf{i}x + \mathbf{j}y + \mathbf{k}z \in \mathbb{R}^3$,

$$q = w + \mathbf{i}x + \mathbf{j}y + \mathbf{k}z \in \mathbb{R} \oplus \mathbb{R}^3.$$

This phrase connotes that scalars w for the linear space $\mathbb{H}$ are identified with elements $(w, 0, 0, 0)$ in $\mathbb{H}$, and vectors $\mathbf{i}x + \mathbf{j}y + \mathbf{k}z$ with $(0, x, y, z)$. Each quaternion $q = w + \mathbf{i}x + \mathbf{j}y + \mathbf{k}z$ is uniquely expressible in the form $q = \langle q \rangle_0 + \langle q \rangle_1$, where $\langle q \rangle_0 = w \in \mathbb{R}$ and $\langle q \rangle_1 = \mathbf{i}x + \mathbf{j}y + \mathbf{k}z \in \mathbb{R}^3$, $\langle q \rangle_0$ being called the *scalar part* of q and $\langle q \rangle_1$ the *pure part*[5] of q. If the scalar part vanishes, then the quaternion is called *pure*.

A quaternion is a scalar, with vanishing pure part, if and only if it commutes with every quaternion. In other words, $\mathbb{R}$ is the center of the ring $\mathbb{H}$. As a consequence, the ring structure of $\mathbb{H}$ induces the real linear structure in $\mathbb{H}$. A ring automorphism or anti-automorphism of $\mathbb{H}$ is a real linear automorphism of $\mathbb{H}$. A ring automorphism or anti-automorphism of $\mathbb{H}$ is a real algebra automorphism or anti-automorphism of $\mathbb{H}$.

In contrast, the field of complex numbers $\mathbb{C}$ has automorphisms which send nonrational real numbers in $\mathbb{R} \setminus \mathbb{Q}$ to nonreal complex numbers in $\mathbb{C} \setminus \mathbb{R}$.

A nonzero quaternion q is pure if and only if its square is a negative scalar, $q^2 < 0$. The direct sum decomposition of $\mathbb{H}$ to the subspaces of scalars and pure quaternions $\mathbb{H} = \mathbb{R} \oplus \mathbb{R}^3$ is induced by the ring structure of $\mathbb{H}$.

---

[4] Two typefaces are in use: italics i, j, k to emphasize origin in complex numbers and boldface $\mathbf{i}, \mathbf{j}, \mathbf{k}$ to emphasize association with vectors, that is, $\mathbf{i}x + \mathbf{j}y + \mathbf{k}z \in \mathbb{R}^3$.

[5] The pure part of a quaternion has a dual role: as a vector (generator of translations) and as a bivector (generator of rotations). Hence a nondiscriminating label: pure.

1.3.2 Quaternion product via the scalar and cross products

The product of two quaternions $a = a_0 + \mathbf{a}$ and $b = b_0 + \mathbf{b}$, where $a_0, b_0 \in \mathbb{R}$ and $\mathbf{a}, \mathbf{b} \in \mathbb{R}^3$, can be written as

$$ab = a_0 b_0 - \mathbf{a} \cdot \mathbf{b} + a_0 \mathbf{b} + \mathbf{a} b_0 + \mathbf{a} \times \mathbf{b}.$$

The product of two vectors $\mathbf{a} = \mathbf{i}a_1 + \mathbf{j}a_2 + \mathbf{k}a_3$ and $\mathbf{b} = \mathbf{i}b_1 + \mathbf{j}b_2 + \mathbf{k}b_3$ is a sum of a scalar and a vector,

$$\mathbf{ab} = -\mathbf{a} \cdot \mathbf{b} + \mathbf{a} \times \mathbf{b},$$

where we recognize the scalar product $\mathbf{a} \cdot \mathbf{b} = a_1 b_1 + a_2 b_2 + a_3 b_3$ and the cross product $\mathbf{a} \times \mathbf{b} = \mathbf{i}(a_2 b_3 - a_3 b_2) + \mathbf{j}(a_3 b_1 - a_1 b_3) + \mathbf{k}(a_1 b_2 - a_2 b_1)$.

We may reobtain the scalar/cross product of two vectors $\mathbf{a}, \mathbf{b} \in \mathbb{R}^3$ in terms of the scalar/pure part of their quaternion product:

$$\mathbf{a} \cdot \mathbf{b} = -\langle \mathbf{ab} \rangle_0, \quad \mathbf{a} \times \mathbf{b} = \langle \mathbf{ab} \rangle_1.$$

For all $\mathbf{a}, \mathbf{b} \in \mathbb{R}^3$, $\mathbf{a} \cdot \mathbf{b} = -\frac{1}{2}(\mathbf{ab} + \mathbf{ba})$, $\mathbf{a} \times \mathbf{b} = \frac{1}{2}(\mathbf{ab} - \mathbf{ba})$, and $(\mathbf{a} \times \mathbf{b}) \perp \mathbf{a}$, $(\mathbf{a} \times \mathbf{b}) \perp \mathbf{b}$.

1.3.3 Matrix representation of quaternion multiplication

The product $qu = v$ of two quaternions $q = w + \mathbf{i}x + \mathbf{j}y + \mathbf{k}z$ and $u = u_0 + \mathbf{i}u_1 + \mathbf{j}u_2 + \mathbf{k}u_3$ can be represented by matrix multiplication:

$$\begin{pmatrix} w & -x & -y & -z \\ x & w & -z & y \\ y & z & w & -x \\ z & -y & x & w \end{pmatrix} \begin{pmatrix} u_0 \\ u_1 \\ u_2 \\ u_3 \end{pmatrix} = \begin{pmatrix} v_0 \\ v_1 \\ v_2 \\ v_3 \end{pmatrix}.$$

Swapping the multiplication to the right, that is, $uq = v'$, gives a partially transformed matrix:

$$\begin{pmatrix} w & -x & -y & -z \\ x & w & z & -y \\ y & -z & w & x \\ z & y & -x & w \end{pmatrix} \begin{pmatrix} u_0 \\ u_1 \\ u_2 \\ u_3 \end{pmatrix} = \begin{pmatrix} v'_0 \\ v'_1 \\ v'_2 \\ v'_3 \end{pmatrix}.$$

Let us denote the matrices of $\mathbb{H} \to \mathbb{H}$, $u \to qu$ and $\mathbb{H} \to \mathbb{H}$, $u \to uq$, respectively, by L_q and R_q, that is,

$$L_q(u) = qu \ (= v) \quad \text{and} \quad R_q(u) = uq \ (= v').$$

We find that[6]

$$L_{\mathbf{i}} L_{\mathbf{j}} L_{\mathbf{k}} = -I \quad \text{and} \quad R_{\mathbf{i}} R_{\mathbf{j}} R_{\mathbf{k}} = I.$$

[6]Note that $R_{\mathbf{i}}^{\mathsf{T}} R_{\mathbf{j}}^{\mathsf{T}} R_{\mathbf{k}}^{\mathsf{T}} = -I$.

The sets $\{L_q \in \mathbb{R}(4) \mid q \in \mathbb{H}\}$ and $\{R_q \in \mathbb{R}(4) \mid q \in \mathbb{H}\}$ form two subalgebras of $\mathbb{R}(4)$, both isomorphic to $\mathbb{H}$. For two arbitrary quaternions $a, b \in \mathbb{H}$ these two matrix representatives commute, that is, $L_a R_b = R_b L_a$. Any real 4×4-matrix is a linear combination of matrices of the form $L_a R_b$. The above observations together with $(\dim \mathbb{H})^2 = \dim \mathbb{R}(4)$ can be expressed as

$$\mathbb{R}(4) \simeq \mathbb{H} \otimes \mathbb{H},$$

or more informatively, $\mathbb{R}(4) = \mathbb{H} \otimes \mathbb{H}^*$.

Take a matrix of the form $U = L_a R_b$ in $\mathbb{R}(4)$. Then $U^\top U = |a|^2 |b|^2 I$, but in general $U + U^\top \neq \alpha I$. Take a matrix of the form $V = L_a + R_b$ in $\mathbb{R}(4)$. Then $V + V^\top = 2(\mathrm{Re}(a) + \mathrm{Re}(b))I$, but in general $V^\top V \neq \beta I$. Conversely, if $U \in \mathbb{R}(4)$ is such that $U + U^\top = \alpha I$ and $U^\top U = \beta I$, then the matrix U belongs either to $\mathbb{H}$ or to $\mathbb{H}^*$.

Besides real 4×4-matrices, quaternions can also be represented by complex 2×2-matrices:

$$w + \mathbf{i}x + \mathbf{j}y + \mathbf{k}z \simeq \begin{pmatrix} w - iz & -ix - y \\ -ix + y & w + iz \end{pmatrix}.$$

The orthogonal unit vectors $\mathbf{i}, \mathbf{j}, \mathbf{k}$ are represented by Pauli matrices $\sigma_1, \sigma_2, \sigma_3$ divided by $i = \sqrt{-1}$:

$$\mathbf{i} \simeq \begin{pmatrix} 0 & -i \\ -i & 0 \end{pmatrix}, \quad \mathbf{j} \simeq \begin{pmatrix} 0 & -1 \\ 1 & 0 \end{pmatrix}, \quad \mathbf{k} \simeq \begin{pmatrix} -i & 0 \\ 0 & i \end{pmatrix}.$$

1.3.4 Linear spaces over $\mathbb{H}$

Much of the theory of linear spaces over commutative fields extends to $\mathbb{H}$. Because of the noncommutativity of $\mathbb{H}$ it is, however, necessary to distinguish between two types of linear spaces over $\mathbb{H}$, namely *right* linear spaces and *left* linear spaces.

A *right* linear space over $\mathbb{H}$ consists of an additive group V and a map

$$V \times \mathbb{H} \to V, \quad (\mathbf{x}, \lambda) \to \mathbf{x}\lambda$$

such that the usual distributivity and unity axioms hold and such that, for all λ, $\mu \in \mathbb{H}$ and $\mathbf{x} \in V$,

$$(\mathbf{x}\lambda)\mu = \mathbf{x}(\lambda\mu).$$

A *left* linear space over $\mathbb{H}$ consists of an additive group V and a map

$$\mathbb{H} \times V \to V, \ (\lambda, \mathbf{x}) \to \lambda\mathbf{x}$$

such that the usual distributivity and unity axioms hold and such that, for all λ, $\mu \in \mathbb{H}$ and $\mathbf{x} \in V$,

$$\lambda(\mu\mathbf{x}) = (\lambda\mu)\mathbf{x}.$$

A mapping $L : V \to U$ between two right linear spaces V and U is a *right linear map* if it respects addition and, for all $\mathbf{x} \in V$, $\lambda \in \mathbb{H}$, $L(\mathbf{x}\lambda) = (L(\mathbf{x}))\lambda$.

Comment 2 *In the matrix form the above definition means that*

$$\begin{pmatrix} a & b \\ c & d \end{pmatrix} \left[\begin{pmatrix} x_1 \\ x_2 \end{pmatrix} \lambda \right] = \begin{pmatrix} a & b \\ c & d \end{pmatrix} \begin{pmatrix} x_1\lambda \\ x_2\lambda \end{pmatrix} = \left[\begin{pmatrix} a & b \\ c & d \end{pmatrix} \begin{pmatrix} x_1 \\ x_2 \end{pmatrix} \right] \lambda.$$

Remark 3 *Although there are linear spaces over* $\mathbb{H}$, *there are no algebras over* $\mathbb{H}$, *since noncommutativity of* $\mathbb{H}$ *precludes bilinearity over* $\mathbb{H}$: $\lambda(xy) = (\lambda x)y \neq (x\lambda)y = x(\lambda y) \neq x(y\lambda) = (xy)\lambda$.

1.4 Clifford algebra of the Euclidean space $\mathbb{R}^3$

The 3-dimensional Euclidean space $\mathbb{R}^3$ has a basis consisting of three orthogonal unit vectors $\mathbf{e}_1$, $\mathbf{e}_2$, $\mathbf{e}_3$. The *Clifford algebra* $C\ell_3$ of $\mathbb{R}^3$ is the real associative algebra generated by the set $\{\mathbf{e}_1, \mathbf{e}_2, \mathbf{e}_3\}$ satisfying the relations

$$\mathbf{e}_1^2 = 1, \quad \mathbf{e}_2^2 = 1, \quad \mathbf{e}_3^2 = 1,$$
$$\mathbf{e}_1\mathbf{e}_2 = -\mathbf{e}_2\mathbf{e}_1, \quad \mathbf{e}_1\mathbf{e}_3 = -\mathbf{e}_3\mathbf{e}_1, \quad \mathbf{e}_2\mathbf{e}_3 = -\mathbf{e}_3\mathbf{e}_2.$$

The Clifford algebra $C\ell_3$ is 8-dimensional with the following basis:

1	the scalar
$\mathbf{e}_1$, $\mathbf{e}_2$, $\mathbf{e}_3$	vectors
$\mathbf{e}_1\mathbf{e}_2$, $\mathbf{e}_1\mathbf{e}_3$, $\mathbf{e}_2\mathbf{e}_3$	bivectors
$\mathbf{e}_1\mathbf{e}_2\mathbf{e}_3$	a volume element.

We abbreviate the unit bivectors as $\mathbf{e}_{ij} = \mathbf{e}_i\mathbf{e}_j$, when $i \neq j$, and the unit oriented volume element as $\mathbf{e}_{123} = \mathbf{e}_1\mathbf{e}_2\mathbf{e}_3$. An arbitrary element in $C\ell_3$ is a sum of a scalar, a vector, a bivector and a volume element, and can be written as $\alpha + \mathbf{a} + \mathbf{b}\mathbf{e}_{123} + \beta\mathbf{e}_{123}$, where $\alpha, \beta \in \mathbb{R}$ and $\mathbf{a}, \mathbf{b} \in \mathbb{R}^3$.

1.4.1 Matrix representation of $C\ell_3$

The set of 2×2-matrices with complex numbers as entries is denoted by $\mathbb{C}(2)$. Mostly we shall regard this set as a *real* algebra with scalar multiplication taken over the real numbers in $\mathbb{R}$ although the matrix entries are in the complex field $\mathbb{C}$. The Pauli matrices

$$\sigma_1 \simeq \begin{pmatrix} 0 & 1 \\ 1 & 0 \end{pmatrix}, \quad \sigma_2 \simeq \begin{pmatrix} 0 & -i \\ i & 0 \end{pmatrix}, \quad \sigma_3 \simeq \begin{pmatrix} 1 & 0 \\ 0 & -1 \end{pmatrix}$$

satisfy the multiplication rules

$$\sigma_1^2 = \sigma_2^2 = \sigma_3^2 = I \quad \text{and}$$
$$\sigma_1\sigma_2 = i\sigma_3 = -\sigma_2\sigma_1,$$
$$\sigma_3\sigma_1 = i\sigma_2 = -\sigma_1\sigma_3,$$
$$\sigma_2\sigma_3 = i\sigma_1 = -\sigma_3\sigma_2.$$

They also generate the real algebra $\mathbb{C}(2)$. The correspondences $\mathbf{e}_1 \simeq \sigma_1, \mathbf{e}_2 \simeq \sigma_2$, $\mathbf{e}_3 \simeq \sigma_3$ establish an isomorphism between the real algebras, $C\ell_3 \simeq \mathbb{C}(2)$, with the following correspondences of the basis elements:

$\mathbb{C}(2)$	$C\ell_3$
I	1
$\sigma_1,\ \sigma_2,\ \sigma_3$	$\mathbf{e}_1,\ \mathbf{e}_2,\ \mathbf{e}_3$
$\sigma_1\sigma_2,\ \sigma_1\sigma_3,\ \sigma_2\sigma_3$	$\mathbf{e}_{12},\ \mathbf{e}_{13},\ \mathbf{e}_{23}$
$\sigma_1\sigma_2\sigma_3$	$\mathbf{e}_{123}$

Note that $\mathbf{e}_{ij} = -\mathbf{e}_{ji}$ for $i \neq j$. The essential difference between the Clifford algebra $C\ell_3$ and its matrix image $\mathbb{C}(2)$ is that in the Clifford algebra $C\ell_3$ we will, in its definition, distinguish a particular subspace, the vector space $\mathbb{R}^3$, in which the square of a vector equals its length squared, that is, $\mathbf{r}^2 = |\mathbf{r}|^2$. No such distinguished subspace has been singled out in the definition of the matrix algebra $\mathbb{C}(2)$. Instead, we have chosen the traceless Hermitian matrices to represent $\mathbb{R}^3$, and thereby added extra structure to $\mathbb{C}(2)$.[7]

1.4.2 *The center of $C\ell_3$*

The element $\mathbf{e}_{123}$ commutes with all the vectors $\mathbf{e}_1, \mathbf{e}_2, \mathbf{e}_3$ and therefore with every element of $C\ell_3$. In other words, elements of the form

$$x + y\mathbf{e}_{123} \simeq \begin{pmatrix} x+iy & 0 \\ 0 & x+iy \end{pmatrix}$$

commute with all the elements in $C\ell_3$. The subalgebra spanned by scalars and 3-vectors

$$\mathbb{R} \oplus \bigwedge^3 \mathbb{R}^3 = \{x + y\mathbf{e}_{123} \mid x,\, y \in \mathbb{R}\}$$

consists of those elements of $C\ell_3$, which commute with every element of $C\ell_3$, that is, it is the *center* $\operatorname{Cen}(C\ell_3)$ of $C\ell_3$. Note that $\sigma_1\sigma_2\sigma_3 = iI$. Since $\mathbf{e}_{123}^2 = -1$, the center of $C\ell_3$ is isomorphic to the complex field $\mathbb{C}$, that is,

$$\operatorname{Cen}(C\ell_3) = \mathbb{R} \oplus \bigwedge^3 \mathbb{R}^3 \simeq \mathbb{C}.$$

---

[7] We could also have chosen, for the representatives of the anticommuting (and therefore orthogonal) unit vectors in $\mathbb{R}^3$, the following matrices:

$$u_1 = \frac{1}{4}\begin{pmatrix} 3i & 5 \\ 5 & -3i \end{pmatrix}, \quad u_2 = \begin{pmatrix} 0 & -i \\ i & 0 \end{pmatrix}, \quad u_3 = \frac{1}{4}\begin{pmatrix} 5 & -3i \\ -3i & -5 \end{pmatrix},$$

that is, $u_1 = \frac{1}{4}(5\sigma_1 + 3\sigma_1\sigma_2)$, $u_2 = \sigma_2$, $u_3 = \frac{1}{4}(5\sigma_3 - 3\sigma_2\sigma_3)$. These matrices are non-Hermitian and satisfy $u_ju_k + u_ku_j = 2\delta_{jk}I$.

1.4.3 *The even subalgebra $C\ell_3^0$*

The elements 1 and $\mathbf{e}_{12} = \mathbf{e}_1\mathbf{e}_2$, $\mathbf{e}_{13} = \mathbf{e}_1\mathbf{e}_3$, $\mathbf{e}_{23} = \mathbf{e}_2\mathbf{e}_3$ are called *even*, because they are products of an even number of vectors. The even elements are represented by the following matrices:

$$w + x\mathbf{e}_{23} + y\mathbf{e}_{31} + z\mathbf{e}_{12} \simeq \begin{pmatrix} w+iz & ix+y \\ ix-y & w-iz \end{pmatrix}.$$

The even elements form a real subspace

$$\begin{aligned} \mathbb{R} \oplus \bigwedge^2 \mathbb{R}^3 &= \{w + x\mathbf{e}_{23} + y\mathbf{e}_{31} + z\mathbf{e}_{12} \mid w, x, y, z \in \mathbb{R}\} \\ &\simeq \{wI + xi\sigma_1 + yi\sigma_2 + zi\sigma_3 \mid w, x, y, z \in \mathbb{R}\} \end{aligned}$$

which is closed under multiplication. Thus, the subspace $\mathbb{R} \oplus \bigwedge^2 \mathbb{R}^3$ is a subalgebra, called the *even subalgebra* of $C\ell_3$. We will denote the even subalgebra by even$(C\ell_3)$ or for short by $C\ell_3^0$. The even subalgebra is isomorphic to the division ring of quaternions $\mathbb{H}$, as can be seen by the following correspondences:

$\mathbb{H}$	$C\ell_3^0$
i	$-\mathbf{e}_{23}$
j	$-\mathbf{e}_{31}$
k	$-\mathbf{e}_{12}$

Remark 4 *The Clifford algebra $C\ell_3$ contains two subalgebras, isomorphic to $\mathbb{C}$ [the center] and $\mathbb{H}$ [the even subalgebra], in such a way that*

1. $ab = ba$ *for* $a \in \mathbb{C}$ *and* $b \in \mathbb{H}$,
2. $C\ell_3$ *is generated as a real algebra by* $\mathbb{C}$ *and* $\mathbb{H}$,
3. $(\dim \mathbb{C})(\dim \mathbb{H}) = \dim C\ell_3$.

These three observations can be expressed as

$$\mathbb{C} \otimes \mathbb{H} \simeq C\ell_3.$$

1.4.4 *Involutions of $C\ell_3$*

The Clifford algebra $C\ell_3$ has three involutions similar to complex conjugation. Take an arbitrary element

$$u = \langle u \rangle_0 + \langle u \rangle_1 + \langle u \rangle_2 + \langle u \rangle_3 \quad \text{in} \quad C\ell_3,$$

written as a sum of a scalar $\langle u \rangle_0$, a vector $\langle u \rangle_1$, a bivector $\langle u \rangle_2$ and a volume element $\langle u \rangle_3$. We introduce the following involutions:

$$\begin{aligned} \hat{u} &= \langle u \rangle_0 - \langle u \rangle_1 + \langle u \rangle_2 - \langle u \rangle_3, && \text{grade involution,} \\ \tilde{u} &= \langle u \rangle_0 + \langle u \rangle_1 - \langle u \rangle_2 - \langle u \rangle_3, && \text{reversion,} \\ \bar{u} &= \langle u \rangle_0 - \langle u \rangle_1 - \langle u \rangle_2 + \langle u \rangle_3, && \text{Clifford conjugation.} \end{aligned}$$

Clifford conjugation is a composition of the two other involutions: $\bar{u} = \hat{u}^{\tilde{}} = \tilde{u}^{\hat{}}$.

The correspondences $\sigma_1 \simeq \mathbf{e}_1, \sigma_2 \simeq \mathbf{e}_2, \sigma_3 \simeq \mathbf{e}_3$ fix the following representations for the involutions:

$$u \simeq \begin{pmatrix} a & b \\ c & d \end{pmatrix}, \qquad a, b, c, d \in \mathbb{C},$$

$$\hat{u} \simeq \begin{pmatrix} d^* & -c^* \\ -b^* & a^* \end{pmatrix}, \qquad \tilde{u} \simeq \begin{pmatrix} a^* & c^* \\ b^* & d^* \end{pmatrix}, \qquad \bar{u} \simeq \begin{pmatrix} d & -b \\ -c & a \end{pmatrix},$$

where the asterisk denotes complex conjugation. We recognize that the reverse $\tilde{u}$ is represented by the Hermitian conjugate $u^\dagger$ and the Clifford-conjugate $\bar{u}$ by the matrix $u^{-1} \det u \in \mathbb{R}(2)$ [for an invertible u].

The grade involution is an automorphism, that is,

$$\widehat{uv} = \hat{u}\hat{v},$$

while the reversion and the conjugation are anti-automorphisms, that is,

$$\widetilde{uv} = \tilde{v}\tilde{u} \quad \text{and} \quad \overline{uv} = \bar{v}\bar{u}.$$

The grade involution induces the even-odd grading of $C\ell_3 = C\ell_3^0 \oplus C\ell_3^1$.

1.5 The electron spin in a magnetic field

In classical mechanics kinetic energy $\frac{1}{2}mv^2 = \frac{p^2}{2m}$, $\vec{p} = m\vec{v}$, and potential energy $W = W(\vec{r})$ sum up to the total energy [8]

$$E = \frac{p^2}{2m} + W.$$

Replace the total energy and momentum by their differential operators,

$$E = i\hbar\frac{\partial}{\partial t} \quad \text{and} \quad \vec{p} = -i\hbar\nabla,$$

operating on a complex valued wave function, $\psi(\vec{r}, t) \in \mathbb{C}$. The result is a wave equation, known as the *Schrödinger equation*,[9]

$$i\hbar\frac{\partial\psi}{\partial t} = -\frac{\hbar^2}{2m}\nabla^2\psi + W\psi, \quad \psi : \mathbb{R}^3 \times \mathbb{R} \to \mathbb{C},$$

a quantum mechanical description of the electron. The Schrödinger equation explains all atomic phenomena except those involving magnetism and relativity.

---

[8] This holds in a conservative force field $\vec{F}$, where $\oint_C \vec{F} \cdot d\vec{s} = 0$ for all loops C.

[9] Light has both wave and particle properties. This prompted de Broglie to propose that perhaps particles also have wave properties such as interference and diffraction.

In 1922 the Stern–Gerlach experiment passed a beam of silver atoms through an inhomogeneous magnetic field. On the screen there appeared, instead of a smear of silver, two distinct traces: the beam split in two in the direction of the magnetic field. In 1925 Uhlenbeck & Goudsmit proposed that the electron has an intrinsic angular momentum, the *spin*. The electron spin can take only two orientations relative to the magnetic field: either parallel or opposite. In the Stern & Gerlach experiment the electron spin made atoms into magnetic dipoles which interacted with the magnetic field: atoms went up or down according as the spin was parallel or opposite to the vertical magnetic field.

1.5.1 Gauge transformations

The probability density of the electron, $|\psi(\vec{r},t)|^2$, [10] does not change under a phase transition, that is, $|\psi(\vec{r},t)|^2 = |\psi(\vec{r},t)e^{i\alpha(\vec{r},t)}|^2$. This suggests that *gauge transformations*,

$$\psi(\vec{r},t) \to \psi(\vec{r},t)e^{i\alpha(\vec{r},t)}, \quad e^{i\alpha(\vec{r},t)} \in U(1),$$

should preserve the form of the Schrödinger equation. By the Leibniz rule

$$\begin{aligned} i\hbar\frac{\partial}{\partial t}(\psi e^{i\alpha}) &= (i\hbar\frac{\partial\psi}{\partial t})e^{i\alpha} - \hbar\psi e^{i\alpha}\frac{\partial\alpha}{\partial t}, \\ -i\hbar\nabla(\psi e^{i\alpha}) &= (-i\hbar\nabla\psi)e^{i\alpha} + \hbar\psi e^{i\alpha}\nabla\alpha, \end{aligned}$$

and differentiating the latter twice,

$$\begin{aligned} (-i\hbar\nabla)^2(\psi e^{i\alpha}) &= (-i\hbar\nabla)\cdot[(-i\hbar\nabla\psi)e^{i\alpha} + \hbar\psi e^{i\alpha}\nabla\alpha] \\ &= [-\hbar^2\nabla^2\psi - i2\hbar^2(\nabla\psi)\cdot(\nabla\alpha) + \hbar^2\psi(\nabla\alpha)^2 - i\hbar^2\psi(\nabla^2\alpha)]e^{i\alpha} \\ &= [-\hbar^2\nabla^2\psi + i2\hbar e(\nabla\psi)\cdot\vec{A} + e^2\vec{A}^2\psi + i\hbar e(\nabla\cdot\vec{A})\psi]e^{i\alpha}, \end{aligned}$$

where we have abbreviated $e\vec{A} = -\hbar\nabla\alpha$. The extra terms containing $\vec{A}$ can be given a physical interpretation by recalling the gauge invariance of the electromagnetic field.

1.5.2 The electron in an electromagnetic field

Electromagnetic interaction is brought in via potentials $V, \vec{A}$ of the fields

$$\vec{E} = -\nabla V - \frac{\partial\vec{A}}{\partial t}, \quad \vec{B} = \nabla\times\vec{A},$$

provided that one agrees on a physical interpretation of the $U(1)$-gauge:

$$eV = -\hbar\frac{\partial\alpha}{\partial t}, \quad e\vec{A} = -\hbar\nabla\alpha.$$

[10] The probability of finding an electron in a space region is the integral of $|\psi(\vec{r},t)|^2$ over that region. This is the Born interpretation of QM.

By comparison we see that $i\hbar\frac{\partial}{\partial t}(\psi e^{i\alpha})$ and $(-\hbar^2\nabla^2)(\psi e^{i\alpha})$ agree, respectively, up to a phase, with

$$\begin{aligned}&(i\hbar\frac{\partial}{\partial t}+eV)\psi, \quad \text{and}\\ &[(-i\hbar\nabla-e\vec{A})\cdot(-i\hbar\nabla-e\vec{A})]\psi\\ &\qquad=-\hbar^2\nabla^2\psi+i\hbar e\nabla\cdot(\vec{A}\psi)+ie\hbar\vec{A}\cdot(\nabla\psi)+e^2\vec{A}^2\psi\\ &\qquad=-\hbar^2\nabla^2\psi+i\hbar e(\nabla\cdot\vec{A})\psi+i2\hbar e\vec{A}\cdot(\nabla\psi)+e^2\vec{A}^2\psi,\end{aligned}$$

where in the last row we have evaluated $\nabla\cdot(\vec{A}\psi)$ by regarding ψ as a scalar (although complex scalar). In the case of $W=0$ and electromagnetic potentials $V,\vec{A}$ (or in the case of $W=-eV$ and a magnetic potential $\vec{A}$):

$$i\hbar\frac{\partial\psi}{\partial t}=\frac{1}{2m}[(-i\hbar\nabla-e\vec{A})\cdot(-i\hbar\nabla-e\vec{A})]\psi-eV\psi, \tag{1}$$

abbreviated as, in terms of the generalized momentum $\vec{\pi}=-i\hbar\nabla-e\vec{A}$,

$$i\hbar\frac{\partial\psi}{\partial t}=\frac{1}{2m}[\vec{\pi}\cdot\vec{\pi}]\psi-eV\psi, \quad \psi:\mathbb{R}^3\times\mathbb{R}\to\mathbb{C}.$$

This equation does not yet involve the spin of the electron.

1.5.3 Interaction of the spin and a magnetic field

Applying the anticommutation relations of the Pauli matrices σ_k and the commutation relations[11]

$$\pi_1\pi_2-\pi_2\pi_1=i\hbar eB_3 \quad \text{(permute 1, 2, 3 cyclically)}$$

of the components $\pi_k=p_k-eA_k$ of the generalized momentum $\vec{\pi}=\vec{p}-e\vec{A}$ results in [12] [13]

$$(\vec{\sigma}\cdot\vec{\pi})^2=(\vec{\pi}\cdot\vec{\pi})I-\hbar e(\vec{\sigma}\cdot\vec{B}).$$

Pauli replaced $\vec{\pi}\cdot\vec{\pi}$ by $(\vec{\sigma}\cdot\vec{\pi})^2$ in equation (1):

$$i\hbar\frac{\partial\psi}{\partial t}=\frac{1}{2m}[(\vec{\pi}\cdot\vec{\pi})I-\hbar e(\vec{\sigma}\cdot\vec{B})]\psi-eV\psi, \quad \psi:\mathbb{R}^3\times\mathbb{R}\to\mathbb{C}^2.$$

This *Pauli equation* describes the spin by virtue of the term

$$\pm\frac{\hbar e}{2m}(\vec{\sigma}\cdot\vec{B}).$$

---

[11] $\pi_1\pi_2\psi=(-i\hbar\frac{\partial}{\partial x}-eA_1)(-i\hbar\frac{\partial}{\partial y}-eA_2)\psi=i\hbar e(\frac{\partial}{\partial x}(A_2\psi)+A_1\frac{\partial\psi}{\partial y});$
$(\pi_1\pi_2-\pi_2\pi_1)\psi=i\hbar e(\frac{\partial A_2}{\partial x}-\frac{\partial A_1}{\partial y})\psi=i\hbar eB_3\psi.$

[12] The notation $\vec{\sigma}\cdot\vec{B}=\begin{pmatrix}0&1\\1&0\end{pmatrix}B_1+\begin{pmatrix}0&-i\\i&0\end{pmatrix}B_2+\begin{pmatrix}1&0\\0&-1\end{pmatrix}B_3$ is not a scalar product of $\vec{\sigma}$ and $\vec{B}$; it is only a temporary abbreviation, used for historical reasons.

[13] $(\pi_1\sigma_1)(\pi_2\sigma_2)+(\pi_2\sigma_2)(\pi_1\sigma_1)=(\pi_1\pi_2-\pi_2\pi_1)(\sigma_1\sigma_2)=(i\hbar eB_3)(i\sigma_3)=-\hbar eB_3\sigma_3.$

1.5.4 Pauli's extra term as a byproduct in $C\ell_3$

In terms of the Clifford algebra $C\ell_3$ Pauli replaced

$$\vec{\pi}\cdot\vec{\pi} = (-i\hbar\nabla - e\vec{A})\cdot(-i\hbar\nabla - e\vec{A}) \quad \text{by} \quad \vec{\pi}^2 = (-i\hbar\nabla - e\vec{A})^2.$$

In so doing Pauli actually invoked the Clifford algebra $C\ell_3$ of the Euclidean space $\mathbb{R}^3$. Pauli observed that, although $e^2(\vec{A}\wedge\vec{A}) = 0$ and $-\hbar^2(\nabla\wedge\nabla) = 0$,

$$\begin{aligned}(\vec{\pi}\wedge\vec{\pi})\psi &= [(-i\hbar\nabla - e\vec{A})\wedge(-i\hbar\nabla - e\vec{A})]\psi \\ &= i\hbar e\nabla\wedge(\vec{A}\psi) + i\hbar e\vec{A}\wedge(\nabla\psi) \\ &= i\hbar e[(\nabla\wedge\vec{A})\psi + (\nabla\psi)\wedge\vec{A} + \vec{A}\wedge(\nabla\psi)] \\ &= i\hbar e[\mathbf{e}_{123}(\nabla\times\vec{A})]\psi = -\hbar e(\nabla\times\vec{A})\psi = -\hbar e\vec{B}\psi,\end{aligned}$$

where in the last row we used the identification $\mathbf{e}_{123} = i$.[14] Thus, Pauli came across the equation

$$i\hbar\frac{\partial\psi}{\partial t} = \frac{1}{2m}(\vec{\pi}\cdot\vec{\pi} + \vec{\pi}\wedge\vec{\pi})\psi - eV\psi$$

where[15]

$$\vec{\pi}\cdot\vec{\pi} + \vec{\pi}\wedge\vec{\pi} = \vec{\pi}\cdot\vec{\pi} - \hbar e\vec{B}$$

where the last term is responsible for the spin. Of course, $\vec{\pi}^2 = \vec{\pi}\cdot\vec{\pi} + \vec{\pi}\wedge\vec{\pi}$. Indeed,

$$\begin{aligned}\vec{\pi}^2\psi &= (-i\hbar\nabla - e\vec{A})(-i\hbar\nabla - e\vec{A})\psi \\ &= -\hbar^2\nabla^2\psi + i\hbar e\nabla(\vec{A}\psi) + i\hbar e\vec{A}\nabla\psi + e^2\vec{A}^2\psi \\ &= -\hbar^2\nabla^2\psi + i\hbar e(\nabla\vec{A})\psi + i\hbar e\dot{\nabla}\vec{A}\dot{\psi} + i\hbar e\vec{A}\nabla\psi + e^2\vec{A}^2\psi \\ &= -\hbar^2\nabla^2\psi + i\hbar e(\nabla\vec{A})\psi - i\hbar e\vec{A}\nabla\psi + 2i\hbar e(\vec{A}\cdot\nabla)\psi + i\hbar e\vec{A}\nabla\psi + e^2\vec{A}^2\psi \\ &= -\hbar^2\nabla^2\psi + i\hbar e(\nabla\cdot\vec{A})\psi + i\hbar e(\nabla\wedge\vec{A})\psi + 2i\hbar e(\vec{A}\cdot\nabla)\psi + e^2\vec{A}^2\psi \\ &= [-\hbar^2\nabla^2 + e^2\vec{A}^2 + i\hbar e(\nabla\cdot\vec{A} + 2\vec{A}\cdot\nabla)]\psi + i\hbar e\,\mathbf{e}_{123}(\nabla\times\vec{A})\psi, \\ &= [-\hbar^2\nabla^2 + e^2\vec{A}^2 + i\hbar e(\nabla\cdot\vec{A} + 2\vec{A}\cdot\nabla)]\psi - \hbar e\vec{B}\psi,\end{aligned}$$

where this time we evaluated $\nabla(\vec{A}\psi)$ by regarding ψ as an element of $C\ell_3$. We see that $\vec{\pi}^2\psi = (\vec{\pi}\cdot\vec{\pi} + \vec{\pi}\wedge\vec{\pi})\psi$.

---

[14] Note that there is no metric involved in $i\hbar e(\nabla\wedge\vec{A})$, before operation on ψ, and no orientation without identification $i = \mathbf{e}_{123}$.

[15] The terms mean $(\vec{\pi}\cdot\vec{\pi})\psi = \vec{\pi}\cdot(\vec{\pi}\psi)$ and $(\vec{\pi}\wedge\vec{\pi})\psi = \vec{\pi}\wedge(\vec{\pi}\psi)$. Compare this to $\vec{\pi}^2\psi = \vec{\pi}(\vec{\pi}\psi)$.

1.6 From column spinors to spinor operators

1.6.1 Square matrix spinors

In the nonrelativistic theory of the spinning electron one considers the complex linear space of the two-component column matrices, the column spinors

$$\psi = \begin{pmatrix} \psi_1 \\ \psi_2 \end{pmatrix} \in \mathbb{C}^2 \quad \text{where} \quad \psi_1, \psi_2 \in \mathbb{C}.$$

An isomorphic complex linear space is obtained if one replaces column spinors by *square matrix spinors*

$$\psi = \begin{pmatrix} \psi_1 & 0 \\ \psi_2 & 0 \end{pmatrix}$$

where only the first column is nonzero. This replacement brings all elements, vectors and spinors, inside a single algebra, $\mathbb{C}(2)$. If we multiply a square matrix spinor ψ on the left by an arbitrary element $u \in \mathbb{C}(2)$, then the result is also of the same type, another square matrix spinor:

$$\begin{pmatrix} u_{11} & u_{12} \\ u_{21} & u_{22} \end{pmatrix} \begin{pmatrix} \psi_1 & 0 \\ \psi_2 & 0 \end{pmatrix} = \begin{pmatrix} \varphi_1 & 0 \\ \varphi_2 & 0 \end{pmatrix}.$$

Such matrices, with only the first column being nonzero, form a subspace S of $\mathbb{C}(2)$. The subspace S is a *left ideal* of $\mathbb{C}(2)$, that is,

$$u\psi \in S \quad \text{for all} \quad u \in \mathbb{C}(2) \quad \text{and} \quad \psi \in S.$$

This left ideal S of $\mathbb{C}(2)$ contains no left ideals other than S itself and the zero ideal $\{0\}$. Such a left ideal is called *minimal* in $\mathbb{C}(2)$. The fact that ψ is in a minimal left ideal, or that only its first column is nonzero, can be expressed as

$$\psi \in \mathbb{C}(2)f = \{Mf \mid M \in \mathbb{C}(2)\}$$

where

$$f = \tfrac{1}{2}(I + \sigma_3) = \begin{pmatrix} 1 & 0 \\ 0 & 0 \end{pmatrix}.$$

The element f is an *idempotent*, that is, $f^2 = f$.

1.6.2 Ideal spinors

We shall regard the correspondences $\mathbf{e}_1 \simeq \sigma_1$, $\mathbf{e}_2 \simeq \sigma_2$, $\mathbf{e}_3 \simeq \sigma_3$ as an identification between $C\ell_3$ and $\mathbb{C}(2)$. This allows us to use the same notations for elements in $C\ell_3$ and its matrix image $\mathbb{C}(2)$.

As a real linear space, S has a basis (f_0, f_1, f_2, f_3) where

$$\begin{aligned} f_0 = f = \tfrac{1}{2}(1+\mathbf{e}_3) &\simeq \begin{pmatrix} 1 & 0 \\ 0 & 0 \end{pmatrix}, \\ f_1 = \mathbf{e}_{23} f = \tfrac{1}{2}(\mathbf{e}_{23}+\mathbf{e}_2) &\simeq \begin{pmatrix} 0 & 0 \\ i & 0 \end{pmatrix}, \\ f_2 = \mathbf{e}_{31} f = \tfrac{1}{2}(\mathbf{e}_{31}-\mathbf{e}_1) &\simeq \begin{pmatrix} 0 & 0 \\ -1 & 0 \end{pmatrix}, \\ f_3 = \mathbf{e}_{12} f = \tfrac{1}{2}(\mathbf{e}_{12}+\mathbf{e}_{123}) &\simeq \begin{pmatrix} i & 0 \\ 0 & 0 \end{pmatrix}. \end{aligned}$$

As a complex linear space, S has a basis (g_1, g_2), where

$$g_1 = f = \tfrac{1}{2}(1+\mathbf{e}_3) \simeq \begin{pmatrix} 1 & 0 \\ 0 & 0 \end{pmatrix}, \quad g_2 = \mathbf{e}_1 f = \tfrac{1}{2}(\mathbf{e}_1+\mathbf{e}_{13}) \simeq \begin{pmatrix} 0 & 0 \\ 1 & 0 \end{pmatrix}.$$

The square matrix spinor in $\mathbb{C}(2)f$

$$\psi = \psi_1 g_1 + \psi_2 g_2 = \begin{pmatrix} \psi_1 & 0 \\ \psi_2 & 0 \end{pmatrix} \quad \text{where} \quad \psi_1, \psi_2 \in \mathbb{C}$$

corresponds, in terms of Clifford algebra, to an *ideal spinor* in $C\ell_3 f$

$$\psi = \psi_1 g_1 + \psi_2 g_2 \quad \text{where} \quad \psi_k = \mathrm{Re}(\psi_k) + \mathrm{Im}(\psi_k)\mathbf{e}_{123}.$$

1.6.3 Complex linear structure on the spinor space

In this section we present a complex structure for ideal spinors in $S = C\ell_3 f$, $f = \frac{1}{2}(1+\mathbf{e}_3)$, without any reference to complex numbers in $\mathbb{C}(2)$. The subset

$$\mathbb{F} = f C\ell_3 f \simeq \left\{ \begin{pmatrix} c & 0 \\ 0 & 0 \end{pmatrix} \middle| \, c \in \mathbb{C} \right\}$$

of $C\ell_3$ is a subring with unity f, that is, $af = fa$ for $a \in \mathbb{F}$. None of the elements of $\mathbb{F}$ is invertible as an element of $C\ell_3$, but for each nonzero $a \in \mathbb{F}$ there is a unique $b \in \mathbb{F}$ such that $ab = f$. Thus, $\mathbb{F}$ is a *division ring* with unity f. [16] As a 2-dimensional real division algebra $\mathbb{F}$ must be isomorphic to $\mathbb{C}$. The isomorphism $\mathbb{F} \simeq \mathbb{C}$ is seen by the equation $f_3^2 = -f_0$ relating the basis elements f_0, f_3 of the real algebra $\mathbb{F}$.

Comment 5 *The multiplication of an element ψ of the real linear space S on the left by an arbitrary even element $u \in C\ell_3^+$, expressed in coordinate form in the basis (f_0, f_1, f_2, f_3),*

$$u\psi = (u_0 + u_1\mathbf{e}_{23} + u_2\mathbf{e}_{31} + u_3\mathbf{e}_{23})(\psi_0 f_0 + \psi_1 f_1 + \psi_2 f_2 + \psi_3 f_3),$$

[16]This follows from the idempotent f being primitive in $C\ell_3$. An idempotent is primitive if it is not the sum of two nonzero idempotents which annihilate each other.

corresponds to the matrix multiplication

$$u\psi \simeq \begin{pmatrix} u_0 & -u_1 & -u_2 & -u_3 \\ u_1 & u_0 & u_3 & -u_2 \\ u_2 & -u_3 & u_0 & u_1 \\ u_3 & u_2 & -u_1 & u_0 \end{pmatrix} \begin{pmatrix} \psi_0 \\ \psi_1 \\ \psi_2 \\ \psi_3 \end{pmatrix}.$$

The square matrices corresponding to the left multiplication by even elements constitute a subring of $\mathbb{R}(4)$; *this subring is an isomorphic image of the quaternion ring* $\mathbb{H}$.

The minimal left ideal

$$S = C\ell_3 f \simeq \left\{ \begin{pmatrix} \psi_1 & 0 \\ \psi_2 & 0 \end{pmatrix} \middle| \ \psi_1, \psi_2 \in \mathbb{C} \right\}$$

has a natural right $\mathbb{F}$-linear structure defined by

$$S \times \mathbb{F} \to S, \ (\psi, \lambda) \to \psi\lambda.$$

We shall provide the minimal left ideal S with this right $\mathbb{F}$-linear structure, and call it a *spinor space.*[17]

The map $C\ell_3 \to \mathrm{End}_{\mathbb{F}} S$, $u \to \tau(u)$, where $\tau(u)$ is defined by the relation $\tau(u)\psi = u\psi$, is a real algebra isomorphism. Employing the basis $f_0, -f_2$ for the $\mathbb{F}$-linear space S, the elements $\tau(\mathbf{e}_1), \tau(\mathbf{e}_2), \tau(\mathbf{e}_3)$ will be represented by the matrices $\sigma_1, \sigma_2, \sigma_3$. In this way the Pauli matrices are reproduced.

1.6.4 The two scalar products of spinors

There is a natural way to introduce scalar products on the spinor space $S = C\ell_3 f \simeq \mathbb{C}(2)f$. First, note that for all $\psi, \varphi \in S$ the product

$$\tilde{\psi}\varphi \simeq \begin{pmatrix} \psi_1^* & \psi_2^* \\ 0 & 0 \end{pmatrix} \begin{pmatrix} \varphi_1 & 0 \\ \varphi_2 & 0 \end{pmatrix} = \begin{pmatrix} \psi_1^*\varphi_1 + \psi_2^*\varphi_2 & 0 \\ 0 & 0 \end{pmatrix}$$

falls in the division ring $\mathbb{F}$ ($z \to z^*$ means complex conjugation). To show that the mapping

$$S \times S \to \mathbb{F}, \ (\psi, \varphi) \to \tilde{\psi}\varphi$$

defines a scalar product we only have to verify that the reversion $\psi \to \tilde{\psi}$ is a right-to-left $\mathbb{F}$-semilinear map. For all $\psi \in S$, $\lambda \in \mathbb{F}$ we have $(\psi\lambda)^{\tilde{}} = \bar{\lambda}\tilde{\psi}$ where the map $\lambda \to \bar{\lambda}$ is an anti-involution of the division algebra $\mathbb{F}$ (actually complex conjugation).

---

[17]Note that multiplying a matrix ψ in $S = \mathbb{C}(2)f$, a left ideal, on the *left* by $\lambda \in \mathbb{F}$ does not result in a left $\mathbb{F}$-linear structure. Also note that in the right $\mathbb{F}$-linear space $S = C\ell_3 f$, $\psi = g_1\psi_1 + g_2\psi_2$, where $\psi_k = \mathrm{Re}(\psi_k) + \mathrm{Im}(\psi_k)\mathbf{e}_{12}$ multiplies from the *right* the basis elements $g_1 = \frac{1}{2}(1 + \mathbf{e}_3)$, $g_2 = \frac{1}{2}(\mathbf{e}_1 + \mathbf{e}_{13})$.

Multiplying a spinor $\psi \in S = C\ell_3 f$ by an element $u \in C\ell_3$ is a right $\mathbb{F}$-linear transformation $S \to S$, $\psi \to u\psi$. The automorphism group of the scalar product is formed by those right $\mathbb{F}$-linear transformations which preserve the scalar product, that is,

$$(u\psi)^{\tilde{}}(u\varphi) = \tilde{\psi}\varphi \quad \text{for all} \quad \psi, \varphi \in S.$$

The automorphism group of the scalar product $\tilde{\psi}\varphi$ is seen to be the group $\{u \in C\ell_3 \mid \tilde{u}u = 1\}$ which is isomorphic to the group of unitary 2×2-matrices,

$$U(2) = \{u \in \mathbb{C}(2) \mid u^\dagger u = I\}.$$

We can also use the Clifford conjugate $u \to \bar{u}$ of $C\ell_3$ to introduce a scalar product for spinors. In this case, the element

$$\bar{\psi}\varphi \simeq \begin{pmatrix} 0 & 0 \\ -\psi_2 & \psi_1 \end{pmatrix} \begin{pmatrix} \varphi_1 & 0 \\ \varphi_2 & 0 \end{pmatrix} = \begin{pmatrix} 0 & 0 \\ \psi_1\varphi_2 - \psi_2\varphi_1 & 0 \end{pmatrix}$$

does not appear in the division ring $\mathbb{F} = fC\ell_3 f$. However, we can find an invertible element $a \in C\ell_3$ so that $a\bar{\psi}\varphi \in \mathbb{F}$, e.g., $a = \mathbf{e}_1$ or $a = \mathbf{e}_{31}$. The map

$$S \times S \to \mathbb{F}, \ (\psi, \varphi) \to a\bar{\psi}\varphi$$

defines a scalar product. Writing

$$J = \begin{pmatrix} 0 & 1 \\ -1 & 0 \end{pmatrix}$$

we find that $a\bar{\psi}\varphi \simeq \tau(\psi)^\top J \tau(\varphi)$. Hence, the automorphism group

$$\{u \in C\ell_3 \mid \bar{u}u = 1\}$$

of the scalar product $a\bar{\psi}\varphi$ is the group of symplectic 2×2-matrices,

$$Sp(2, \mathbb{C}) = \{u \in \mathbb{C}(2) \mid u^\top J u = J\}.$$

1.6.5 Operator spinors in the even subalgebra $C\ell_3^+$

Up until now spinors have been objects which have been operated upon. Next we will replace such passive spinors by more active operators. Instead of ideal spinors $\psi \in C\ell_3 f$, with matrix images

$$\psi = \begin{pmatrix} \psi_1 & 0 \\ \psi_2 & 0 \end{pmatrix} \in \mathbb{C}(2)f,$$

in minimal left ideals we will consider the following even elements:

$$\Psi = 2\,\text{even}(\psi) \simeq \begin{pmatrix} \psi_1 & -\psi_2^* \\ \psi_2 & \psi_1^* \end{pmatrix},$$

also computed as $\Psi = \psi + \hat{\psi} \in C\ell_3^+$ for $\psi \in C\ell_3 f$. Note that this *even spinor* $\Psi \in C\ell_3^+$ carries the same amount of information as the original column spinor $\psi \in \mathbb{C}^2$ or the ideal spinor $\psi \in C\ell_3 f$: no information was lost in projecting out the even part.

Classically, the expectation values of the components of the spin vector $\mathbf{s} = (s_1, s_2, s_3)$ have been determined in terms of the column spinor $\psi \in \mathbb{C}^2$ by computing the following three real numbers:

$$s_1 = \psi^\dagger \sigma_1 \psi, \quad s_2 = \psi^\dagger \sigma_2 \psi, \quad s_3 = \psi^\dagger \sigma_3 \psi.$$

In terms of the ideal spinor $\psi \in C\ell_3 f$ there is an analogous computation of the components:

$$s_1 = 2\langle \tilde{\psi} \mathbf{e}_1 \psi \rangle_0, \quad s_2 = 2\langle \tilde{\psi} \mathbf{e}_2 \psi \rangle_0, \quad s_3 = 2\langle \tilde{\psi} \mathbf{e}_3 \psi \rangle_0.$$

More conveniently, the spin vector $\mathbf{s} = s_1\mathbf{e}_1 + s_2\mathbf{e}_2 + s_3\mathbf{e}_3$ can be computed, benefiting the multivector structure of $C\ell_3$, directly as

$$\mathbf{s} = 2\langle \psi \mathbf{e}_3 \tilde{\psi} \rangle_1, \quad \psi \in C\ell_3 f.$$

It should be emphasized that not only did we get all the components of the spin vector $\mathbf{s}$ at one stroke, but we also got the entity $\mathbf{s}$ as a whole.

In terms of the even spinor $\Psi \in C\ell_3^+$ we may compute $\mathbf{s} = s_1\mathbf{e}_1 + s_2\mathbf{e}_2 + s_3\mathbf{e}_3$ directly as

$$\mathbf{s} = \Psi \mathbf{e}_3 \tilde{\Psi}.$$

More importantly, the spin vector $\mathbf{s}$ is obtained without using any projection operators: this form of $\mathbf{s}$ is applicable in further computations. Since the even spinor Ψ acts here like an operator, it is also referred to as an *operator spinor*.

Remark 6 *The mapping $C\ell_3^+ \to \mathbb{R}^3$, $\Psi \to \Psi \mathbf{e}_3 \tilde{\Psi}$ $(= \Psi \sigma_3 \Psi^\dagger)$ is the KS-transformation (introduced by Kustaanheimo & Stiefel 1965) for spinor regularization of Kepler motion, and its restriction to norm-one spinor operators Ψ satisfying $\Psi\tilde{\Psi} = 1$ (or equivalently $\Psi\Psi^\dagger = I$) results in a Hopf fibration $S^1 \to S^3 \to S^2$ (the matrix $\Psi\sigma_3\Psi^\dagger$ is both unitary and involutory and represents a reflection of the spinor space with axis ψ).*

The above mapping should not be confused with the 'Cartan map', see Cartan 1966 p. 41 and Keller & Rodríguez-Romo 1991 p. 1591. A 'Cartan map' $S \times S \to C\ell_3$, $(\psi, \varphi) \to 2\psi \mathbf{e}_1 \bar{\varphi}$, where $S = C\ell_3 f$, sends a pair of ideal spinors to a complex paravector $x_0 + \mathbf{x}$ where

$$x_0 = -(\psi_1\varphi_2 - \psi_2\varphi_1),$$
$$\mathbf{x} = \mathbf{e}_1(\psi_1\varphi_1 - \psi_2\varphi_2) + \mathbf{e}_2(i(\psi_1\varphi_1 + \psi_2\varphi_2)) + \mathbf{e}_3(-(\psi_1\varphi_2 + \psi_2\varphi_1))$$

(complex means coefficients in $\{x + y\mathbf{e}_{123} \mid x, y \in \mathbb{R}\}$). When $\psi = \varphi$, the complex vector $\mathbf{x}$ is null, $\mathbf{x}^2 = 0$.

Note also that $\operatorname{trace}(\psi\psi^\dagger) = 2\langle\psi\tilde{\psi}\rangle_0 = \Psi\bar{\Psi}$ which equals $\Psi\bar{\Psi} = \det(\Psi)$.

In operator form the Pauli equation

$$\boxed{i\hbar\frac{\partial\Psi}{\partial t} = \frac{1}{2m}(\vec{\pi}\cdot\vec{\pi})\Psi - \frac{\hbar e}{2m}\vec{B}\Psi\mathbf{e}_3 - eV\Psi, \quad \Psi : \mathbb{R}^3 \times \mathbb{R} \to C\ell_3^+}$$

shows explicitly the quantization direction $\mathbf{e}_3$ of the spin. The occurrence of $\mathbf{e}_3$ is due to the injection $\mathbb{C}^2 \to \mathbb{C}(2)f$, $f = \frac{1}{2}(I + \sigma_3)$; made explicit in the mapping $C\ell_3 f \to C\ell_3^+$, $2\,\mathrm{even}(\vec{B}\psi) = \vec{B}\Psi\mathbf{e}_3$ with a fixed $\vec{B} \in \mathbb{R}^3$. If we rotate the system 90° about the y-axis, counterclockwise as seen from the positive y-axis, then vectors and operator spinors transform to

$$\vec{B}' = u\vec{B}u^{-1} \quad \text{and} \quad \Psi' = u\Psi \quad \text{where} \quad u = \exp(\frac{\pi}{4}\mathbf{e}_{13}),$$

and the Pauli equation transforms to

$$i\hbar\frac{\partial\Psi'}{\partial t} = \frac{1}{2m}(\vec{\pi}'\cdot\vec{\pi}')\Psi' - \frac{\hbar e}{2m}\vec{B}'\Psi'\mathbf{e}_3 - eV\Psi'.$$

If this equation is multiplied on the right by u^{-1}, then $\mathbf{e}_3$ goes to $\mathbf{e}_1 = u\mathbf{e}_3u^{-1}$, and the equation looks like

$$i\hbar\frac{\partial\Psi''}{\partial t} = \frac{1}{2m}(\vec{\pi}'\cdot\vec{\pi}')\Psi'' - \frac{\hbar e}{2m}\vec{B}'\Psi''\mathbf{e}_1 - eV\Psi'',$$

where $\Psi'' = u\Psi u^{-1}$. Both the transformation laws give the same values for observables, that is, $\Psi'\mathbf{e}_3\tilde{\Psi}' = \Psi''\mathbf{e}_1\tilde{\Psi}''$.

1.7 In 4D: Clifford algebra $C\ell_4$ of $\mathbb{R}^4$

The Clifford algebra $C\ell_4$ of $\mathbb{R}^4$ with an orthonormal basis $(\mathbf{e}_1, \mathbf{e}_2, \mathbf{e}_3, \mathbf{e}_4)$ is generated by the relations

$$\mathbf{e}_1^2 = \mathbf{e}_2^2 = \mathbf{e}_3^2 = \mathbf{e}_4^2 = 1 \quad \text{and} \quad \mathbf{e}_i\mathbf{e}_j = -\mathbf{e}_j\mathbf{e}_i \quad \text{for} \quad i \neq j.$$

It is a 16-dimensional algebra with basis consisting of

scalar	1
vectors	$\mathbf{e}_1, \mathbf{e}_2, \mathbf{e}_3, \mathbf{e}_4$
bivectors	$\mathbf{e}_{12}, \mathbf{e}_{13}, \mathbf{e}_{14}, \mathbf{e}_{23}, \mathbf{e}_{24}, \mathbf{e}_{34}$
3-vectors	$\mathbf{e}_{123}, \mathbf{e}_{124}, \mathbf{e}_{134}, \mathbf{e}_{234}$
volume element	$\mathbf{e}_{1234}$

where $\mathbf{e}_{ij} = \mathbf{e}_i\mathbf{e}_j$ for $i \neq j$ and $\mathbf{e}_{1234} = \mathbf{e}_1\mathbf{e}_2\mathbf{e}_3\mathbf{e}_4$.

The Clifford algebra $C\ell_4$ is isomorphic to the real algebra of 2×2-matrices $\mathbb{H}(2)$ with quaternions as entries,

$$\mathbf{e}_1 \simeq \begin{pmatrix} 0 & -i \\ i & 0 \end{pmatrix}, \mathbf{e}_2 \simeq \begin{pmatrix} 0 & -j \\ j & 0 \end{pmatrix}, \mathbf{e}_3 \simeq \begin{pmatrix} 0 & -k \\ k & 0 \end{pmatrix}, \mathbf{e}_4 \simeq \begin{pmatrix} 0 & 1 \\ 1 & 0 \end{pmatrix}.$$

The even subalgebra $C\ell_4^0$ is isomorphic to the direct sum $\mathbb{H} \oplus \mathbb{H}$, where addition and multiplication is defined componentwise.

1.7.1 Bivectors in $\bigwedge^2 \mathbb{R}^4 \subset C\ell_4$

The essential difference between 3D and 4D spaces is that bivectors are no longer products of two vectors. Instead, bivectors are sums of products of two vectors in $\mathbb{R}^4$. In the 3-dimensional space $\mathbb{R}^3$ there are only *simple* bivectors, that is, all the bivectors represent a plane. In the 4-dimensional space $\mathbb{R}^4$ this is not the case any more.

Example 7 *The bivector* $\mathbf{B} = \mathbf{e}_{12} + \mathbf{e}_{34} \in \bigwedge^2 \mathbb{R}^4$ *is not simple. For all simple elements the square is real, but* $\mathbf{B}^2 = -2 + 2\mathbf{e}_{1234} \notin \mathbb{R}$.

If the square of a bivector is real, then it is simple.[18]

Usually a bivector in $\bigwedge^2 \mathbb{R}^4$ can be uniquely written as a sum of two simple bivectors, which represent completely orthogonal planes. There is an exception to this uniqueness, crucial to the study of four dimensions: If the simple components of a bivector have equal squares, that is equal norms, then the decomposition to a sum of simple components is not unique.

Example 8 *The bivector* $\mathbf{e}_1\mathbf{e}_2 + \mathbf{e}_3\mathbf{e}_4$ *can also be decomposed into a sum of two completely orthogonal bivectors as follows:*

$$\mathbf{e}_1\mathbf{e}_2 + \mathbf{e}_3\mathbf{e}_4 = \tfrac{1}{2}(\mathbf{e}_1 + \mathbf{e}_3)(\mathbf{e}_2 + \mathbf{e}_4) + \tfrac{1}{2}(\mathbf{e}_1 - \mathbf{e}_3)(\mathbf{e}_2 - \mathbf{e}_4).$$

1.8 Clifford algebra of Minkowski spacetime

The Clifford algebra $C\ell_{3,1}$ of $\mathbb{R}^{3,1}$ with an orthonormal basis $(\mathbf{e}_1, \mathbf{e}_2, \mathbf{e}_3, \mathbf{e}_4)$ is generated by the relations

$$\mathbf{e}_1^2 = \mathbf{e}_2^2 = \mathbf{e}_3^2 = 1, \quad \mathbf{e}_4^2 = -1 \quad \text{and} \quad \mathbf{e}_i\mathbf{e}_j = -\mathbf{e}_j\mathbf{e}_i \quad \text{for} \quad i \neq j.$$

It is a 16-dimensional algebra isomorphic to the real algebra of 4×4-matrices $\mathbb{R}(4)$.

---

[18] Even when the square of a 3-vector is real, this does not mean that it is simple (or decomposable). For instance, $\mathbf{V} = \mathbf{e}_{123} + \mathbf{e}_{456} \in \bigwedge^3 \mathbb{R}^6$ is not simple although its square is real. This can be seen by computing $\mathbf{V}\mathbf{e}_i\mathbf{V}^{-1}$, $i = 1, 2, \ldots, 6$, (note that $\mathbf{V}^2 = -2$ hence $\mathbf{V}^{-1} = -\frac{1}{2}\mathbf{V}$) and observing that none of them is a vector: all are 5-vectors. [This footnote has been corrected by the Editors.]

The even subalgebra $C\ell_{3,1}^0$ of $C\ell_{3,1}$ is isomorphic to $\mathbb{C}(2)$. The even elements $u \in C\ell_{3,1}^0$ of unit norm $u\tilde{u} = 1$ form the subgroup $\mathbf{Spin}(3,1)$, which is a 2-fold covering group of the Lorentz group $SO(3,1)$.

1.9 The exterior algebra and contractions

The exterior algebra $\bigwedge V$ of a linear space V, with basis $(\mathbf{e}_1, \mathbf{e}_2, \ldots, \mathbf{e}_n)$, has a basis consisting of

$$\begin{array}{rl} 1 & \text{scalar} \\ \mathbf{e}_1, \mathbf{e}_2, \ldots, \mathbf{e}_n & \text{vectors} \\ \mathbf{e}_1 \wedge \mathbf{e}_2,\ \mathbf{e}_1 \wedge \mathbf{e}_3,\ \ldots,\ \mathbf{e}_{n-1} \wedge \mathbf{e}_n & \text{bivectors} \\ \vdots & \vdots \\ \mathbf{e}_1 \wedge \mathbf{e}_2 \wedge \ldots \wedge \mathbf{e}_n & \text{volume element.} \end{array}$$

The multiplication rules are

$$\mathbf{e}_i \wedge \mathbf{e}_j = -\mathbf{e}_j \wedge \mathbf{e}_i$$

together with associativity and the unity 1. A scalar product on V can be extended to the homogeneous parts $\bigwedge^k V$ by

$$<\mathbf{x}_1 \wedge \mathbf{x}_2 \wedge \ldots \wedge \mathbf{x}_k, \mathbf{y}_1 \wedge \mathbf{y}_2 \wedge \ldots \wedge \mathbf{y}_k> = \det(\mathbf{x}_i \cdot \mathbf{y}_j)$$

and further by orthogonality to all of $\bigwedge V$. This scalar valued product can be used to define the **contraction** $\bigwedge V \times \bigwedge V \to \bigwedge V$, $(u, v) \to u \,\lrcorner\, v$ by

$$<u \,\lrcorner\, v, w> = <v, \tilde{u} \wedge w> \quad \text{for all} \quad w \in \bigwedge V.$$

The contraction could also be defined by its characteristic properties

$$\begin{aligned} &\mathbf{x} \,\lrcorner\, \mathbf{y} = \mathbf{x} \cdot \mathbf{y}, \\ &\mathbf{x} \,\lrcorner\, (u \wedge v) = (\mathbf{x} \,\lrcorner\, u) \wedge v + \hat{u} \wedge (\mathbf{x} \wedge v), \\ &(u \wedge v) \,\lrcorner\, w = u \,\lrcorner\, (v \,\lrcorner\, w) \end{aligned}$$

which hold for all $\mathbf{x}, \mathbf{y} \in V$ and $u, v, w \in \bigwedge V$.

The contraction could also be introduced via the Clifford product as

$$u \,\lrcorner\, v = (u \wedge (v\mathbf{e}_{12\ldots n}))\mathbf{e}_{12\ldots n}^{-1}.$$

This can be proved by observing $<u\mathbf{e}_{12\ldots n}, w> = <\mathbf{e}_{12\ldots n}, \tilde{u} \wedge w>$ and computing

$$\begin{aligned} &<(u \wedge (v\mathbf{e}_{12\ldots n}))\mathbf{e}_{12\ldots n}^{-1}, w> \\ &\quad = <\mathbf{e}_{12\ldots n}^{-1}, (u \wedge (v\mathbf{e}_{12\ldots}))^{\sim} \wedge w> \\ &\quad = <\mathbf{e}_{12\ldots n}^{-1}, (v\mathbf{e}_{12\ldots n})^{\sim} \wedge \tilde{u} \wedge w> \\ &\quad = <v\mathbf{e}_{12\ldots n}\mathbf{e}_{12\ldots}^{-1}, \tilde{u} \wedge w> = <v, \tilde{u} \wedge w> = <u \,\lrcorner\, v, w>. \end{aligned}$$

Geometrically, for decomposable homogeneous multivectors $\mathbf{A}$ and $\mathbf{B}$, not orthogonal to each other, the contraction $\mathbf{A} \rfloor \mathbf{B}$ is the largest subspace of $\mathbf{B}$ orthogonal to $\mathbf{A}$.[19]

1.10 The Grassmann–Cayley algebra and shuffle products

In the Grassmann–Cayley algebra, we have the exterior product and the shuffle product defined (in terms of the Clifford product) by

$$u \vee v = ((u\mathbf{e}_{12\ldots n}^{-1}) \wedge (v\mathbf{e}_{12\ldots n}^{-1}))\mathbf{e}_{12\ldots n}.$$

The shuffle product is associative and metric independent, but depends on orientation, while its unity is $\mathbf{e}_{12\ldots n}$. The shuffle product is like the exterior product upside down, lowering degrees. The shuffle inverse is given by

$$u^{\vee(-1)} = ((u\mathbf{e}_{12\ldots n}^{-1})^{\wedge(-1)})\mathbf{e}_{12\ldots n}.$$

Under a linear transformation L of V the shuffle product transforms as

$$\bigwedge L(u \vee v) = \bigwedge L(u) \vee \bigwedge L(v) / \det(L).$$

In geometry, joins of subspaces are dealt with the exterior product and meets with the shuffle product. It should be noted that the definition of the shuffle product in terms of the Clifford product makes the Grassmann–Cayley algebra unnecessary.

1.11 Alternative definitions of the Clifford algebra

We work over a field $\mathbb{F}$ (although the generalization to commutative rings is apparent). Let us denote by $\mathbb{F}$-alg the category of associative algebras over $\mathbb{F}$: its objects are unital associative $\mathbb{F}$-algebras and its morphisms are $\mathbb{F}$-algebra homomorphisms preserving the unit elements. Let V be a vector space over $\mathbb{F}$ equipped with a quadratic form Q, i.e., $Q : V \to \mathbb{F}$ satisfies

(i) $Q(\lambda \mathbf{x}) = \lambda^2 Q(\mathbf{x})$ for all $\lambda \in \mathbb{F}, \mathbf{x} \in V$;
(ii) the map $(\mathbf{x}, \mathbf{y}) \to Q(\mathbf{x}+\mathbf{y}) - Q(\mathbf{x}) - Q(\mathbf{y})$ is $\mathbb{F}$-bilinear.

---

[19] This sentence needs an explanation. Recall that the exterior product of k linearly independent vectors of V is called a *decomposable (homogeneous)* or *simple* multivector of degree k, and that the subspace of V spanned by these k vectors is called the *subspace* of the decomposable multivector. Scalars are considered as decomposable elements of degree 0, with the associated subspace reduced to $\{0\}$. The contraction $\mathbf{A} \rfloor \mathbf{B}$ of two decomposable multivectors $\mathbf{A}$ and $\mathbf{B}$ is still a decomposable multivector, unless it vanishes. Geometrically, if S and T are the subspaces of $\mathbf{A}$ and $\mathbf{B}$, respectively, the subspace of $\mathbf{A} \rfloor \mathbf{B}$ is the largest subspace of T orthogonal to S provided that no vector of S (except 0) is orthogonal to T. In fact, existence of such a nonzero vector is a sufficient and necessary condition to assure that $\mathbf{A} \rfloor \mathbf{B} = 0$. [*Editors*]

Definition 9 (of category Q-alg)*: The objects are pairs (A, f), where A is an object of $\mathbb{F}$-alg and $f : V \to A$ is an $\mathbb{F}$-linear map with the property that $f(\mathbf{x})^2 = Q(\mathbf{x})1_A$ for all $\mathbf{x} \in V$ (1_A being the unit element of A); the morphisms from (A, f) to (B, g) are $\mathbb{F}$-alg morphisms h from A to B with the property that $g = h \circ f$.*

The initial object in Q-alg is called the **Clifford algebra** $C\ell(Q)$ of Q. The terminal object of Q-alg is $(\{0\}, f)$, where $\{0\}$ is the trivial ring and f is the zero map.

Definition 10 (of category Q-geoalg)*: The objects A are objects in $\mathbb{F}$-alg such that V is a subspace of A and $\mathbf{x}^2 = Q(\mathbf{x})$ for all $\mathbf{x} \in V$; the morphisms from A to B are $\mathbb{F}$-alg morphisms h such that $h|V$ is an isometry of Q on V.*

The initial object in Q-geoalg is called the **Clifford algebra** $C\ell(Q)$ of Q. This category Q-geoalg has no terminal object.

Definition 11 (of Clifford algebra by generators and relations)*: An associative algebra over $\mathbb{F}$ with unity is the Clifford algebra $C\ell(Q)$ of a nondegenerate Q if it contains V and $\mathbb{F}$ as distinct subspaces so that*

1) $\mathbf{x}^2 = Q(\mathbf{x})$ for all $\mathbf{x}$ in V,
2) V generates $C\ell(Q)$ as an algebra over $\mathbb{F}$,
3) $C\ell(Q)$ is not generated by any proper subspace of V.

Let B be a bilinear form, not necessary symmetric, on V. The contraction is a bilinear product $\bigwedge V \times \bigwedge V \to \bigwedge V$, $(u, v) \to u \,\lrcorner\, v$ determined by

1) $\mathbf{x} \,\lrcorner\, \mathbf{y} = B(\mathbf{x}, \mathbf{y})$,
2) $\mathbf{x} \,\lrcorner\, (u \wedge v) = (\mathbf{x} \,\lrcorner\, u) \wedge v + \hat{u} \wedge (\mathbf{x} \,\lrcorner\, v)$,
3) $(u \wedge v) \,\lrcorner\, w = u \,\lrcorner\, (v \,\lrcorner\, w)$,

for all $\mathbf{x}, \mathbf{y}$ in V and u, v, w in $\bigwedge V$.

Definition 12 (of Clifford algebra $C\ell(B)$, B nondegenerate, but not necessarily symmetric)*: Introduce the Clifford product $\mathbf{x}u = \mathbf{x} \,\lrcorner\, u + \mathbf{x} \wedge u$ for all $\mathbf{x}$ in V and u in $\bigwedge V$ and extend the Clifford product by linearity and associativity to all of $\bigwedge V$. The Clifford algebra $C\ell(B)$ of B is the pair ($\bigwedge V$, the Clifford product).*

1.11.1 Discussion on the definitions

The category Q-alg is bigger than Q-geoalg, which consists of those objects of Q-alg, which contain V.

Example 13 *1. Consider a 2-dimensional real linear space V with a vanishing quadratic form $Q = 0$. Then $C\ell(Q) = \bigwedge V$, the exterior algebra of V. Recall that $\bigwedge V = \mathbb{R} \oplus V \oplus \bigwedge^2 V$. The exterior algebra $\bigwedge V$ has the following two-sided* *ideals*[20] *of dimension respectively* $0,\ 1,\ 2,\ 3,\ 4:$

$$\{0\}, \quad \bigwedge^2 V, \quad L \oplus \bigwedge^2 V, \quad V \oplus \bigwedge^2 V, \quad \bigwedge V;$$

the third one is the ideal generated by L, *that is any line through the origin in* V. *The corresponding factor algebras, of dimension* $4,\ 3,\ 2,\ 1,\ 0$, *are*

$$\bigwedge V, \quad \mathbb{R} \oplus V, \quad \mathbb{R} \oplus L', \quad \mathbb{R}, \quad \{0\},$$

where L' *is any line supplementary to* L *in* V, *so that there exists a surjective algebra morphism from* $\bigwedge V$ *onto* $\mathbb{R} \oplus L'$ *with kernel* $L \oplus \bigwedge^2 V$; *this algebra morphism maps every element of* V *to its parallel projection onto* L' *with respect to* L. *The lattice of factor algebras is*

$$\bigwedge V \to \mathbb{R} \oplus V \to \mathbb{R} \oplus L' \to \mathbb{R} \to \{0\};$$

the third object $\mathbb{R} \oplus L'$ *is one among infinitely many analogous factor algebras. The category Q-alg contains all these five algebras, whereas Q-geoalg contains only the first two.*

Example 14 *Consider* $C\ell_{0,3} \simeq \mathbb{H} \oplus \mathbb{H}$, *which contains two nontrivial ideals, both isomorphic to* $\mathbb{H}$, *but only in the category* $\mathbb{R}$*-alg, not in the category Q-alg. The ideals* $\frac{1}{2}(1 \pm \mathbf{e}_{123})C\ell_{0,3}$ *have unit elements* $\frac{1}{2}(1 \pm \mathbf{e}_{123})$ *(denoted by 1) and three other basis elements*

$$\begin{aligned} i' &\simeq \tfrac{1}{2}(1+\mathbf{e}_{123})\mathbf{e}_1, & j' &\simeq \tfrac{1}{2}(1+\mathbf{e}_{123})\mathbf{e}_2, & k' &\simeq \tfrac{1}{2}(1+\mathbf{e}_{123})\mathbf{e}_3, \\ i &\simeq \tfrac{1}{2}(1-\mathbf{e}_{123})\mathbf{e}_1, & j &\simeq \tfrac{1}{2}(1-\mathbf{e}_{123})\mathbf{e}_2, & k &\simeq \tfrac{1}{2}(1-\mathbf{e}_{123})\mathbf{e}_3. \end{aligned}$$

In the former ideal $i'j'k' = 1$ *and in the latter* $ijk = -1$. *The algebras* $\mathbb{H}$ *and* $\mathbb{H}'$ *with bases* $(1, i, j, k)$ *and* $(1, i', j', k')$ *are not isomorphic in the category Q-alg (the identity mapping on* V *does not extend to an Q-alg isomorphism from* $\mathbb{H}$ *to* $\mathbb{H}'$*). The lattice of ideals is*

$$\begin{array}{ccc} & C\ell_{0,3} & \\ \swarrow & & \searrow \\ \mathbb{H} & & \mathbb{H}' \\ \searrow & & \swarrow \\ & \{0\} & \end{array}$$

where the last object does not belong to Q-geoalg.

The Clifford algebra $C\ell(B)$ is just $C\ell(Q)$, where $Q(\mathbf{x}) = B(\mathbf{x}, \mathbf{x})$ for all $\mathbf{x}$ in V. However, $C\ell(Q)$ has the same multivector structure as $C\ell(B)$ only for a symmetric B. In particular, the bivector spaces (and all multivector spaces of

---

[20] This list of ideals was completed by the Editors by adding an extra 2-dimensional ideal $L \oplus \bigwedge^2 V$. The subsequent list of factor algebras has been completed accordingly.

higher grade) are different in $C\ell(Q)$ and $C\ell(B)$ for a nonsymmetric B. A consequence is that the reversions are different in $C\ell(Q)$ and $C\ell(B)$, if defined so that their invariant spaces are of homogeneous grade. Alternatively, and more appropriately, the reversion should be defined as an anti-involution fixing scalars and vectors.

The Clifford algebra $C\ell(B)$ of an antisymmetric bilinear form B is just the exterior algebra $\bigwedge V$. One should not confuse such a $C\ell(B)$ with the symplectic Clifford algebras (also of antisymmetric bilinear forms). In characteristic 0, the symplectic Clifford algebra of $B = 0$ is just the algebra of symmetric tensors, which is of infinite dimension.

All the above Clifford algebras are algebras of quadratic forms and bilinear forms. One could also study Clifford algebras linearizing polynomial forms f of degrees $d > 2$ on n-dimensional spaces V. For $n \geq 2$ and $d \geq 3$, such Clifford algebras $C\ell(V, f)$ are infinite dimensional.

1.12 REFERENCES

[1] R. Abłamowicz, P. Lounesto, On Clifford algebras of a bilinear form with an antisymmetric part, in *Clifford Algebras with Numeric and Symbolic Computations*. Eds. R. Abłamowicz, P. Lounesto, J.M. Parra, Birkhäuser, Boston, 1996, pp. 167–188.

[2] R. Abłamowicz, Clifford algebra computations with Maple, *Proc. Clifford (Geometric) Algebras*, Banff, Alberta Canada, 1995. Ed. W. E. Baylis, Birkhäuser, Boston, 1996, pp. 463–501.

[3] J. Helmstetter, Algèbres de Clifford et algèbres de Weyl, *Cahiers Math.* **25**, Montpellier, 1982.

[4] P. Lounesto, *Clifford Algebras and Spinors*, Cambridge University Press, Cambridge, 1997, 2001.

[5] W. Pauli, Zur Quantenmekanik des magnetischen Elektrons. *Z. Physik* **43** (1928), 601–623.

[6] G. Sobczyk, The generalized spectral decomposition of a linear operator, *The College Mathematics Journal* **28**:1 (1997), 27–38.

[7] D. Sturmfels, *Algorithms af Invariant Theory*, Springer, Vienna, 1993.

Received: January 31, 2002; Revised: March 15, 2003 (Editors).

2

Mathematical Structure of Clifford Algebras

Ian Porteous

ABSTRACT The first part of this chapter is mainly concerned with the construction of Clifford algebras for real and complex nondegenerate quadratic spaces of arbitrary rank and signature, these being presented as matrix algebras over $\mathbb{R}$, $\mathbb{C}$, $\mathbb{H}$, ${}^2\mathbb{R}$, ${}^2\mathbb{C}$ or ${}^2\mathbb{H}$. In each case the algebra has an anti-involution known as conjugation, and the second part is concerned with determining products, or equivalently correlations, on the spinor space for which the induced adjoint anti-involution on the matrix algebra is conjugation. Applications of the classification are to the description of the Spin groups and conformal groups for quadratic spaces of low dimension.

2.1 Clifford algebras

2.1.1 Construction of the algebras $C\ell_{p,q}$

We have seen in the previous lecture how well adapted the algebra of quaternions $\mathbb{H}$ is to the study of the groups $SO(3)$ and $SO(4)$. The center of interest is a finite-dimensional vector space X over the real field $\mathbb{R}$, furnished with a quadratic form, in the one case $\mathbb{R}^3$ and in the other case $\mathbb{R}^4$. In either case the real associative algebra of quaternions $\mathbb{H}$ contains both $\mathbb{R}$ and X as linear subspaces, there being an anti-involution, namely conjugation, of the algebra, such that, for all $x \in X$,

$$\overline{x}\,x = x^{(2)} = x \cdot x.$$

In the former case, when $\mathbb{R}^3$ is identified with the subspace of pure quaternions, this formula can also be written in the simpler form

$$x^2 = -x^{(2)} = -x \cdot x.$$

In an analogous, but more elementary way, the real algebra of complex numbers $\mathbb{C}$ may be used in the study of the group $SO(2)$.

This lecture was presented at "Lecture Series on Clifford Algebras and their Applications", May 18 and 19, 2002, as part of the 6th International Conference on Clifford Algebras and their Applications in Mathematical Physics, Cookeville, TN, May 20–25, 2002.

AMS Subject Classification: 15A66, 17B37, 20C30, 81R25.

Keywords: Clifford algebra, spinor, Radon–Hurwitz number, correlation, conjugation, conformal transformation, Vahlen matrix.

Our aim is to put these rather special cases into a wider context. To keep the algebra simple, the emphasis is laid at first on generalizing the second of the two displayed formulae.

Let X be a finite-dimensional real *quadratic space*, that is a finite-dimensional vector space assigned a possibly degenerate quadratic form. As follows from one of the definitions in Lecture 1, the *Clifford Algebra* for X is a real associative algebra, $C\ell(X)$, with unit element 1, containing isomorphic copies of $\mathbb{R}$ and X as linear subspaces, and generated as a real associative algebra by them in such a way that, for all $x \in X$, $x^2 = -x^{(2)}$. The dimension of the algebra is $2^{\dim X}$.

Superficially, it would seem to be simpler to arrange things so that for all $x \in X$, $x^2 = x^{(2)}$, as was the case in Lecture 1. But there is a good reason for adopting the 'negative' convention, as we shall see in a moment. To simplify notations, in practice $\mathbb{R}$ and X are identified with their copies in $C\ell(X)$.

The Clifford algebra has an anti-involution, known as *conjugation*

$$C\ell(X) \to C\ell(X);\ a \mapsto a^-$$

such that, for all $x \in X$, $x^- = -x$. Then, for all $x \in X$, $x \cdot x = x^- x$.

The vector space $\mathbb{R} \oplus X$ is called the space of *paravectors* of the Clifford algebra. It becomes a real quadratic space on being assigned the quadratic form $\lambda + x \mapsto \lambda^2 + x \cdot x = \lambda^2 - x^2 = (\lambda - x)(\lambda + x) = (\lambda + x)^-(\lambda + x)$. Here is where we get a positive payoff for adhering to the negative convention! In particular, if X is a positive-definite quadratic space of dimension n then the space of paravectors $\mathbb{R} \oplus X$ is a positive-definite quadratic space of dimension $n+1$.

One almost invariably works with an orthonormal basis for the vector space X. We denote by $\mathbb{R}^{p,q,r}$ the real vector space $\mathbb{R}^{p+q+r}$ such that the first p vectors of the standard basis have scalar square -1, the next q have scalar square $+1$, and the final r have scalar square 0, these basis vectors being mutually orthogonal. The Clifford algebra generated by these basis vectors is then denoted by $C\ell_{p,q,r}$, the first p basis vectors having square $+1$, the next q vectors having square -1 and the final r basis vectors having square 0. In the Clifford algebra the basis vectors *anticommute*, the equation

$$0 = x \cdot y = \tfrac{1}{2}((x+y) \cdot (x+y) - x \cdot x - y \cdot y)$$

becoming in the Clifford algebra

$$0 = -xy - yx = -(x+y)^2 + x^2 + y^2.$$

The quadratic space $\mathbb{R}^{p,q,0}$ is also denoted by $\mathbb{R}^{p,q}$, the associated Clifford algebra being denoted by $C\ell_{p,q}$. In the same spirit, the quadratic space $\mathbb{R}^{0,n}$ is also denoted by $\mathbb{R}^n$, the associated Clifford algebra being denoted by $C\ell_{0,n}$ (not $C\ell_n$, to avoid confusion with the conventions of Lecture 1). A quadratic space is said to be *neutral* if it is nondegenerate, that is $r = 0$ and the number of positive squares is equal to the number of negative squares, that is $p = q$.

It is generally helpful to have an explicit construction of the Clifford algebras for quadratic spaces of arbitrary rank, signature and nullity.

Proposition 1 *Let W be a linear subspace of a finite-dimensional real quadratic space X, assigned the induced quadratic form with Clifford algebra $C\ell(X)$. Then the subalgebra of $C\ell(X)$ generated by W may be identified with $C\ell(W)$.*

By this proposition the existence of a Clifford algebra for an arbitrary n-dimensional quadratic space X is implied by the existence of a Clifford algebra for the neutral nondegenerate space $\mathbb{R}^{n,n}$. Such an algebra is constructed below.

The starting point for all our work is the observation that the three matrices

$$\begin{pmatrix} 1 & 0 \\ 0 & -1 \end{pmatrix}, \quad \begin{pmatrix} 0 & 1 \\ 1 & 0 \end{pmatrix} \quad \text{and} \quad \begin{pmatrix} 0 & -1 \\ 1 & 0 \end{pmatrix}$$

in $\mathbb{R}(2)$ are mutually anticommutative and satisfy the equations

$$\begin{pmatrix} 1 & 0 \\ 0 & -1 \end{pmatrix}^2 = 1, \quad \begin{pmatrix} 0 & 1 \\ 1 & 0 \end{pmatrix}^2 = 1 \quad \text{and} \quad \begin{pmatrix} 0 & -1 \\ 1 & 0 \end{pmatrix}^2 = -1,$$

the product of any two of them being plus or minus the third.

We select the second and third of these to represent an orthonormal set of vectors of the quadratic space $\mathbb{R}^{1,1}$, embedded as a linear subspace of the matrix algebra $\mathbb{R}(2)$ of 2×2 matrices over the field of real numbers $\mathbb{R}$. These vectors generate the algebra, the identity matrix and the three matrices together forming a basis for the algebra as a four-dimensional linear space. With these choices, the algebra represents the real Clifford algebra $C\ell_{1,1}$ of the quadratic space $\mathbb{R}^{1,1}$.

It is now very easy to construct a representation of the Clifford algebra $C\ell_{2,2}$ of the quadratic space $\mathbb{R}^{2,2}$. Let a and b be the generators of $C\ell_{1,1}$. The algebra we require is none other than $\mathbb{R}(4)$, thought of as $\mathbb{R}(2)(2)$, taking as a set of four mutually anticommuting matrices the matrices

$$\begin{pmatrix} a & 0 \\ 0 & -a \end{pmatrix}, \quad \begin{pmatrix} b & 0 \\ 0 & -b \end{pmatrix}, \quad \begin{pmatrix} 0 & -1 \\ 1 & 0 \end{pmatrix} \quad \text{and} \quad \begin{pmatrix} 0 & 1 \\ 1 & 0 \end{pmatrix}$$

with 0 the 2×2 matrix $\left(\begin{smallmatrix} 0 & 0 \\ 0 & 0 \end{smallmatrix}\right)$ and 1 the 2×2 matrix $\left(\begin{smallmatrix} 1 & 0 \\ 0 & 1 \end{smallmatrix}\right)$. Clearly two of these have square -1 and two have square $+1$, and together they generate the sixteen-dimensional real algebra $\mathbb{R}(4)$. Continuing in the same way, one proves by induction that the real algebra $\mathbb{R}(2^n)$ represents the real Clifford algebra $C\ell_{n,n}$ of the real quadratic space $\mathbb{R}^{n,n}$.

Now consider the case of the quadratic space $\mathbb{R}^{p,q}$ with an orthonormal basis, the first p vectors of which have scalar square -1, so Clifford square $+1$, while the remaining q have scalar square 1, so Clifford square -1. Clearly its Clifford algebra $C\ell_{p,q}$ is representable as a subalgebra of the real algebra $C\ell_{n,n}$ where n is the larger of p and q.

We can even construct in this way Clifford algebras for degenerate quadratic spaces, where some of the generating vectors have square 0. For in any of the algebras constructed so far if one chooses one basis vector a, say of square -1, and another b, say of square $+1$, then because a and b anticommute, $a + b$ and

$a - b$ both have square 0. In particular the *Grassmann algebra* of the vector space $\mathbb{R}^n$, all of whose basis vectors anticommute and have square 0, is representable as a real subalgebra of the real algebra $C\ell_{n,n} \cong \mathbb{R}(2^n)$.

The main theorem of this section is that for any p, q the Clifford algebra $C\ell_{p,q}$ of the *nondegenerate* real quadratic space $\mathbb{R}^{p,q}$ is representable as a full matrix algebra with entries in one of the real algebras $\mathbb{R}, \mathbb{C}, \mathbb{H}, {}^2\mathbb{R}$ or ${}^2\mathbb{H}$. ${}^2\mathbb{R}$ is shorthand notation for the real algebra of diagonal 2×2 real matrices, ${}^2\mathbb{H}$ is shorthand notation for the real algebra of diagonal 2×2 quaternionic matrices, and ${}^2\mathbb{C}$ is a shorthand notation for the algebra of diagonal 2×2 complex matrices, regarded either as a real or as a complex algebra according to the context. For example $C\ell_{0,1} \cong \mathbb{C}$, with generator i, $C\ell_{0,2} \cong \mathbb{H}$, with generators i and k, while $C\ell_{0,3} \cong {}^2\mathbb{H}$, with generators $\begin{pmatrix} \mathrm{i} & 0 \\ 0 & -\mathrm{i} \end{pmatrix}$, $\begin{pmatrix} \mathrm{j} & 0 \\ 0 & -\mathrm{j} \end{pmatrix}$ and $\begin{pmatrix} \mathrm{k} & 0 \\ 0 & -\mathrm{k} \end{pmatrix}$.

This last algebra is to be preferred to the algebra $\mathbb{H}$ as a Clifford algebra for $\mathbb{R}^{0,3}$, since although i, j and k mutually anticommute, and each has square -1, the product ij $=$ k, and they generate only the four-dimensional real algebra $\mathbb{H}$, and not an eight-dimensional algebra. (In Porteous 1995 $\mathbb{H}$ is referred to as a *nonuniversal* Clifford algebra for $\mathbb{R}^{0,3}$.)

One then obtains $\mathbb{H}(2)$ as a representation of the Clifford algebra $\mathbb{R}_{0,4}$ by taking as the fourth generator the matrix $\begin{pmatrix} 0 & -1 \\ 1 & 0 \end{pmatrix}$, which clearly anticommutes with the previous three and also has square -1.

The main result of this section is the following theorem:

Theorem 1 *For* $0 \leq p, q < 7$, *matrix representations of the Clifford algebras* $C\ell_{p,q}$ *are exhibited in the following table.*

Table 1.1

p ↓ \ q →							
	$\mathbb{R}$	$\mathbb{C}$	$\mathbb{H}$	${}^2\mathbb{H}$	$\mathbb{H}(2)$	$\mathbb{C}(4)$	$\mathbb{R}(8)$	${}^2\mathbb{R}(8)$
${}^2\mathbb{R}$	$\mathbb{R}(2)$	$\mathbb{C}(2)$	$\mathbb{H}(2)$	${}^2\mathbb{H}(2)$	$\mathbb{H}(4)$	$\mathbb{C}(8)$	$\mathbb{R}(16)$
$\mathbb{R}(2)$	${}^2\mathbb{R}(2)$	$\mathbb{R}(4)$	$\mathbb{C}(4)$	$\mathbb{H}(4)$	${}^2\mathbb{H}(4)$	$\mathbb{H}(8)$	$\mathbb{C}(16)$
$\mathbb{C}(2)$	$\mathbb{R}(4)$	${}^2\mathbb{R}(4)$	$\mathbb{R}(8)$	$\mathbb{C}(8)$	$\mathbb{H}(8)$	${}^2\mathbb{H}(8)$	$\mathbb{H}(16)$
$\mathbb{H}(2)$	$\mathbb{C}(4)$	$\mathbb{R}(8)$	${}^2\mathbb{R}(8)$	$\mathbb{R}(16)$	$\mathbb{C}(16)$	$\mathbb{H}(16)$	${}^2\mathbb{H}(16)$
${}^2\mathbb{H}(2)$	$\mathbb{H}(4)$	$\mathbb{C}(8)$	$\mathbb{R}(16)$	${}^2\mathbb{R}(16)$	$\mathbb{R}(32)$	$\mathbb{C}(32)$	$\mathbb{H}(32)$
$\mathbb{H}(4)$	${}^2\mathbb{H}(4)$	$\mathbb{H}(8)$	$\mathbb{C}(16)$	$\mathbb{R}(32)$	${}^2\mathbb{R}(32)$	$\mathbb{R}(64)$	$\mathbb{C}(64)$
$\mathbb{C}(8)$	$\mathbb{H}(8)$	${}^2\mathbb{H}(8)$	$\mathbb{H}(16)$	$\mathbb{C}(32)$	$\mathbb{R}(64)$	${}^2\mathbb{R}(64)$	$\mathbb{R}(128)$

the table extending to higher p, q *with 'period'* 8.

Most of this follows at once from what we have already done. For example, for any p, q, $C\ell_{p+1,q+1} \cong C\ell_{p,q}(2)$. Also, clearly, $C\ell_{1,0} \cong {}^2\mathbb{R}$. What is missing are the remarks, first that, if S is an orthonormal subset of type $(p+1, q)$, generating a real associative algebra A, then, for any $a \in S$ with $a^2 = 1$, the set

$$\{ba : b \in S \backslash \{a\}\} \cup \{a\}$$

is an orthonormal subset of type $(q+1, p)$ generating A. From this it follows at once that the Clifford algebras $\mathbb{R}_{p+1,q}$ and $\mathbb{R}_{q+1,p}$ are isomorphic. The second remark requires first that we introduce the concept of the *tensor product* of two real algebras.

A tensor product decomposition of a real algebra is somewhat analogous to a direct sum decomposition of a vector space, but involves the multiplicative structure rather than the additive structure. Suppose that B and C are subalgebras of a finite-dimensional algebra A over $\mathbb{K}$, the algebra being associative and with unit element, such that (i) for any $b \in B$, $c \in C$, $cb = bc$, (ii) A is generated as an algebra by B and C, and (iii) $\dim A = \dim B \dim C$.

Then we say that A is the *tensor product* $B \otimes_{\mathbb{K}} C$ over $\mathbb{K}$, the abbreviation $B \otimes C$ being used when the field $\mathbb{K}$ is not in doubt.

Let B and C be subalgebras of a finite-dimensional algebra A over $\mathbb{K}$, such that $A = B \otimes C$, the algebra A being associative and with unit element. Then $B \cap C = \mathbb{K}$ (the field $\mathbb{K}$ is identified with the set of scalar multiples of the unit element $1_{(A)}$).

For our present purposes the important facts are that

$$\mathbb{R} \otimes \mathbb{R} = \mathbb{R}, \qquad \mathbb{C} \otimes \mathbb{R} = \mathbb{C}, \qquad \mathbb{H} \otimes \mathbb{R} = \mathbb{H},$$
$$\mathbb{C} \otimes \mathbb{C} \cong {}^2\mathbb{C}, \quad \mathbb{H} \otimes \mathbb{C} \cong \mathbb{C}(2) \quad \mathbb{H} \otimes \mathbb{H} \cong \mathbb{R}(4).$$

The first three of these statements are obvious. The two mutually commuting copies of $\mathbb{C}$ in ${}^2\mathbb{C}$ are multiples of the identity matrix and the matrix $\left(\begin{smallmatrix} 1 & 0 \\ 0 & -1 \end{smallmatrix}\right)$. The mutually commuting copies of $\mathbb{C}$ and $\mathbb{H}$ in $\mathbb{C}(2)$ are multiples of the identity and the standard representation of quaternions as 2×2 complex matrices. The two mutually commuting copies of $\mathbb{H}$ in $\mathbb{R}(4)$ are the representations of quaternions, first by themselves multiplying other quaternions on the left, and secondly by their conjugates multiplying other quaternions on the right, the commutativity in this case following from the associativity of the quaternions. Moreover, for any m, n, $\mathbb{R}(m) \otimes \mathbb{R}(n) \cong \mathbb{R}(mn)$. We leave it as an exercise to work out why.

The other result that we need is that, for all p, q,

$$C\ell_{p,q+4} \cong C\ell_{p,q} \otimes C\ell_{0,4}$$

bearing in mind that $C\ell_{0,4} \cong \mathbb{H}(2)$.

To prove this, let a, b, c, d be the last four generators of $C\ell_{p,q+4}$, and let the others be e_i. Then $(abcd)^2 = 1$. Let $f_i = abcde_i$, for each i. It is easily verified that the f_i anticommute, and that $f_i^2 = e_i^2$, for all i. The f_i then generate a copy of $C\ell_{p,q}$ in $C\ell_{p,q+4}$ that commutes with the copy of $C\ell_{0,4}$, generated by a, b, c, d. Hence the result.

For example

$$\begin{aligned} C\ell_{0,5} &\cong \mathbb{C} \otimes \mathbb{H}(2) \cong \mathbb{C}(4), \\ C\ell_{0,6} &\cong \mathbb{H} \otimes \mathbb{H}(2) \cong \mathbb{R}(8), \\ C\ell_{0,7} &\cong {}^2\mathbb{H} \otimes \mathbb{H}(2) \cong {}^2\mathbb{R}(8), \\ \text{and} \quad C\ell_{0,8} &\cong \mathbb{H}(2) \otimes \mathbb{H}(2) \cong \mathbb{R}(16). \end{aligned}$$

The *negative* of a real quadratic space is that obtained by replacing the quadratic form by the same form multiplied by -1. In particular the negative of the quadratic space $\mathbb{R}^{p,q}$ is isomorphic to the quadratic space $\mathbb{R}^{q,p}$. An important, and perhaps surprising feature of Table 1.1 is that the Clifford algebras of a quadratic vector space and of its negative are quite different.

2.1.2 *The even Clifford algebras* $C\ell^0_{p,q}$

A map from a vector space or algebra to itself that respects the vector space or algebra structure is said to be an *involution* if its square is the identity. Such a map from an algebra to itself is an *anti-involution* if the order of products is reversed. For example, conjugation on the real algebra of quaternions $\mathbb{H}$ is an anti-involution, since, for any $q, q' \in \mathbb{H}$, $\overline{qq'} = \overline{q'}\,\overline{q}$.

Let $C\ell(X)$ be the Clifford algebra of a finite-dimensional quadratic space X. We have already mentioned *conjugation* as the unique anti-involution of $C\ell(X)$ that sends each vector to its negative. In fact, the linear involution $X \to X : x \mapsto -x$ also extends uniquely to an involution of $C\ell(X)$, known as the *main involution* of the algebra, while the identity on X extends to an anti-involution of $C\ell(X)$, known as *reversion*. The main involution will be denoted by $a \mapsto \hat{a}$, while reversion will be denoted by $a \mapsto a^{\sim}$. The two anti-involutions and the involution commute with each other, each the composite of the other two.

It is not difficult to prove that the main involution leaves invariant a subalgebra of $C\ell(X)$, spanned by all even products of basis elements of X, and known as the *even Clifford algebra* of X. The even subalgebra of the Clifford algebra $C\ell_{p,q}$ will be denoted here by $C\ell^0_{p,q}$

Proposition 2 *For any finite* p, q,

$$C\ell^0_{p,q+1} \cong C\ell_{p,q}, \qquad C\ell^0_{p+1,q} \cong C\ell_{q,p}.$$

We give the proof in the particular case that $p = 0$, $q = n + 1$. Let $e_1, \ldots,$ e_{n+1} be the standard orthonormal basis for $\mathbb{R}^{n+1}$. Each has square equal to -1 in $C\ell_{0,n+1}$, and, of course, they anticommute. Then the n elements $e_1e_{n+1}, \ldots,$ e_ne_{n+1} each have square -1, and these also anticommute, while together with $\mathbb{R}$ they generate $C\ell^0_{0,n+1}$ as a real algebra, thus being an algebra isomorphic to $C\ell_{0,n}$.

Similar arguments hold in the other cases.

It follows, in particular, that the table of the even Clifford algebras $C\ell^0_{p,q}$ is the same as Table 1.1, except that there is an additional line of entries down the left-hand side matching the existing line of entries across the top row. The symmetry about the main diagonal in the table of even Clifford algebras expresses the fact that the *even* Clifford algebras of a finite-dimensional nondegenerate quadratic space, and of its negative, are mutually isomorphic.

2.1.3 *Complex Clifford algebras*

So far we have only considered real Clifford algebras of *real* quadratic spaces. Everything that we have done before can be carried through for the complex quadratic spaces $\mathbb{C}^n$. We denote this Clifford algebra by $C\ell_n(\mathbb{C})$. This algebra may be considered as the tensor product over $\mathbb{R}$ or $\mathbb{C}$ and any of the real Clifford algebras $C\ell_{p,q}$, where $p+q=n$.

Theorem 2 *For any finite k,*

$$\mathbb{C}\otimes C\ell_{0,2k}\cong\mathbb{C}(2^k)\ \textit{and}\ \mathbb{C}\otimes C\ell_{0,2k+1}\cong{}^2\mathbb{C}(2^k).$$

In defining the conjugation anti-involution for the algebra $\mathbb{C}\otimes C\ell_{p,q}$, we have to decide whether or not to conjugate the complex *coefficients*. It turns out, as we shall see in a later section, that if we do not conjugate the coefficients, then the only invariant is the rank n of the relevant real quadratic space, but that if we conjugate the coefficients, then signature remains important. We also will have occasion to consider briefly the tensor products of $C\ell_{p,q}$ by ${}^2\mathbb{R}^\sigma$ and of $\mathbb{C}_n$ by ${}^2\mathbb{C}^\sigma$, where σ denotes the involution that consists in *swapping* the components of ${}^2\mathbb{R}$ and ${}^2\mathbb{C}$, respectively, this swap to be invoked when conjugation is being defined.

2.1.4 Spinors

Table 1.1 exhibits each of the Clifford algebras $C\ell_{p,q}$ as the real algebra of endomorphisms of a right $\mathbb{A}$-linear space of the form $\mathbb{A}^m$, where $\mathbb{A}=\mathbb{R}$, $\mathbb{C}$, $\mathbb{H}$, ${}^2\mathbb{R}$ or ${}^2\mathbb{H}$. This space is called the *(real) spinor space* or *space of (real) spinors* of the quadratic space $\mathbb{R}^{p,q}$. It is identifiable with a minimal left ideal of the algebra, namely the space of matrices with every column except the first nonzero. However as a minimal left ideal it is nonunique.

Physicists concerned with space-time have to choose between the Clifford algebras $C\ell_{1,3}\cong\mathbb{H}(2)$ and $C\ell_{3,1}\cong\mathbb{R}(4)$, with $C\ell_{1,3}^0\cong C\ell_{3,1}^0\cong\mathbb{C}(2)$. Roughly speaking, in the former case the spinor space $\mathbb{H}^2$ is known as the space of *Dirac spinors*, though these, as originally defined, consisted of quadruplets of complex numbers rather than pairs of quaternions, while in the latter case the spinor space $\mathbb{R}^4$ is known as the space of *Majorana spinors*. The complexification of each of these algebras is isomorphic to $\mathbb{C}(4)$. This algebra is known as the *Weyl algebra*, the complex spinor space $\mathbb{C}^4$ being known as the space of *Weyl spinors*.

Minimal left ideals of a matrix algebra are generated by *primitive idempotents*. An idempotent of an algebra is an element y such that $y^2=y$. It is *primitive* if it cannot be expressed as the sum of two idempotents whose product is zero. The simplest example in a matrix algebra is the matrix consisting entirely of zeros, except for a single entry of 1 somewhere in the main diagonal. The minimal ideal generated by such an idempotent then consists of matrices all of whose columns except one consist of zeros. The easiest idempotents to construct are of the form $\frac{1}{2}(1+x)$ where $x^2=1$, but *not* $x^2=-1$. Of course, $\frac{1}{2}(1-x)$ also is an idempotent, so that spinor spaces constructed in this way come naturally in pairs. However, these are not necessarily primitive. They are when the matrix algebra consists of 2×2 matrices over $\mathbb{R}$, $\mathbb{C}$ or $\mathbb{H}$, but in the case of 4×4 matrix algebras, the primitive idempotents are products of commuting pairs of such idempotents.

Although as minimal left ideals of matrix algebras any two spinor spaces are equivalent, they may behave differently when the Clifford algebra structure of the matrix algebra is taken into account, and so may have possibly different physical interpretations. Therefore, in applications the terms Majorana, Dirac and Weyl spinors may be reserved for specific minimal left ideals of the Clifford algebra.

For example, suppose that the algebra $C\ell_{1,3} \cong \mathbb{H}(2)$ is generated by mutually anti-commuting vectors $\gamma_0, \gamma_1, \gamma_2, \gamma_3$, where $\gamma_0^2 = 1, \gamma_1^2 = \gamma_2^2 = \gamma_3^2 = -1$. Then the spinor space generated by the primitive idempotent $\frac{1}{2}(1 + \gamma_0\gamma_1)$ consists of Dirac spinors, while the spinor space of $\mathbb{C} \otimes \mathbb{H}(2) \cong \mathbb{C}(4)$ generated by the primitive idempotent $\frac{1}{4}(1 + \gamma_0\gamma_1)(1 + \mathrm{i}\gamma_0\gamma_1\gamma_2\gamma_3)$ consists of Weyl spinors.

Likewise, suppose that the algebra $C\ell_{3,1} \cong \mathbb{R}(4)$ is generated by mutually anti-commuting vectors e_0, e_1, e_2, e_3, where $e_0^2 = -1, e_1^2 = e_2^2 = e_3^2 = 1$. Then the spinor space generated by the primitive idempotent $\frac{1}{4}(1 + e_1)(1 + e_0e_2)$ consists of Majorana spinors.

2.1.5 Groups of motions

The Clifford algebra $C\ell_{0,n}$ plays a central role in studying the group of *isometries* or *motions* of $\mathbb{R}^n$, not only the rotations and anti-rotations (orientation-reversing) of $\mathbb{R}^n$, but also translations of $\mathbb{R}^n$. Indeed, we can go further and include conformal transformations of $\mathbb{R}^n$ in our study, though we defer that discussion to a later section.

The starting point of all this is the observation that if v is any nonzero (and therefore invertible) vector of $\mathbb{R}^n$, then the map

$$\rho_v : \mathbb{R}^n \to \mathbb{R}^n; x \mapsto vx\hat{v}^{-1} = -vxv^{-1}$$

is reflection of $\mathbb{R}^n$ in the orthogonal complement in $\mathbb{R}^n$ of the line through the origin spanned by the vector v since any real multiple of v maps to minus itself, while any vector anti-commuting in the algebra with v maps to itself. Now any rotation of $\mathbb{R}^n$ is representable as the composite of an *even* number of hyperplane reflections and so is representable as conjugation of $\mathbb{R}^n$ by that composite, that is by an element of the *even* Clifford algebra $C\ell_{0,n}^0$. Explicitly one has the following proposition.

Proposition 3 *Let g be a product of nonzero vectors in $C\ell_{0,n}$. Then the map*

$$\rho_g : x \mapsto gx\hat{g}^{-1}$$

is an orthogonal automorphism of $\mathbb{R}^n$, being a rotation if the number of factors is even, in which case $\hat{g} = g$.

The group of products of invertible vectors in $C\ell_{0,n}$ is known as the *Clifford* or *Lipschitz* group $\Gamma(n)$ of $\mathbb{R}^n$. The subgroup, each of whose elements is the product of an even number of nonzero vectors, is the even Clifford group, and is denoted by $\Gamma^0(n)$.

The *quadratic norm* of any element $g \in \Gamma(n)$ is defined to be the necessarily nonnegative real number $g^- g$, since each vector in $\mathbb{R}^n$ has quadratic norm equal to $v \cdot v$. The subgroup of $\Gamma^0(n)$, consisting of all elements of quadratic norm equal to 1, is known as the group $\mathbf{Spin}(n)$. Clearly any element g of $\mathbf{Spin}(n)$ induces the same rotation as its negative, and it turns out that this is the only ambiguity – the group $\mathbf{Spin}(n)$ *doubly covers* the *special orthogonal group* $SO(n)$ of *rotations* of $\mathbb{R}^n$.

Of course, the original space $\mathbb{R}^n$ does not belong to the even Clifford algebra $C\ell^0_{0,n}$. We can get around this by multiplying the elements of $\mathbb{R}^n$ by the last basis vector, the one that we used to identify $C\ell^0_n$ with $C\ell_{0,n-1}$. Then the space $\mathbb{R}^n$ maps to the space of *paravectors* in $C\ell_{0,n-1}$.

Proposition 3 therefore implies:

Proposition 4 *Let g be an element of* $\mathbf{Spin}(n)$. *Then, since $\hat{g} = g$, the map*

$$\rho_g : y \mapsto gyg^-,$$

is a rotation of $\mathbb{R}^n$ and any rotation of that space may be so induced, the only ambiguity being one of the sign of g.

If we work with paravectors this is replaced by

Proposition 5 *Let g be an element of* $\mathbf{Spin}(n)$, *regarded as a subset of $C\ell_{0,n-1}$. Then the map*

$$\rho_g : y \mapsto gy\tilde{g},$$

where $y = \lambda + x$, with $\lambda \in \mathbb{R}$, and $x \in \mathbb{R}^{n-1}$, is a rotation *of the space of paravectors $\mathbb{R} \oplus \mathbb{R}^{n-1}$, and any rotation of that space may be so induced, the only ambiguity being one of the sign of g.*

Still remaining with the positive-definite case, we have

Theorem 3 *Let $\begin{pmatrix} a & b \\ b & a \end{pmatrix}$ in $C\ell_{0,n}(2)$ represent an element of the even Clifford group $\Gamma^0(n+1)$ in $C\ell_{0,n+1}$ with $a \in \Gamma^0(n)$. Then the map $\mathbb{R}^n \to \mathbb{R}^n$, $x \mapsto axa^{-1} + ba^{-1}$ is an orientation-preserving isometry of $\mathbb{R}^n$ and any isometry of $\mathbb{R}^n$ may be so represented, the representation being unique up to nonzero real multiples of a and b.*

Strictly speaking what is involved here is the subgroup of $C\ell_{0,n}(2)$ consisting of all matrices of the form $\begin{pmatrix} a & b \\ 0 & a \end{pmatrix}$, with $a \in \Gamma^0(n)$ and $b = a\,p$, where $p \in \mathbb{R}^n$. In the particular case that $n = 3$, $\mathbf{Spin}(4)$ is most frequently identified with the group $S^3 \times S^3 \subset {}^2\mathbb{H}$. An alternative to ${}^2\mathbb{H}$ consists of the matrices of $\mathbb{H}(2)$ of the form $\begin{pmatrix} a & b \\ b & a \end{pmatrix}$. The subalgebra of $\mathbb{H}(2)$ consisting of all matrices of the form $\begin{pmatrix} a & b \\ 0 & a \end{pmatrix}$ is known as Clifford's algebra (1873) of *biquaternions*. Elements of it are all of the form $a + b\epsilon$, where a and b are quaternions and $\epsilon^2 = 0$, ϵ being represented in $\mathbb{H}(2)$ by the matrix $\begin{pmatrix} 0 & 1 \\ 0 & 0 \end{pmatrix}$. This is $C\ell^0_{0,3,1}$.

Much of what we have said about isometries of the positive-definite spaces $\mathbb{R}^n$ extends to the indefinite spaces $\mathbb{R}^{p,q}$, especially the important cases for physics where either $p = 3, q = 1$, or $p = 1, q = 3$. We return to these in the next section, where we also show how Clifford algebras may be used to handle conformal transformations efficiently.

2.2 Conjugation

2.2.1 *Symmetric and skew products and anti-involutions*

In order to have a clear understanding of the various **Spin** groups, we begin by classifying the conjugation anti-involutions of Clifford algebras of nondegenerate real or complex quadratic spaces. According to Theorem 1 and Theorem 2, any such Clifford algebra is representable as a matrix algebra $\mathbb{A}(m)$, for some number m, where $\mathbb{A} = \mathbb{K}$ or ${}^2\mathbb{K}$ and $\mathbb{K} = \mathbb{R}$, $\mathbb{C}$ or $\mathbb{H}$.

A *correlation* on an n-dimensional vector space over $\mathbb{K} = \mathbb{R}, \mathbb{C}$ or $\mathbb{H}$ is just a *real* linear map from the vector space to its dual, this being *nondegenerate* if its kernel is zero, in which case it is a linear isomorphism. Let $\xi : v \mapsto v^\xi$ be such a nondegenerate correlation. It induces a $\mathbb{K}$-valued bilinear product $(a, b) \mapsto a^\xi b$.

It is an important fact, which here we take without proof, that any anti-involution of the matrix algebra $\mathbb{K}(n)$ is induced by either a symmetric or skew correlation on the vector space $\mathbb{K}^n$, in the sense discussed below.

The most familiar example is the map from $\mathbb{R}^n$ to its dual, which sends a column vector to the row vector which is its transpose. The induced product on $\mathbb{R}^n$ is then the standard positive-definite scalar product, *symmetric* since, for any $x, y \in \mathbb{R}^n, x \cdot y = y \cdot x$. The induced anti-involution on the real algebra $\mathbb{R}(n)$ of linear maps from $\mathbb{R}^n$ to itself, known as the *adjoint* anti-involution, is in this case a transposition. Let us denote the transpose of an element a of this algebra by a^τ. Then the elements a such that $a^\tau a = 1$ form the *orthogonal group* $O(n)$.

The product just discussed is *positive-definite*, since the square $x^{(2)} = x \cdot x$ of any element x of the vector space $\mathbb{R}^n$ is greater than or equal to zero, being equal to zero only when $x = 0$. Besides this, one has products of signature (p, q), where p of the basis vectors have scalar square -1 and q have scalar square $+1$. The associated groups are the groups $O(p, q)$, with $O(0, n) = O(n)$. The determinant of any element of $O(p, q)$ is equal to $+1$ or -1. The elements with determinant equal to $+1$ form the *special orthogonal group* $SO(p, q)$.

Another example of a correlation, this time on $\mathbb{R}^2$, is the map $(x, y) \mapsto (y, -x)^\tau$ inducing a *skew* product

$$(x', y') \cdot (x, y) = y'x - x'y,$$

skew, since $(x, y) \cdot (x', y') = -(x', y') \cdot (x, y)$.

The induced anti-involution of $\mathbb{R}(2)$ is the map

$$\begin{pmatrix} a & c \\ b & d \end{pmatrix} \mapsto \begin{pmatrix} d & -c \\ -b & a \end{pmatrix}.$$

This product is known as the *standard symplectic product* on $\mathbb{R}^2$, the induced group consisting of all 2×2 real matrices with determinant 1. The notation used here for this group is $\mathrm{Sp}(2;\mathbb{R})$, or when it is necessary to save space in tables, by $\mathrm{Sp}_2(\mathbb{R})$.

An analogous nondegenerate symplectic product may be defined on each even-dimensional real vector space. The associated classical groups are the *real symplectic groups* $\mathrm{Sp}(2n;\mathbb{R})$. There are complex analogues of these products and groups, the associated classical groups being $O(n;\mathbb{C})$ and $\mathrm{Sp}(2n;\mathbb{C})$.

Elements of $O(n;\mathbb{C})$ have determinant equal to $+1$ or to -1. Those of determinant equal to $+1$ form the *special complex orthogonal group* $SO(n;\mathbb{C})$. All elements of both $\mathrm{Sp}(2n;\mathbb{R})$ and $\mathrm{Sp}(2n;\mathbb{C})$ have determinant equal to $+1$.

Also of importance is the real linear, but complex *semilinear* map sending a column vector of $\mathbb{C}^n$ to its *conjugate transpose*. This real linear map induces a *sesquilinear* (that is, one-and-a-half times linear) product on $\mathbb{C}^n$ which is symmetric in the sense that if one forms the product of two vectors in reverse order, then the resulting product is the *conjugate* of the product of the elements in the given order. The product that is simply the same one multiplied by the scalar i is then a *skew* sesquilinear product. Both give rise to the same anti-involution of the algebra $\mathbb{C}(n)$, namely conjugate transposition, the group of matrices whose inverse is the conjugate transpose forming the *unitary group* $U(n)$. Here again one has analogous products with signature (p,q), and associated groups $U(p,q)$, with $U(0,n) = U(n)$. Elements of $U(p,q)$ have determinant a complex number of modulus 1. The subgroup of those with determinant $+1$ is denoted by $SU(p,q)$.

For quaternionic spaces one has to be careful, since $\mathbb{H}$ is not commutative. The usual convention is to regard $\mathbb{H}^n$ as a *right-vector* space, this implying that the scalars multiply vectors on the *right*. The dual space is then a left-vector space, and any correlation between the two has to be semilinear, involving one of the anti-involutions of $\mathbb{H}$. One of these is of course conjugation. The others correspond to reflections of the three-dimensional space of pure quaternions in a plane, the typical one that we choose being that in which only the quaternion j changes sign. Symmetric correlations of either type turn out to be equivalent to skew correlations of the other type. The groups that arise are denoted by $\mathrm{Sp}(n)$ for the symmetric conjugation case, equivalently for the skew symmetric case, and $O(n;\ \mathbb{H})$, or, more briefly, $O_n(\mathbb{H})$ for the opposite case. In the quaternionic symplectic case, one again has the possibility of other signatures, the groups being $\mathrm{Sp}(p,q)$, with $\mathrm{Sp}(0,n) = \mathrm{Sp}(n)$.

Finally, for each of the spaces ${}^2\mathbb{K}^n$, where $\mathbb{K} = \mathbb{R}, \mathbb{C}$ or $\mathbb{H}$, there are symmetric or equivalent skew correlations involving the involution swap. It is of some interest that in each of these cases the group that arises is isomorphic in an obvious way to the *general linear group* $GL(n;\ \mathbb{K})$, elements of the group consisting of 2×2 matrices with entries in $\mathbb{K}(n)$, the top left entry being any invertible matrix, the bottom right entry some transform of this, depending on the details of the product, the other two off-diagonal entries being zero. The subgroup of $GL(n;\ \mathbb{R})$, consisting of those elements with determinant $+1$, forms the *special linear group* $SL(n;\ \mathbb{R})$. The subgroup of $GL(n;\ \mathbb{C})$, consisting of those elements with determinant $+1$, forms the *special linear group* $SL(n; \mathbb{C})$.

The *determinant* of an $n \times n$ quaternionic matrix is defined to be the square root of the determinant of the matrix regarded as a $2n \times 2n$ complex matrix, the latter necessarily having as determinant a nonnegative real number. The subgroup of $GL(n; \mathbb{H})$ consisting of those elements with determinant $+1$ forms the *special linear group* $SL(n; \mathbb{H})$.

There are altogether, up to fairly obvious equivalences, *ten* families of symmetric or skew correlations to choose from.

Theorem 4 *Any symmetric or skew-symmetric correlation on a right $\mathcal{A}$-linear space of finite dimension > 1 belongs to one of the following ten types, which are mutually exclusive.*

0	*a symmetric $\mathbb{R}$-correlation;*
1	*a symmetric, or equivalently, a skew ${}^2\mathbb{R}^\sigma$-correlation;*
2	*a skew $\mathbb{R}$-correlation;*
3	*a skew $\mathbb{C}$-correlation;*
4	*a skew $\tilde{\mathbb{H}}$- or equivalently a symmetric $\overline{\mathbb{H}}$-correlation;*
5	*a skew, or equivalently a symmetric, ${}^2\overline{\mathbb{H}}^\sigma$-correlation;*
6	*a symmetric, $\tilde{\mathbb{H}}$- or equivalently a skew $\overline{\mathbb{H}}$-correlation;*
7	*a symmetric $\mathbb{C}$-correlation;*
8	*a symmetric, or equivalently a skew, $\overline{\mathbb{C}}$-correlation;*
9	*a symmetric, or equivalently a skew, ${}^2\overline{\mathbb{C}}^\sigma$-correlation.*

The logic behind the numbering of these ten types derives from the order in which most of the cases appear in Table 6.1 below.

The job of identification in any particular case is made easier by the fact that an anti-involution of an algebra is uniquely determined by its restriction to any subset that generates the algebra. This means that in classifying the conjugation anti-involution of a Clifford algebra $C\ell(X)$, all one has to do is to examine the representatives in the algebra of an orthonormal basis for the quadratic space X.

What has to be noted is that among all the products that arise, signature is an invariant for just three, symmetric real products on $\mathbb{R}^n$, unitary products on $\mathbb{C}^n$, and the quaternionic symplectic products on $\mathbb{H}^n$. The corresponding real quadratic spaces are denoted by $\mathbb{R}^{p,q}$, $\overline{\mathbb{C}}^{p,q}$ and $\overline{\mathbb{H}}^{p,q}$. In each of these cases, products of signature (p, p) will be said to be *neutral*.

It is not possible to give all the details here. The interested reader is referred to Porteous (1995) for the full story. Table 6.1 first appeared explicitly in Hampson (1969), was published in Porteous (1969), but was implicit in Wall (1968).

One case that is easily verified is that of the positive-definite quadratic spaces. One has

Theorem 5 *Conjugation on the Clifford algebra $C\ell_{0,n}$ is conjugate transposition of the matrices representing the elements of the algebra.*

We begin by reminding the reader that the Clifford algebras $C\ell_{p,q}$ for $0 \leq$

p, $q < 7$ and $\mathbb{C} \otimes C\ell_{0,n}$ for $0 \leq n < 7$ are as in Table 1.1 and Theorem 2. First, we recall

Table 1.1

$p \downarrow$ / $q \rightarrow$								
± 1	$\mathbb{R}$	$\mathbb{C}$	$\mathbb{H}$	$^2\mathbb{H}$	$\mathbb{H}(2)$	$\mathbb{C}(4)$	$\mathbb{R}(8)$	$^2\mathbb{R}(8)$
$\mathbb{R}$	$^2\mathbb{R}$	$\mathbb{R}(2)$	$\mathbb{C}(2)$	$\mathbb{H}(2)$	$^2\mathbb{H}(2)$	$\mathbb{H}(4)$	$\mathbb{C}(8)$	$\mathbb{R}(16)$
$\mathbb{C}$	$\mathbb{R}(2)$	$^2\mathbb{R}(2)$	$\mathbb{R}(4)$	$\mathbb{C}(4)$	$\mathbb{H}(4)$	$^2\mathbb{H}(4)$	$\mathbb{H}(8)$	$\mathbb{C}(16)$
$\mathbb{H}$	$\mathbb{C}(2)$	$\mathbb{R}(4)$	$^2\mathbb{R}(4)$	$\mathbb{R}(8)$	$\mathbb{C}(8)$	$\mathbb{H}(8)$	$^2\mathbb{H}(8)$	$\mathbb{H}(16)$
$^2\mathbb{H}$	$\mathbb{H}(2)$	$\mathbb{C}(4)$	$\mathbb{R}(8)$	$^2\mathbb{R}(8)$	$\mathbb{R}(16)$	$\mathbb{C}(16)$	$\mathbb{H}(16)$	$^2\mathbb{H}(16)$
$\mathbb{H}(2)$	$^2\mathbb{H}(2)$	$\mathbb{H}(4)$	$\mathbb{C}(8)$	$\mathbb{R}(16)$	$^2\mathbb{R}(16)$	$\mathbb{R}(32)$	$\mathbb{C}(32)$	$\mathbb{H}(32)$
$\mathbb{C}(4)$	$\mathbb{H}(4)$	$^2\mathbb{H}(4)$	$\mathbb{H}(8)$	$\mathbb{C}(16)$	$\mathbb{R}(32)$	$^2\mathbb{R}(32)$	$\mathbb{R}(64)$	$\mathbb{C}(64)$
$\mathbb{R}(8)$	$\mathbb{C}(8)$	$\mathbb{H}(8)$	$^2\mathbb{H}(8)$	$\mathbb{H}(16)$	$\mathbb{C}(32)$	$\mathbb{R}(64)$	$^2\mathbb{R}(64)$	$\mathbb{R}(128)$

extending indefinitely either way with period 8.

Here, and in the tables that follow, the left-hand column is to be added in when the even subalgebras are under consideration.

Theorem 6 *Conjugation types for the algebras $C\ell_{p,q}$ are to be overlaid on Table 1.1.*

Table 6.1

p mod 8 $\downarrow$ / q mod 8 $\rightarrow$								
^{20}	0	8	4	^{24}	4	8	0	^{20}
0	1	2	3	4	5	6	7	0
8	2	^{22}	2	8	6	^{26}	6	8
4	3	2	1	0	7	6	5	4
^{24}	4	8	0	^{20}	0	8	4	^{24}
4	5	6	7	0	1	2	3	4
8	6	^{26}	6	8	2	^{22}	2	8
0	7	6	5	4	3	2	1	0

For the complexifications of $C\ell_{p,q}$ we have two results:

Theorem 7 *Table 7.2 classifies conjugation for the complexification of the algebras $C\ell_{p,q}$ when the complex coefficients are not conjugated. Here signature is not important, the result depending only on $n = p + q$. The table of algebras is*

Table 7.1

n mod 8 $\rightarrow$								
± 1	$\mathbb{C}$	$^2\mathbb{C}$	$\mathbb{C}(2)$	$^2\mathbb{C}(2)$	$\mathbb{C}(4)$	$^2\mathbb{C}(4)$	$\mathbb{C}(8)$	$^2\mathbb{C}(8)$

extending indefinitely with period 8.

The classification then follows by overlaying the following table on it.

Table 7.2

n mod 8 $\rightarrow$								
^{27}	7	9	3	^{23}	3	9	7	^{27}

Theorem 8 *Table 8.2 classifies conjugation for the complexification of the algebras $C\ell_{p,q}$ when the complex coefficients are conjugated. Here signature remains important.*

The table of algebras is

Table 8.1

		q	$\rightarrow$
p	± 1	$\mathbb{C}$	${}^2\mathbb{C}$
$\downarrow$	$\mathbb{C}$	${}^2\mathbb{C}$	$\mathbb{C}(2)$

extending indefinitely with period 2.

The classification then follows by overlaying the following table on it.

Table 8.2

		$q \bmod 2$	$\rightarrow$
$p \bmod 2$	^{28}	8	^{28}
$\downarrow$	8	8	8

For the codes 0, 4 and 8 in Tables 6.1 and 7.2, there is a further classification by signature. The choice along the top row or down the extra column on the left is the positive-definite one. Elsewhere, the choice is the neutral one.

For completeness we also have the following two results:

Theorem 9 *For the tensor product of $C\ell_{p,q}$ by ${}^2\mathbb{R}$, with conjugation swapping the components of the coefficients, we have*

Table 9.1

$-p+q \bmod 8$ $\rightarrow$								
^{21}	1	9	5	^{25}	5	9	1	^{21}

Theorem 10 *For the tensor product of $C\ell_{p,q}$ by ${}^2\mathbb{C}$, with conjugation swapping the components of the coefficients, we have, with $n = p + q$,*

Table 10.1

$n \bmod 2$ $\rightarrow$		
^{29}	9	^{29}

Tables 6.1, 7.2, 8.2, 9.1 and 10.1 may be more appreciated if the various code numbers are replaced by the classical groups that preserve the sesquilinear forms on the spinor spaces. We give them here for $0 \leq p+q < 8$ in Tables 6.1′ to 10.1′. To save space, we abbreviate the notations slightly in obvious ways.

Table 6.1′

$p \downarrow$ \ $q \rightarrow$								
	O_1	U_1	Sp_1	${}^2Sp_1$	Sp_2	U_4	O_8	${}^2O_8$
	$GL_1(\mathbb{R})$	$Sp_2(\mathbb{R})$	$Sp_2(\mathbb{C})$	$Sp_{1,1}$	$GL_2(\mathbb{H})$	$O_4(\mathbb{H})$	$O_8(\mathbb{C})$	
	$Sp_2(\mathbb{R})$	${}^2Sp_2(\mathbb{R})$	$Sp_4(\mathbb{R})$	$U_{2,2}$	$O_4(\mathbb{H})$	${}^2O_4(\mathbb{H})$		
	$Sp_2(\mathbb{C})$	$Sp_4(\mathbb{R})$	$GL_4(\mathbb{R})$	$O_{4,4}$	$O_8(\mathbb{C})$			
	$Sp_{1,1}$	$U_{2,2}$	$O_{4,4}$	${}^2O_{4,4}$				
	$GL_2(\mathbb{H})$	$O_4(\mathbb{H})$	$O_8(\mathbb{C})$					
	$O_4(\mathbb{H})$	${}^2O_4(\mathbb{H})$						
	$O_8(\mathbb{C})$							

Table 7.1′

$n \rightarrow$

$O_1(\mathbb{C})$ $GL_1(\mathbb{C})$ $Sp_2(\mathbb{C})$ ${}^2Sp_2(\mathbb{C})$ $Sp_4(\mathbb{C})$ $GL_4(\mathbb{C})$ $O_8(\mathbb{C})$ ${}^2O_8(\mathbb{C})$

Table 8.1′

$p \downarrow$ \ $q \rightarrow$								
	U_1	${}^2U_1$	U_2	${}^2U_2$	U_4	${}^2U_4$	U_8	${}^2U_8$
	$GL_1(\mathbb{C})$	U_2	$GL_2(\mathbb{C})$	U_4	$GL_4(\mathbb{C})$	U_8	$GL_8(\mathbb{C})$	
	U_2	${}^2U_2$	U_4	${}^2U_4$	U_8	${}^2U_8$		
	$GL_2(\mathbb{C})$	U_4	$GL_4(\mathbb{C})$	U_8	$GL_8(\mathbb{C})$			
	U_4	${}^2U_4$	U_8	${}^2U_8$				
	$GL_4(\mathbb{C})$	U_8	$GL_8(\mathbb{C})$					
	U_8	${}^2U_8$						
	$GL_8(\mathbb{C})$							

Table 9.1′

$p \downarrow$ \ $q \rightarrow$								
	$GL_1(\mathbb{R})$	$GL_1(\mathbb{C})$	$GL_1(\mathbb{H})$	${}^2GL_1(\mathbb{H})$	$GL_2(\mathbb{H})$	$GL_4(\mathbb{C})$	$GL_8(\mathbb{R})$	${}^2GL_8(\mathbb{R})$
	${}^2GL_1(\mathbb{R})$	$GL_2(\mathbb{R})$	$GL_2(\mathbb{C})$	$GL_2(\mathbb{H})$	${}^2GL_2(\mathbb{H})$	$GL_4(\mathbb{H})$	$GL_8(\mathbb{C})$	$GL_{16}(\mathbb{R})$
	etc.							

Table 10.1′

$n \rightarrow$

$GL_1(\mathbb{C})$ ${}^2GL_1(\mathbb{C})$ $GL_2(\mathbb{C})$ ${}^2GL_2(\mathbb{C})$ $GL_4(\mathbb{C})$ ${}^2GL_4(\mathbb{C})$ $GL_8(\mathbb{C})$ ${}^2GL_8(\mathbb{C})$

The *dimension* of the classical group consisting of matrices g such that $g^- g = 1$ is the dimension of its *Lie algebra*, the real vector space of matrices g such that $g + g^- = 0$.

The dimensions of the groups in Table 6.1′ are shown in Table 6.1″.

Table 6.1″

0	1	3	6	10	16	28	56
1	3	6	10	16	28	56	
3	6	10	16	28	56		
6	10	16	28	56			
10	16	28	56				
16	28	56					
28	56						
56							

These depend only on the rank $n = p + q$ and not on the index p, q. The dimensions of the groups in Table 7.1′ are twice those of the groups in Table 6.1′.

2.2.2 *Tables of spin groups*

For each n, the quadratic norm $g^{-}g$ of any element g of the group $\mathbf{Spin}(n)$ is equal to $+1$, the group being a subgroup of the *even* classical group associated in Table 6.1′ to the signature $(0, n)$ or $(n, 0)$. That group, being the part of the classical group that lies in the even Clifford algebra for the given signature, lies in the table either in the position $p = 0$, $q = n - 1$ or in the position $p = n$, $q = -1$ (in an extra column on the left that matches the first row). For any p, q, with neither p nor q equal to 0, the quadratic norm of an element of $\mathbf{Spin}(p, q)$ may be equal either to $+1$ or to -1. It is then the subgroup $\mathbf{Spin}^{+}(p, q)$, consisting of those elements of $\mathbf{Spin}(p, q)$ with quadratic norm $+1$, that is a subgroup of the even classical group for the signature, namely the classical group in the position p, $q - 1$. The group $SO^{+}(p,q)$, with neither p nor q equal to 0, which it covers twice, is the group of *orthochronous* isometries of $\mathbb{R}^{p,q}$ that preserve not only the orientations of $\mathbb{R}^{p,q}$ but also its *semi-orientations*, the important case for physics being when p, $q = 3$, 1 or 1, 3.

The group $\mathbf{Spin}(n)$ or $\mathbf{Spin}^{+}(p, q)$, with $n = p+q$, has dimension $\frac{1}{2}n(n-1)$. It is the whole group in Table 6.1′, for $n = p + q \leq 5$, but is of dimension one less than this, namely of dimension 15, rather than 16, for $n = p + q = 6$. In this case, each algebra has a real-valued determinant, and lowering the dimension by 1 corresponds to taking the determinant equal to 1.

Theorem 11 *The groups* $\mathbf{Spin}(n)$ *for* $n \leq 6$, *as well as the groups* $\mathbf{Spin}^{+}(p, q)$ *for* $p + q \leq 6$, *where both* p *and* q *are nonzero, are shown in Table 11.1.*

Table 11.1

	$q \rightarrow$						
p	± 1	O_1	U_1	Sp_1	${}^2Sp_1$	Sp_2	SU_4
$\downarrow$	O_1	$GL_1(\mathbb{R})$	$Sp_2(\mathbb{R})$	$Sp_2(\mathbb{C})$	$Sp_{1,1}$	$SL_2(\mathbb{H})$	
	U_1	$Sp_2(\mathbb{R})$	${}^2Sp_2(\mathbb{R})$	$Sp_4(\mathbb{R})$	$SU_{2,2}$		
	Sp_1	$Sp_2(\mathbb{C})$	$Sp_4(\mathbb{R})$	$SL_4(\mathbb{R})$			
	${}^2Sp_1$	$Sp_{1,1}$	$SU_{2,2}$				
	Sp_2	$SL_2(\mathbb{H})$					
	SU_4						

The groups $\mathbf{Spin}(n;\ \mathbb{C})$, *for* $n \leq 6$, *are shown in Table 11.2.*

Table 11.2

$n \rightarrow$						
± 1	$O_1(\mathbb{C})$	$GL_1(\mathbb{C})$	$Sp_2(\mathbb{C})$	${}^2Sp_2(\mathbb{C})$	$Sp_4(\mathbb{C})$	$SL_4(\mathbb{C})$

As a matter of fact it is enough to prove that $\mathbf{Spin}(6;\ \mathbb{C}) \cong SL(4;\ \mathbb{C})$. All the real cases then follow by restriction. Many writers do not give the entirety of these tables and some are in error. Note in particular that $\mathbf{Spin}^+(3,3) \cong SL(4;\ \mathbb{R})$. Of course, for physics, an important case is

$$\mathbf{Spin}^+(3,1) \cong \mathbf{Spin}^+(1,3) \cong Sp_2(\mathbb{C}),$$

lying in the even Clifford algebra

$$C\ell^0_{3,1} \cong C\ell^0_{1,3} \cong \mathbb{C}(2),$$

this group just being the six-dimensional group of 2×2 complex matrices of determinant 1.

2.2.3 *The Radon–Hurwitz numbers*

Because of the overarching position of the positive-definite quadratic spaces, the top row of Table 1.1 is especially important.

One application is to the construction of linear subspaces of the groups $GL(s;\ \mathbb{R})$, for finite s, a *linear subspace* of $GL(s;\ \mathbb{R})$ being, by definition, a linear subspace of the real matrix algebra $\mathbb{R}(s)$ all of whose elements, with the exception of the origin, are invertible.

For example, the standard copy of $\mathbb{C}$ in $\mathbb{R}(2)$ is a linear subspace of $GL(2;\ \mathbb{R})$ of dimension 2, while either of the standard copies of $\mathbb{H}$ in $\mathbb{R}(4)$ is a linear subspace of $GL(4;\ \mathbb{R})$ of dimension 4. On the other hand, when s is odd, there is no linear subspace of $GL(s;\ \mathbb{R})$ of dimension greater than 1. For if there were such a space of dimension greater than 1, then there would exist linearly independent elements a and b of $GL(s;\ \mathbb{R})$ such that, for all $\lambda \in \mathbb{R}$, $a + \lambda b \in GL(s;\ \mathbb{R})$ and therefore such that $c + \lambda 1 \in GL(s;\ \mathbb{R})$, where $c = b^{-1}a$. However, by the fundamental theorem of algebra, there is a real number λ such that $\det(c + \lambda 1) = 0$,

$\mathbb{R} \to \mathbb{R};\ \lambda \mapsto \det(c + \lambda 1)$ being a polynomial map of odd degree. This provides a contradiction.

Proposition 6 *Let $\mathbb{K}(m)$ be a possibly nonuniversal Clifford algebra for the positive-definite quadratic space $\mathbb{R}^n$, for any positive integer n. Then $\mathbb{R} \oplus \mathbb{R}^n$ is a linear subspace of $GL(m;\ \mathbb{K})$ and therefore of one of the groups $GL(m;\ \mathbb{R})$, $GL(2m;\ \mathbb{R})$ or $GL(4m;\ \mathbb{R})$, according as $\mathbb{K} = \mathbb{R}$, $\mathbb{C}$ or $\mathbb{H}$. Moreover, the conjugate of any element of $\mathbb{R} \oplus \mathbb{R}^n$ is the conjugate transpose of the representative in $GL(m;\ \mathbb{K})$ or, equivalently, the transpose of its representative in $GL(m;\ \mathbb{R})$, $GL(2m;\ \mathbb{R})$ or $GL(4m;\ \mathbb{R})$.*

The following follows from the top line of Table 1.1.

Proposition 7 *Let $\{\chi(k)\}$ be the sequence of positive integers defined by $\chi(8p + q) = 4p + j$, where $j = 0$ for $q = 0$, 1 for $q = 1$, 2 for $q = 2$ or 3 and 3 for $q = 4$, 5, 6 or 7. Then if $2^{\chi(k)}$ divides s, there exists a k-dimensional linear subspace X of $GL(s;\ \mathbb{R})$ such that*

1. *for each $x \in X$, $x^{\tau} = -x$, $x^{\tau}x = -x^2$, being a nonnegative real multiple of ${}^s1$, and zero only if $x = 0$,*

2. *$\mathbb{R} \oplus X$ is a $(k + 1)$-dimensional linear subspace of $GL(s;\ \mathbb{R})$.*

The sequence χ is called the *Radon–Hurwitz sequence* (Radon (1923) and Hurwitz (1923)). It can be proved that there is no linear subspace of $GL(s;\ \mathbb{R}))$ of dimension greater than that asserted here.

As a particular case, there is an eight-dimensional linear subspace of $GL(8; \mathbb{R})$, since $\mathbb{R}(8)$ is a (nonuniversal) Clifford algebra for $\mathbb{R}^7$. This remark provides a route into the study of the algebra of *Cayley numbers*, also known as the *octonion* algebra.

2.2.4 *Vahlen matrices and conformal transformations*

Consider the positive-definite quadratic space $\mathbb{R}^n$, with Clifford algebra $C\ell_{0,n}$ and Clifford group Γ. It follows from earlier work that the real algebra $C\ell_{0,n}(2)$ of 2×2-matrices with entries in $C\ell_{0,n}$ is isomorphic to $C\ell_{1,n+1}$, where elements of the vector space $\mathbb{R}^{1,n+1}$ are represented by matrices of the form $\left(\begin{smallmatrix} x & \nu \\ \mu & -x \end{smallmatrix}\right)$, where $x \in \mathbb{R}^n$ and μ, $\nu \in \mathbb{R}$, such matrices being referred to below as *vectors* in $C\ell_{0,n}(2)$. Let $\Gamma(2)$ denote the Clifford group of $C\ell_{0,n}(2)$. For many applications, one would like to characterize the elements of $\Gamma(2)$ in terms of $\mathbb{R}^n$ and Γ. Such a characterization was given by Vahlen (1902), extending the work of Clifford on biquaternions, and his work was re-presented in a series of papers by Ahlfors in the early 80s, for example (1985), (1986). The reader is referred to a whole series of papers that have appeared during the last 15 years, for example, the paper of Jan Cnops (1994), developed from earlier work of Maks (1989) and Fillmore and Springer (1990). For a parallel account, involving paravectors, see Elstrodt,

Grunewald and Mennicke (1987). See also Waterman (1993), Porteous (1995) and Pozo and Sobczyk (2002).

Here we limit ourselves to discussing the positive-definite case. We begin by describing conjugation and reversion on $C\ell_{0,n}(2)$.

Proposition 8 *For any element of* $C\ell_{0,n}(2)$

$$\begin{pmatrix} a & c \\ b & d \end{pmatrix}^{-} = \begin{pmatrix} d^{\sim} & -c^{\sim} \\ -b^{\sim} & a^{\sim} \end{pmatrix} \quad \textit{and} \quad \begin{pmatrix} a & c \\ b & d \end{pmatrix}^{\sim} = \begin{pmatrix} d^{-} & c^{-} \\ b^{-} & a^{-} \end{pmatrix}$$

These hold for vectors in $C\ell_{0,n}(2)$ and so for the whole of $C\ell_{0,n}(2)$.

The next theorem describes the Clifford group $\Gamma(2)$ of $C\ell_{0,n}(2)$.

Theorem 12 *Let* Γ *be the Clifford group of* $\mathbb{R}^n$, *and let* G *be the set of all matrices* $\begin{pmatrix} a & c \\ b & d \end{pmatrix}$ *of* $C\ell_{0,n}(2)$ *such that*

$$(a)\ a,\ b,\ c,\ d \in \Gamma; \quad (b)\ a b^{\sim},\ c d^{\sim},\ a^{\sim} c,\ b^{\sim} d \in \mathbb{R}^n; \quad (c)\ \Delta = a d^{\sim} - c b^{\sim} \in \mathbb{R}^*.$$

Then G *is the Clifford group* $\Gamma(2)$.

The number Δ is known as the *pseudo-determinant* of the matrix.

The indefinite case is somewhat trickier to handle. One important difference is that the four entries in a Vahlen matrix need not belong to Γ. Each must be a finite product of vectors, but some of these may be null. Of course, the pseudo-determinant must still be a nonzero real number.

Conformal maps are maps that preserve angle. One's first encounter with conformality is probably in a course on functions of one complex variable, where it is proved that any holomorphic map is conformal. However, it was proved long ago by Liouville (1850) that conformality for transformations of $\mathbb{R}^3$ is much more restrictive. He proved that the image of any plane or sphere must be either a plane or sphere. It then follows from a theorem of Möbius that any such map is representable as the composite of a finite number of orthogonal maps, translations, or inversions of $\mathbb{R}^3$ in spheres. The simplest such inversion is the inversion in the sphere, with centre the origin, namely the map $\mathbb{R}^3 \to \mathbb{R}^3 : x \mapsto x/|x|^2$, defined everywhere except at the origin. The obvious analogue of this theorem holds for positive-definite quadratic spaces of any finite dimension greater than 3. The analogous statement for indefinite quadratic spaces is also true by a theorem of Haantjes (1938).

We show here how Clifford algebras may be used to handle such Möbius maps. We limit ourselves to the case of positive-definite quadratic spaces, explicitly $\mathbb{R}^n$, for any positive n.

The trick is first to map the space $\mathbb{R}^n$ to the unit sphere S^n in $\mathbb{R}^{n+1}$ by stereographic projection from the South Pole $(0, \ldots, 0, 1)$, but factoring this through the null-cone in $\mathbb{R}^{1,n+1}$, as follows:

$$x \to (\tfrac{1}{2}(1 + x \cdot x),\ \tfrac{1}{2}(1 - x \cdot x),\ x) \to (\frac{1 - x \cdot x}{1 + x \cdot x}, \frac{2x}{1 + x \cdot x}).$$

It is not difficult to prove that any orthogonal transformation of $\mathbb{R}^{1,n+1}$ not only preserves the null-cone, but also induces a Möbius transformation of $\mathbb{R}^n$, and moreover any Möbius transformation of $\mathbb{R}^n$ may be so induced.

The *Möbius group* $M(0,n)$ is the connected component of the identity of the group of Möbius transformations of $\mathbb{R}^n$. One can save a dimension by identifying $\mathbb{R}^n$ with the space of paravectors in $\mathbb{R}^{0,n-1}$, and then representing such a paravector x by the matrix

$$\tfrac{1}{2}(1+x\cdot x)\begin{pmatrix}0 & 1\\ 1 & 0\end{pmatrix} + \tfrac{1}{2}(1-x\cdot x)\begin{pmatrix}0 & -1\\ 1 & 0\end{pmatrix} + \begin{pmatrix}x & 0\\ 0 & x^{-}\end{pmatrix}$$

$$= \begin{pmatrix}x & x\,x^{-}\\ 1 & x^{-}\end{pmatrix} = \begin{pmatrix}x\\ 1\end{pmatrix}\begin{pmatrix}1 & x^{-}\end{pmatrix}.$$

Consider now an element of $\mathbf{Spin}^{+}(1, n+1)$ represented by an element of $C\ell_{0,n-1}(2)$ of the form

$$\begin{pmatrix}a & c\\ b & d\end{pmatrix}.$$

By Proposition 5 and Proposition 8, it maps the paravector representing the vector x to

$$\begin{pmatrix}a & c\\ b & d\end{pmatrix}\begin{pmatrix}x & x\,x^{-}\\ 1 & x^{-}\end{pmatrix}\begin{pmatrix}d^{-} & c^{-}\\ b^{-} & a^{-}\end{pmatrix} = \lambda\begin{pmatrix}x' & x'\,x'^{-}\\ 1 & x'\end{pmatrix},$$

where $x' = (ax+c)(bx+d)^{-1}$ and λ is the real number $(bx+d)(bx+d)^{-}$.

For example, the *translation* $x \mapsto x+c$ is represented by the matrix

$$\begin{pmatrix}1 & c\\ 0 & 1\end{pmatrix}$$

and *inflation* by the positive scalar ρ by the matrix

$$\begin{pmatrix}\sqrt{\rho} & 0\\ 0 & \sqrt{\rho}^{-1}\end{pmatrix},$$

while inversion in the unit quasi-sphere composed with the hyperplane reflection $x \mapsto -x^{-}$ is represented by the matrix

$$\begin{pmatrix}0 & -1\\ 1 & 0\end{pmatrix}.$$

Representations of the Möbius groups $M(0,n) = M(\mathbb{R}^n)$ for $n \leq 4$ are given in the following theorem.

Theorem 13

$$\begin{aligned}
M(0,1) &\cong Sp(2,\mathbb{R})/\{1,-1\}\\
M(0,2) &\cong Sp(2,\mathbb{C})/\{1,-1\}\\
M(0,3) &\cong Sp(1,1)/\{1,-1\}\\
M(0,4) &\cong SL(2,\mathbb{H})/\{1,-1\}
\end{aligned}$$

The indefinite cases follow the same route, except for some detail. In particular, for pq odd, the group $\mathbf{Spin}^+(p+1, q+1)$ covers the Möbius group $M(p,q)$ four times and not twice. In particular, for the case p, $q = 1$, 3, important for physics,

$$M(1,3) = SU(2,2)/\{1, \mathrm{i}, -1, -\mathrm{i}\}.$$

2.3 REFERENCES

[1] Ahlfors, L. (1985), Möbius transformations and Clifford numbers, in I. Chavel, H.M. Parkas (eds) *Differential Geometry and Complex Analysis*. Dedicated to H.E. Rauch, Springer-Verlag, Berlin, pp. 65–73.

[2] Ahlfors, L. (1986), Möbius transformations in $\mathbb{R}^n$ expressed through 2×2 matrices of Clifford numbers, *Complex Variables* **5**, 215–224.

[3] Clifford, W.K. (1876), Preliminary sketch of Biquaternions, *Proc. London Math. Soc.* **4**, 381–395.

[4] Cnops, J. (1994), *Hurwitz Pairs and Applications of Möbius Transformations*, Thesis, Universiteit Gent.

[5] Elstrodt, J., Grunewald, F. and Mennicke, J. (1987), Vahlen's group of Clifford matrices and spin groups, *Math. Z.* **196**, 369–390.

[6] Fillmore, J. and Springer, A. (1990), Möbius groups over general fields using Clifford algebras associated with spheres, *Int. J. Theo. Phys.* **29**, 225–246.

[7] Grassmann, H. (1844), *Die Wissenschaft der extensiven Grösse oder die Ausdehnungslehre, eine neue mathematishcen Disclipin*, Leipzig.

[8] Haantjes, J. (1937), Conformal representations of an n-dimensional euclidean space with a nondefinite fundamental form on itself, *Proc. Ned. Akad. Wet. (Math)* **40**, 700–705.

[9] Hamilton, W.R. (1844), On quaternions, or on a new system of imaginaries in algebra, *Phil. Mag.* **25**, 489–495, reprinted in *The Mathematical papers of Sir William Rowan Hamilton*, Vol. III, *Algebra*, Cambridge University Press, London (1967), pp. 106–110.

[10] Hampson, A. (1969), *Clifford Algebras*, MSc. Thesis, University of Liverpool.

[11] Hurwitz, A. (1923), Über die Komposition der quadratischen Formen, *Math. Ann.* **88**, 294–298.

[12] Liouville, J. (1850), Appendix to Monge, G. *Application de l'analyse à la geométrie*, 5 éd. par Liouville.

[13] Lipschitz, R. (1884), *Üntersuchungen über die Summen von Quadraten*, Bonn.

[14] Maks, J. (1989), *Modulo* $(1,1)$ *periodicity of Clifford algebras and the generalized (anti-)Möbius transformations*, Thesis, Technische Universiteit Delft.

[15] Maks, J. (1992), Clifford algebras and Möbius transformations, in A. Micali et al. (eds) *Clifford Algebras and their Applications in Mathematical Physics*, Kluwer Acad. Publ., Dordrecht.

[16] Porteous, I.R. (1969), *Topological Geometry*, 1st Edition, Van Nostrand Reinhold.

[17] Porteous, I.R. (1995), *Clifford Algebras and the Classical Groups*, Cambridge University Press, Cambridge.

[18] Pozo, J.M. and Sobczyk, G (2002), Geometric algebra in linear algebra and geometry, *Acta Applicandae Mathematicae* **71**, 207–244.

[19] Radon, J. (1923), Lineare Scharen orthogonaler Matrizen, *Abh. math. Semin. Univ. Hamburg* 1, 1–14.

[20] Vahlen, K.Th. (1902), Über Bewegungen und complexe Zahlen, *Math. Ann.* **55**, 585–593.

[21] Wall, C.T.C. (1968), Graded algebras, antiinvolutions, simple groups and symmetric spaces, *Bull. Am. Math. Soc.* **74**, 198–202.

[22] Waterman, P.L. (1993), Möbius transformations in several dimensions, *Adv. in Math.* **101**, 87–113.

Ian R. Porteous
Department of Mathematical Sciences
Division of Pure Mathematics
University of Liverpool
Liverpool L69 3BX, U.K.
E-mail: porteous@liverpool.ac.uk

Received: January 31, 2002; Revised: March 15, 2003.

3

Clifford Analysis

John Ryan

ABSTRACT We introduce the basic concepts of Clifford analysis. This analysis started many years ago as an attempt to generalize one variable complex analysis to higher dimensions. Most of the basic analysis was initially developed over the quaternions which are a division algebra. However, it was soon realized that virtually all of this analysis extends to all dimensions using Clifford algebras. Here we introduce a generalized Cauchy–Riemann operator, often called a Dirac operator, and the analogues of holomorphic functions. These functions are called Clifford holomorphic functions or monogenic functions. We give a generalization of Cauchy's theorem and Cauchy's integral formula. Using Cauchy's theorem, we can establish the Möbius invariance of monogenic functions. We will also introduce the Plemelj formulas and operators, and Hardy spaces.

3.1 Introduction

In this chapter we regard Clifford algebras as natural generalizations of the complex number system. First, note that if z is a complex number, then $\overline{z}z = \|z\|^2$. For a quaternion q, we also have $\overline{q}q = \|q\|^2$. Quaternions in this way may be regarded as a generalization of the complex number system. It seems natural to ask if one can extend basic results of one complex variable analysis on holomorphic function theory to four dimensions using quaternions. The answer is yes. This was developed by the Swiss mathematician Rudolph Fueter in the 1930s and 1940s and also by Moisil and Theodorescu [29]. See for instance [12]. An excellent review of this work is given in the survey article "Quaternionic analysis" by Sudbery, see [47]. There is also earlier work of Dixon [11]. However, in previous lectures we have seen that for a vector $x \in \mathbb{R}^n$, when we consider $\mathbb{R}^n$ embedded in the Clifford algebra $C\ell_n$, then $x^2 = -\|x\|^2$. So it is reasonable to ask if all that is known in the quaternionic setting further extends to the Clifford algebra setting. Again the answer is yes. The earlier aspects of this study were developed

---

This lecture was presented at "Lecture Series on Clifford Algebras and their Applications", May 18 and 19, 2002, as part of the 6th International Conference on Clifford Algebras and their Applications in Mathematical Physics, Cookeville, TN, May 20–25, 2002.

AMS Subject Classification: 30G35.

Keywords: Cauchy–Riemann operator, monogenic functions, Cauchy's theorem, Cauchy's integral, Plemelj formula, Hardy space.

by among others, Richard Delanghe [9], Viorel Iftimie [16] and David Hestenes [15]. The subject that has grown from these works is now called Clifford analysis.

In more recent times, Clifford analysis has found a wealth of unexpected applications in a number of branches of mathematical analysis, particularly classical harmonic analysis. See, for instance, the work of Alan McIntosh and his collaborators [21, 22], Marius Mitrea [27, 28] and papers in [37]. Links to representation theory and several complex variables may be found in [14, 34–36] and elsewhere.

The purpose of this paper is to review the basic aspects of Clifford analysis.

Alternative accounts of much of this work, together with other related results, can be found in [5, 10, 13, 14, 20, 31, 37, 38].

3.2 Foundations of Clifford analysis

We start by replacing the vector $x = x_1e_1+\ldots+x_ne_n$ by the differential operator $D = \sum_{j=1}^{n} e_j \frac{\partial}{\partial x_j}$. One basic, but interesting, property of D is that $D^2 = -\Delta_n$, the Laplacian $\sum_{j=1}^{n} \frac{\partial^2}{\partial x_j^2}$ in $\mathbb{R}^n$. The differential operator D is called a Dirac operator.

This is because the classical Dirac operator constructed over four-dimensional Minkowski space squares to give the wave operator.

Definition 1. *Suppose that U is a domain in $\mathbb{R}^n$ and f and g are C^1-functions defined on U and taking values in $C\ell_n$. Then f is called a left monogenic function if $Df = 0$ on U, while g is called a right monogenic function on U if $gD = 0$, where $gD = \sum_{j=1}^{n} \frac{\partial g}{\partial x_j} e_j$.*

Left monogenic functions are also called left regular functions and, perhaps most appropriately, left Clifford holomorphic functions. The term Clifford holomorphic functions, or Clifford analytic functions appears to be due to Semmes, see [41] and elsewhere. We shall most often use the term Clifford holomorphic functions.

Examples of such functions include the gradients of real valued harmonic functions on U. If h is harmonic on U, and if it is also real valued, then Dh is a vector valued left monogenic function. It is also a right monogenic function. Such a function is commonly referred to as a conjugate harmonic function, or a harmonic 1-form. See for instance [46]. An example of such a function is $G(x) = \frac{x}{\|x\|^n}$.

It should be noted that if f and g are left monogenic functions then, due to the lack of commutativity of the Clifford algebra, it is not in general true that their product $f(x)g(x)$ is left monogenic.

To introduce other examples of left monogenic functions, suppose that μ is a $C\ell_n$ valued measure with compact support $[\mu]$ in $\mathbb{R}^n$. Then the convolution

$$\int_{[\mu]} G(x-y)d\mu(y)$$

defines a left monogenic function on the maximal domain lying in $\mathbb{R}^n \backslash [\mu]$. The previously defined integral is the Cauchy transform of the measure $[\mu]$.

Another way of constructing examples of left monogenic functions was introduced by Littlewood and Gay in [23], for the case $n = 3$, and independently re-introduced for all n by Sommen [43]. Suppose U' is a domain in $\mathbb{R}^{n-1}$, spanned by $e_2, \ldots, e_n$. Suppose also that $f'(x')$ is a $C\ell_n$-valued function such that at each point $x' \in U'$ there is a multiple series expansion in $x_2, \ldots, x_n$ that converges uniformly on some neighborhood of x' in U' to f'. Such a function is called a real analytic function. The series

$$\sum_{k=0}^{\infty} \frac{1}{k!} x_1^k (-e_1 D')^k f'(x') = \exp(-x_1 e_1 D') f'(x'),$$

where $D' = \sum_{j=2}^{n} e_j \frac{\partial}{\partial x_j}$, defines a left monogenic function f in some neighborhood $U(f')$ in $\mathbb{R}^n$ of U'. The left monogenic function f is the Cauchy–Kowalewska extension of f'.

It should be noted that if f is a left monogenic function, then $\overline{f}$ and $\tilde{f}$ are both right monogenic functions[1].

We now turn to analogues of Cauchy's Theorem and Cauchy's integral formula.

Theorem 1 (Clifford–Cauchy Theorem). *Suppose that f is a left Clifford holomorphic function on U, and g is a right Clifford holomorphic function on U. Suppose also that V is a bounded subdomain of U with piecewise differentiable boundary S lying in U. Then*

$$\int_S g(x) n(x) f(x)\, d\sigma(x) = 0 \tag{2.1}$$

where $n(x)$ is the outward pointing normal vector to S at x and σ is the Lebesgue measure on S.

Proof. The proof follows directly from Stokes' Theorem. One important point to keep in mind is that, since $C\ell_n$ is not a commutative algebra, the order of the quantities g, $n(x)$ and f must be maintained. One then has that

$$\int_S g(x) n(x) f(x)\, d\sigma(x) = \int_V ((g(x)D) f(x) + g(x)(Df(x)))\, dx^n = 0.$$

Suppose that g is the gradient of a real valued harmonic function and $f = 1$. Then the real part of Equation 1 gives the following well-known integral formula:

$$\int < \text{grad}\, g(x), n(x) > d\sigma(x) = 0.$$

[1] Here, $\overline{f}$ denotes the Clifford conjugate of f while $\tilde{f}$ is the reversion of a $C\ell_n$-valued function f.

We now turn to the analogue of a Cauchy integral formula.

Theorem 2 (Clifford–Cauchy Integral Formula). *Suppose that U, V, S, f and g are all as in Theorem 1 and that $y \in V$. Then*

$$f(y) = \frac{1}{\omega_n} \int_S G(x-y)n(x)f(x)\, d\sigma(x)$$

and

$$g(y) = \frac{1}{\omega_n} \int_S g(x)n(x)G(x-y)\, d\sigma(x)$$

where ω_n is the surface area of the unit sphere in $\mathbb{R}^n$.

Proof. The proof follows very similar lines to the argument in one variable complex analysis. We shall establish the formula for $f(y)$, the proof being similar for $g(y)$. First, let us take a sphere $S^{n-1}(y,r)$ centered at y and of radius r. The radius r is chosen sufficiently small so that the closed disc with boundary $S^{n-1}(y,r)$ lies in V. Then, by the Clifford–Cauchy theorem,

$$\int_S G(x-y)n(x)f(x)\, d\sigma(x) = \int_{S^{n-1}(y,r)} G(x-y)n(x)f(x)\, d\sigma(x).$$

However, on $S^{n-1}(y,r)$ the vector $n(x) = \frac{y-x}{\|x-y\|}$. So $G(x-y)n(x) = \frac{1}{r^{n-1}}$ and

$$\begin{aligned}&\int_{S^{n-1}(y,r)} G(x-y)n(x)f(x)\, d\sigma(x)\\ &\quad = \int_{S^{n-1}(y,r)} \frac{1}{r^{n-1}}(f(x)-f(y))\, d\sigma(x) + \int_{S^{n-1}(y,r)} \frac{f(y)}{r^{n-1}}\, d\sigma(x).\end{aligned}$$

The right side of the previous expression reduces to

$$\int_{S^{n-1}(y,r)} \frac{(f(x)-f(y))}{r^{n-1}}\, d\sigma(x) + f(y)\int_{S^{n-1}} d\sigma(x).$$

Now $\int_{S^{n-1}} d\sigma(x) = \omega_n$, and by continuity

$$\lim_{r\to 0} \int_{S^{n-1}(y,r)} \frac{(f(x)-f(y)}{r^{n-1}}\, d\sigma(x) = 0.$$

The result follows. □

One important feature to note is that Kelvin inversion, $x^{-1} = \frac{-x}{\|x\|^2}$ whenever x is nonzero, plays a fundamental role in this proof. Moreover, the proof is almost exactly the same as the proof of Cauchy's Integral Formula for piecewise C^1-curves in one variable complex analysis.

Having obtained a Cauchy Integral Formula in $\mathbb{R}^n$, a number of basic results that one might see in a first course in one variable complex analysis carry over more or less automatically to the context described here. This includes Liouville's Theorem and Weierstrass' Convergence Theorem. We leave it to the interested reader to set up and establish the Clifford analysis analogues of these results. Their statements and proofs can be found in [5].

Theorems 1 and 2 show us that the individual components of the equations $Df = 0$ and $gD = 0$ comprise generalized Cauchy–Riemann equations. In the particular case that f is vector valued, $f = \sum_{j=1}^{n} f_j e_j$, the generalized Cauchy–Riemann equations become $\frac{\partial f_j}{\partial x_i} = \frac{\partial f_i}{\partial x_j}$, whenever $i \neq j$, and $\sum_{j=1}^{n} \frac{\partial f_j}{\partial x_j} = 0$. This system of equations is often referred to as the Riesz system.

Having obtained an analogue of Cauchy's integral formula in Euclidean space, we now exploit this result to show how many consequences of the classical Cauchy integral carry over to the context described here. We begin with the Mean Value Theorem.

Theorem 3 (The Mean Value Theorem). *Suppose that $D(y, R)$ is a closed disc centered at y, of radius R and lying in U. Then, for each left Clifford holomorphic function f on U*

$$f(y) = \frac{1}{R\omega_n} \int_{D(y,R)} \frac{f(x)}{\|x - y\|^{n-1}} \, dx^n.$$

Proof. We have already seen that for each $r \in (0, R)$,

$$f(y) = \frac{1}{\omega_n} \int_{S^{n-1}(y,r)} \frac{f(x)}{\|x - y\|^{n-1}} \, d\sigma(x),$$

where $S^{n-1}(y, r)$ is the $(n-1)$-dimensional sphere centered at y and of radius r. We obtain the result by integrating both sides of this expression with respect to the variable r, and dividing throughout by R. □

Let us now explore the real analyticity properties of Clifford holomorphic functions. First note that when n is even,

$$G(x - y) = (-1)^{\frac{n-2}{2}} (x - y)^{-n+1}.$$

Also

$$(x-y)^{-1} = x^{-1}(1-yx^{-1})^{-1} = (1-x^{-1}y)^{-1}x^{-1}, \ \|x^{-1}y\| = \|yx^{-1}\| = \frac{\|y\|}{\|x\|}.$$

So for, $\|y\| < \|x\|$,

$$\begin{aligned}(x - y)^{-1} &= x^{-1}(1 + yx^{-1} + \ldots + yx^{-1} \ldots yx^{-1} + \ldots) \\ &= (1 + x^{-1}y + \ldots + x^{-1}y \ldots x^{-1}y + \ldots)x^{-1}.\end{aligned}$$

Hence, these two sequences converge uniformly to $(x - y)^{-1}$ provided

$$\|y\| \leq r < \|x\|,$$

and they converge pointwise to $(x-y)^{-1}$ provided $\|y\| < \|x\|$. One now takes $(-1)^{\frac{n-2}{2}}$ times the $(n-1)$-fold product of the series expansions of $(x-y)^{-1}$ with itself to obtain a series expansion for $G(x-y)$. In the process of multiplying series together, in order to maintain the same radius of convergence, one needs to group together all linear combinations of monomials in $y_1, \dots, y_n$ that are of the same order. Thus, we have deduced that when n is even, the multiple Taylor series expansion

$$\sum_{j=0}^{\infty}\Big(\sum_{\substack{j_1,\dots,j_n \\ j_1+\dots+j_n=j}} \frac{y_1^{j_1}\dots y_n^{j_n}}{j_1!\dots j_n!} \frac{\partial^j G(x)}{\partial x_1^{j_1}\dots \partial x_n^{j_n}}\Big)$$

converges uniformly to $G(x-y)$ provided $\|y\| < r < \|x\|$, and converges pointwise to $G(x-y)$ provided $\|y\| < \|x\|$.

A similar argument holds when n is odd.

Returning to Cauchy's integral formula, let us suppose that f is a left Clifford holomorphic function defined in a neighborhood of the closure of some ball $B(0,R)$. Then

$$\begin{aligned} f(y) &= \frac{1}{\omega_n}\int_{\partial B(0,R)} G(x-y)n(x)f(x)\,d\sigma(x) \\ &= \frac{1}{\omega_n}\int_{\partial B(0,R)} \sum_{j=0}^{\infty}\Big(\sum_{\substack{j_1\dots j_n \\ j_1+\dots+j_n=j}} \frac{y_1^{j_1}\dots y_n^{j_n}}{j_1!\dots j_n!} \frac{\partial^j G(x)}{\partial x_1^{j_1}\dots \partial x_n^{j_n}}\Big) n(x)f(x)\,d\sigma(x) \end{aligned}$$

provided $\|y\| < \|x\|$. Since this series converges uniformly on each ball $B(0,r)$, for each $r < R$, this last integral can be re-written as

$$\frac{1}{\omega_n}\sum_{j=0}^{\infty}\int_{\partial B(0,R)}\Big(\sum_{\substack{j_1\dots j_n \\ j_1+\dots+j_n=j}} \frac{y_1^{j_1}\dots y_n^{j_n}}{j_1!\dots j_n!} \frac{\partial^j G(x)}{\partial x_1^{j_1}\dots \partial x_n^{j_n}} n(x)f(x)\Big)\,d\sigma(x).$$

Since the summation within the parentheses is a finite summation, this last expression easily reduces to

$$\frac{1}{\omega_n}\sum_{j=0}^{\infty}\Big(\sum_{\substack{j_1\dots j_n \\ j_1+\dots+j_n=j}} \frac{y_1^{j_1}\dots y_n^{j_n}}{j_1!\dots j_n!} \int_{\partial B(0,R)} \frac{\partial^j G(x)}{\partial x_1^{j_1}\dots \partial x_n^{j_n}} n(x)f(x)\Big)\,d\sigma(x).$$

On placing

$$\frac{1}{\omega_n}\int_{\partial B(0,R)} \frac{\partial^j G(x)}{\partial x_1^{j_1}\dots \partial x_n^{j_n}} n(x)f(x)\,d\sigma(x) = a_{j_1\dots j_n}$$

it may be seen that on $B(0,R)$ the series

$$\sum_{j=0}^{\infty}\Big(\sum_{\substack{j_1\dots j_n \\ j_1+\dots+j_n=j}} \frac{x_1^{j_1}\dots x_n^{j_n}}{j_1!\dots j_n!} a_{j_1\dots j_n}\Big)$$

converges pointwise to $f(y)$. Convergence is uniform on each ball $B(0,r)$, provided $r < R$.

Similarly, if g is a right Clifford holomorphic function defined in a neighborhood of the closure of $B(0,R)$, then the series

$$\sum_{j=0}^{\infty}\Bigg(\sum_{\substack{j_1\cdots j_n\\ j_1+\ldots+j_n=j}} b_{j_1\ldots j_n}\frac{y_1^{j_1}\ldots y_n^{j_n}}{j_1!\ldots j_n!}\Bigg)$$

converges pointwise on $B(0,R)$ to $g(y)$ and converges uniformly on $B(0,r)$ for $r < R$, where

$$b_{j_1\ldots j_n} = \frac{1}{\omega_n}\int_{\partial B(0,R)} g(x)n(x)\frac{\partial^j G(x)}{\partial x_1^{j_1}\ldots\partial x_n^{j_n}}\,d\sigma(x).$$

By translating the ball $B(0,R)$ to the ball $B(w,R)$, where

$$w = w_1e_1 + \ldots + w_ne_n,$$

one may readily observe that for any left Clifford holomorphic function f, defined in a neighborhood of the closure of $B(w,R)$, the series

$$\sum_{j=0}^{\infty}\Bigg(\sum_{\substack{j_1\cdots j_n\\ j_1+\ldots j_n=j}} \frac{(y_1-w_1)^{j_1}\ldots(y_n-w_n)^{j_n}}{j_1!\ldots j_n!}a'_{j_1\ldots j_n}\Bigg)$$

converges pointwise on $B(w,R)$ to $f(y)$, where

$$a'_{j_1\ldots j_n} = \frac{1}{\omega_n}\int_{\partial B(w,R)} \frac{\partial^j G(x-w)}{\partial x_1^{j_1}\ldots x_n^{j_n}}n(x)f(x)\,d\sigma(x).$$

Again, the series converges uniformly on $B(w,r)$ for each $r < R$. A similar series may be readily obtained for any right Clifford holomorphic function defined in a neighborhood of the closure of $B(w,R)$.

The types of power series that we have developed for left Clifford holomorphic functions are not entirely satisfactory. In particular, unlike their complex analogues, the homogeneous polynomials

$$\sum_{\substack{j_1\cdots j_n\\ j_1+\ldots+j_n=j}} \frac{x_1^{j_1}\ldots x_n^{j_n}}{j_1!\ldots j_n!}a_{j_1\ldots j_n}$$

are not expressed as a linear combination of left Clifford holomorphic polynomials. To rectify this situation, let us first take a closer look at the Taylor expansion for the Cauchy kernel $G(x-y)$ where all the Taylor coefficients are real. Let us first look at the first order terms in the Taylor expansion. This is the expression

$$y_1\frac{\partial G(x)}{\partial x_1} + \ldots + y_n\frac{\partial G(x)}{\partial x_n}.$$

Since G is a Clifford holomorphic function,

$$\frac{\partial G(x)}{\partial x_1} = -\sum_{j=2}^{n} e_1^{-1} e_j \frac{\partial G(x)}{\partial x_j}.$$

Therefore, the first order terms of the Taylor expansion for $G(x-y)$ can be re-expressed as

$$\sum_{j=2}^{n} (y_j - e_1^{-1} e_j y_1) \frac{\partial G(x)}{\partial x_j}.$$

Moreover, for $2 \leq j \leq n$, the first order polynomial $y_j - e_1^{-1} e_j y_1$ is a left Clifford holomorphic polynomial. Let us now go to second order terms. Again, we replace the operator $\frac{\partial}{\partial x_1}$ by the operator

$$-\sum_{j=2}^{n} e_1^{-1} e_j \frac{\partial}{\partial x_j}$$

whenever it arises. Let us consider the term $\frac{\partial^2 G(x)}{\partial x_i \partial x_j}$, where $i \neq j \neq 1$. We end up with the polynomial

$$\begin{aligned} & y_i y_j - y_i y_1 e_1^{-1} e_j - y_j y_1 e_1^{-1} e_i \\ & \quad = \tfrac{1}{2}((y_i - y_1 e_1^{-1} e_i)(y_j - y_1 e_1^{-1} e_j) + (y_j - y_1 e_1^{-1} e_j)(y_i - y_1 e_1^{-1} e_i). \end{aligned}$$

Similarly, the polynomial attached to the term $\frac{\partial^2 G(x)}{\partial x_i^2}$ is $(y_i - y_1 e_1^{-1} e_i)^2$. Using the Clifford algebra anticommutation relations $e_i e_j + e_j e_i = -2\delta_{ij}$, and on replacing the differential operator $\frac{\partial}{\partial x_1}$ by the operator

$$-\sum_{j=2}^{n} e_1^{-1} e_j \frac{\partial}{\partial x_j},$$

the power series we previously obtained for $G(x-y)$ can be replaced by the series

$$\sum_{j=0}^{\infty} \Big(\sum_{\substack{j_2 \cdots j_n \\ j_2 + \ldots + j_n = j}} P_{j_2 \ldots j_n}(y) \frac{\partial^j G(x)}{\partial x_2^{j_2} \ldots \partial x_n^{j_n}} \Big),$$

where $\|y\| < \|x\|$ and

$$P_{j_2 \ldots j_n}(y) = \frac{1}{j!} \Sigma (y_{\sigma(1)} - y_1 e_1^{-1} e_{\sigma(1)}) \ldots (y_{\sigma(j)} - y_1 e_1^{-1} e_{\sigma(j)}).$$

Here, $\sigma(i) \in \{2, \ldots, n\}$ and the previous summation is taken over all permutations of the monomials $(y_{\sigma(i)} - y_1 e_1^{-1} e_{\sigma(i)})$ without repetition. The quaternionic monogenic analogues for these polynomials were introduced by Fueter [12], while the Clifford analogues, $P_{j_2\ldots j_n}$, were introduced by Delanghe in [9]. It should be noted that each polynomial $P_{j_2\ldots j_n}(y)$ takes its values in the space spanned by $\{1, e_1e_2, \ldots, e_1e_n\}$. Also, each such polynomial is homogeneous of degree j. Similar arguments to those just outlined give

$$G(x-y)=\sum_{j=0}^{\infty}\Big(\sum_{\substack{j_2\ldots j_n\\ j_2+\ldots+j_n=j}}\frac{\partial^j G(x)}{\partial x_2^{j_2}\ldots x_n^{j_n}}\widetilde{P_{j_2\ldots j_n}}(y)\Big)$$

provided $\|y\|<\|x\|$.

Proposition 1. *Each of the polynomials $P_{j_2\ldots j_n}(y)$ is a left Clifford holomorphic polynomial.*

Proof. Calculating

$$DP_{j_2\ldots j_n}(y)=e_1\Big(\frac{\partial}{\partial y_1}+e_1^{-1}\sum_{j=2}^{n}e_j\frac{\partial}{\partial y_j}P_{j_2\ldots j_n}(y)\Big)$$

we then consider the expression

$$(\frac{\partial}{\partial y_1}+\sum_{j=2}^{n}e_1^{-1}e_j\frac{\partial}{\partial y_j})P_{j_2\ldots j_n}(y).$$

This term is equal to

$$\begin{aligned}\Big(\frac{\partial}{\partial y_j}+\sum_{j=2}^{n}e_1^{-1}e_j\frac{\partial}{\partial y_j}\Big)&\Sigma(y_{\sigma(1)}-e_1^{-1}e_{\sigma(1)}y_1)\ldots(y_{\sigma(i-1)}-e_1^{-1}e_{\sigma(i-1)}y_1)\\&\times(y_{\sigma(i)}-e_1^{-1}e_{\sigma(i)}y_1)(y_{\sigma(i+1)}-e_1^{-1}e_{\sigma(i+1)}y_1)\ldots(y_{\sigma(j)}-e_1^{-1}e_{\sigma(j)}y_1).\end{aligned}$$

This is equal to

$$\begin{aligned}&\sum(y_{\sigma(1)}-e_1^{-1}e_{\sigma(1)}y_1)\ldots(y_{\sigma(i-1)}-e_1^{-1}e_{\sigma(i-1)}y_1)(-e_1^{-1}e_{\sigma(i)})\\&\qquad\times(y_{\sigma(i+1)}-e_1^{-1}e_{\sigma(i+1)}y_1)\ldots(y_{\sigma(j)}-e_1^{-1}e_{\sigma(j)}y_1)\\&\quad+\sum e_1^{-1}e_{\sigma(i)}(y_{\sigma(1)}-e_1^{-1}e_{\sigma(1)}y_1)\ldots(y_{\sigma(i-1)}-e_1^{-1}e_{\sigma(i-1)}y_1)\\&\qquad\qquad\times(y_{\sigma(i+1)}-e_1^{-1}e_{\sigma(i+1)}y_1)\ldots(y_{\sigma(j)}-e_1^{-1}e_{\sigma(j)}y_1).\end{aligned}$$

If we multiply the previous term by y_1, and add to it the following term, which is equal to zero,

$$\begin{aligned}&\sum(y_{\sigma(1)}-e_1^{-1}e_{\sigma(1)}y_1)\ldots(y_{\sigma(i-1)}-e_1^{-1}e_{\sigma(i-1)}y_1)(y_{\sigma(i)}-y_{\sigma(i)})\\&\qquad\qquad\times(y_{\sigma(i+1)}-e_1^{-1}e_{\sigma(i+1)}y_1)\ldots(y_{\sigma(j)}-e_1^{-1}e_{\sigma(i)}y_1)\end{aligned}$$

we get, after regrouping terms,

$$\sum(y_{\sigma(1)} - e_1^{-1}e_{\sigma(1)}y_1)\dots(y_{\sigma(i-1)} - e_1^{-1}e_{\sigma(i-1)}y_1)(y_{\sigma(i)} - e_1^{-1}e_{\sigma(i)}y_1)$$
$$\times(y_{\sigma(i+1)} - e_1^{-1}e_{\sigma(i+1)}y_1)\dots(y_{\sigma(j)} - e_1^{-1}e_{\sigma(j)}y_1)$$
$$-\sum(y_{\sigma(i)} - e_1^{-1}e_{\sigma(i)}y_1)(y_{\sigma(1)} - e_1^{-1}e_{\sigma(1)}y_1)\dots(y_{\sigma(i-1)} - e_1^{-1}e_{\sigma(i-1)}y_1)$$
$$\times(y_{\sigma(i+1)} - e_1^{-1}e_{\sigma(i+1)}y_1)\dots(y_{\sigma(j)} - e_1^{-1}e_{\sigma(j)}y_j).$$

Since the summation is taken over all possible permutations, without repetition, the last term vanishes. □

Using Proposition 1 and the results we previously obtained on series expansions, we can obtain the following generalization of Taylor expansions from complex analysis.

Theorem 4 (Taylor Series). *Suppose that f is a left Clifford holomorphic function defined in an open neighborhood of the closure of the ball $B(w,R)$. Then*

$$f(y) = \sum_{j=0}^{\infty}\Big(\sum_{\substack{j_2\cdots j_n \\ j_2+\dots+j_n=j}} P_{j_2\dots j_n}(y-w)a_{j_2\dots j_n}\Big)$$

where

$$a_{j_2\dots j_n} = \frac{1}{\omega_n}\int_{\partial B(w,R)} \frac{\partial^j G(x-w)}{\partial x_2^{j_2}\dots\partial x_n^{j_n}} n(x)f(x)\,d\sigma(x)$$

and $\|y-w\| < R$. Convergence is uniform provided $\|x-w\| < r < R$.

A simple application of Cauchy's theorem tells us that the Taylor series that we obtained for f in the previous theorem remains valid on the largest open ball on which f is defined, and on the largest open ball on which g is defined. Also, the previous identities immediately yield the mutual linear independence of the collection of the left Clifford holomorphic polynomials

$$\{P_{j_2\dots j_n} : j_2 + \dots j_n = j,\ 0 \le j < \infty\}.$$

3.3 Other types of Clifford holomorphic functions

Unlike the classical Cauchy–Riemann operator $\frac{\partial}{\partial \bar z} = \frac{\partial}{\partial x} + i\frac{\partial}{\partial y}$, the generalized Cauchy–Riemann operator D that we have introduced here does not have an identity component. Instead, we could have considered the differential operator

$$D' = \frac{\partial}{\partial x_0} + \sum_{j+1}^{n-1} e_j\frac{\partial}{\partial x_j}.$$

Also for U' a domain in $\mathbb{R} \oplus \mathbb{R}^{n-1}$, spanned by $\{1, e_1, \ldots, e_{n-1}\}$, one can consider $C\ell_{n-1}$-valued differentiable functions f' and g' defined on U' such that $D'f' = 0$ and $g'D' = 0$, where

$$g'D' = \frac{\partial g'}{\partial x_0} + \sum_{j=1}^{n-1} \frac{\partial g'}{\partial x_j} e_j.$$

Traditionally, such functions are also called left monogenic and right monogenic functions. To avoid confusion, we shall call such functions unital left monogenic and unital right monogenic, respectively. In the case where $n = 2$, the operator D' corresponds to the usual Cauchy–Riemann operator, and unital monogenic functions are the usual holomorphic functions studied in one variable complex analysis. The function

$$G'(\underline{x}) = \frac{\overline{\underline{x}}}{\|\underline{x}\|^n} = \underline{x}^{-1}\|\underline{x}\|^{-n+2}$$

is an example of a function which is both unital left monogenic and unital right monogenic. It is a simple matter to observe that f' is unital left monogenic if and only if $\tilde{f}'$ is unital right monogenic. However, $\overline{f}'$ is not unital right monogenic whenever f' is unital left monogenic. Instead, $\overline{f}'$ satisfies the equation $\overline{f}'D' = 0$.

The function theory for unital left monogenic functions is much the same as for left monogenic functions. For instance, if f' is unital left monogenic on U', and g' is unital right monogenic on the same domain, and S' is a piecewise smooth, compact surface lying in U' and bounding a subdomain V', then

$$\int_{S'} g'(\underline{x})n(\underline{x})f'(\underline{x})\, d\sigma(\underline{x}) = 0,$$

where $n(\underline{x})$ is the outward pointing normal vector to S' at $\underline{x}$. Also, for each $\underline{y} \in V'$ there is the following version of Cauchy's integral formula:

$$f'(\underline{y}) = \frac{1}{\omega_n} \int_{S'} G'(\underline{x} - \underline{y})n(\underline{x})f'(\underline{x})\, d\sigma(\underline{x}).$$

To get from the operator D to the operator D', one first rewrites D as

$$e_n\Big(\frac{\partial}{\partial x_n} + \sum_{j=1}^{n-1} e_n^{-1} e_j \frac{\partial}{\partial x_j}\Big).$$

On multiplying on the left by e_n, and changing the variable x_n to x_0, we get the operator

$$D" = \frac{\partial}{\partial x_0} + \sum_{j=1}^{n-1} e_n^{-1} e_j \frac{\partial}{\partial x_j}.$$

This operator takes its values in the even subalgebra $C\ell_n^+$ of $C\ell_n$. Applying the isomorphism

$$\theta : C\ell_{n-1} \to C\ell_n^+, \quad \theta(e_{j_1} \dots e_{j_r}) = e_n^{-1} e_{j_1} \dots e_n^{-1} e_{j_r},$$

it immediately follows that $\theta(D') = D"$. So if f' is unital left monogenic, then $D"\theta(f) = 0$. If we change the variable x_0 of the function $\theta(f(\underline{x}))$ to x_n, we get a left monogenic function, which we denote by $\theta'(f)(x)$, where

$$x = x_1 e_1 + \dots + x_n e_n \in U \subset \mathbb{R}^n$$

if and only if

$$\underline{x} = x_n + x_1 e_1 + \dots + x_{n-1} e_{n-1} \in U' \subset \mathbb{R} \oplus \mathbb{R}^{n-1}.$$

It should be noted that $D'\overline{D'} = \overline{D'}D' = \Delta_n$.

When $n = 3$, the algebra $C\ell_3$ is split by the two projection operators

$$E_\pm = \tfrac{1}{2}(1 \pm e_1 e_2 e_3)$$

into the direct sum

$$C\ell_3 = E_+ C\ell_3 \oplus E_- C\ell_3,$$

and each of these subalgebras is isomorphic to the quaternion algebra $\mathbb{H}$. In this setting, the differential operator $E_\pm D'$ can best be written as

$$E_\pm D' = \frac{\partial}{\partial t} + i\frac{\partial}{\partial x} + j\frac{\partial}{\partial y} + k\frac{\partial}{\partial z},$$

and the operator $E_\pm D$ can best be written as

$$E_\pm D = i\frac{\partial}{\partial x} + j\frac{\partial}{\partial y} + k\frac{\partial}{\partial z}.$$

We shall denote the first of these two operators by $D'_{\mathbb{H}}$ and the second by $D_{\mathbb{H}}$. The operator $D'_{\mathbb{H}}$ is sometimes referred to as the Cauchy–Riemann–Fueter operator. The function theory associated to the differential operators $D_{\mathbb{H}}$ and $D'_{\mathbb{H}}$ is much the same as that associated to the operators D and D'. In fact, historically the starting point for Clifford analysis was to study the function theoretic aspects of the operators $D'_{\mathbb{H}}$ and $D_{\mathbb{H}}$, see for instance [9, 12] and the excellent review article of Sudbery [47].

It is a simple enough matter to set up analogues of Cauchy's theorem and Cauchy's integral formula for the quaternionic valued differentiable functions that are either annihilated by $D'_{\mathbb{H}}$ or $D_{\mathbb{H}}$, either acting on the left or on the right. When dealing with the operator $D'_{\mathbb{H}}$, such functions are called quaternionic monogenic. The quaternionic monogenic Cauchy kernel is the function $q^{-1}\|q\|^{-2}$. Consequently, for each quaternionic left monogenic function $f(q)$ defined on a domain

$U" \subset \mathbb{H}$, and each q_0 lying in a bounded subdomain with piecewise C^1-boundary $S"$,

$$f(q_0) = \frac{1}{\omega_3} \int_{S"} (q - q_0)^{-1} \|q - q_0\|^{-2} n(q) f(q)\, d\sigma(q).$$

Similarly, if g is right quaternionic monogenic on $U"$, then

$$g(q_0) = \frac{1}{\omega_3} \int_{S"} g(q) n(q) (q - q_0)^{-1} \|q - q_0\|^{-2}\, d\sigma(q). \tag{3.2}$$

3.4 The equation $D^k f = 0$

It is reasonably well known that if h is a real valued harmonic function defined on a domain $U \subset \mathbb{R}^n$, then for each $y \in U$ and each compact, piecewise C^1-surface S lying in U such that S bounds a subdomain V of S and $y \in V$,

$$h(y) \\ = \frac{1}{\omega_n} \int_S (H(x-y) < n(x), \text{grad}\, h(x) > - < G(x-y), n(x) > h(x))\, d\sigma(x),$$

where

$$H(x-y) = \frac{1}{(n-2)\|x-y\|^{n-2}}.$$

This is Green's formula for a harmonic function, and it heavily relies on the standard inner product on $\mathbb{R}^n$. Introducing the Clifford algebra $C\ell_n$, the right side of Green's formula is the real part of

$$\frac{1}{\omega_n} \int_S (G(x-y) n(x) h(x) - H(x-y) n(x) D h(x))\, d\sigma(x).$$

Assuming that the function h is C^2, then on applying Stokes' theorem, the previous integral becomes

$$\frac{1}{\omega_n} \int_{S^{n-1}(y, r(y))} \Big(G(x-y) n(x) h(x) - H(x-y) n(x) D h(x) \Big)\, d\sigma(x),$$

where $S^{n-1}(y, r(y))$ is a sphere centered at y, of radius $r(y)$ and lying in V. On letting the radius $r(y)$ tend to zero, the first term of the integral tends to $h(y)$, while the second term tends to zero. Consequently, the Clifford analysis version of Green's formula is

$$h(y) = \frac{1}{\omega_n} \int_S \Big(G(x-y) n(x) h(x) - H(x-y) n(x) D h(x) \Big)\, d\sigma(x).$$

We obtained this formula under the assumption that h is real valued and C^2. The fact that h is real valued can easily be seen to be irrelevant, and so we can assume that h is $C\ell_n$ valued. From now on, we shall assume that all harmonic functions take their values in $C\ell_n$. If h is also a left monogenic function, then the Clifford analysis version of Green's formula becomes Cauchy's integral formula.

Proposition 2. *Suppose that f is a Clifford holomorphic function on some domain U. Then $xf(x)$ is harmonic.*

Proof.

$$Dxf(x) = -nf(x) - \sum_{j=1}^{n} x_j \frac{\partial f(x)}{\partial x_j} - \sum_{\substack{j,k \\ j\neq k}} x_k e_k e_j \frac{\partial f(x)}{\partial x_j}.$$

Now

$$\sum_{\substack{j,k \\ j\neq k}} x_k e_k e_j \frac{\partial f(x)}{\partial x_j} = \sum_{k=1}^{n} \sum_{j\neq k} x_k e_k e_j \frac{\partial f(x)}{\partial x_j}.$$

As f is left monogenic, this last expression simplifies to $\sum_{k=1}^{n} x_k \frac{\partial f(x)}{\partial x_k}$. Moreover, $D(\sum_{j=1}^{n} x_j \frac{\partial f(x)}{\partial x_j}) = 0$. Consequently, $D^2 xf(x) = 0$. □

The previous proof is a generalization of the statement: "if $h(x)$ is a real valued harmonic function, then so is $< x, \text{grad } h(x) >$".

In fact, in the previous proof, we determine that

$$Dxf(x) = -nf(x) - 2\sum_{j=1}^{n} x_j \frac{\partial f(x)}{\partial x_j}.$$

In the special case where $f(x) = P_k(x)$, a left Clifford holomorphic polynomial of order k, this equation simplifies to

$$DxP_k(x) = -(n+2k)P_k(x).$$

Suppose now that $h(x)$ is a harmonic function defined in a neighborhood of the ball $B(0,R)$. Then Dh is a left Clifford holomorphic function, and there is a series $\sum_{l=0}^{\infty} P_l(x)$ of left Clifford holomorphic polynomials with each P_l homogeneous of degree l, such that the series converges locally uniformly on $B(0,R)$ to $Dh(x)$. Now consider the series

$$\sum_{l=0}^{\infty} \frac{-1}{n+2l} P_l(x).$$

Since

$$\frac{1}{n+2l} \|P_l(x)\| < \|P_l(x)\|,$$

this new series converges locally uniformly on $B(0,R)$ to a left Clifford holomorphic function $f_1(x)$. Moreover, $Dxf_1(x) = Dh(x)$ on $B(0,R)$. Consequently, $h(x) - xf_1(x)$ is equal to a left Clifford holomorphic function $f_2(x)$ on $B(0,R)$. We have established:

Proposition 3. *Suppose that h is a harmonic function defined in a neighborhood of $B(0,R)$. Then there are left Clifford holomorphic functions f_1 and f_2 defined on $B(0,R)$ such that $h(x) = xf_1(x) + f_2(x)$ for each $x \in B(0,R)$.*

This result remains invariant under translation. As a consequence, it shows us that all harmonic functions are real analytic functions. So there is no need to specify whether or not a harmonic function is C^2. The result also provides an Almansi type decomposition of harmonic functions in terms of Clifford holomorphic functions over any ball in $\mathbb{R}^n$.

It should be noted that Proposition 3 remains true only if h is real valued.

Proposition 3 gives rise to an alternative proof of the Mean Value Theorem for harmonic functions.

Theorem 5. *For any harmonic function h defined in a neighborhood of a ball $B(a, R)$ and any $r < R$,*

$$h(a) = \frac{1}{\omega_n} \int_{\partial B(a,r)} h(x)\, d\sigma(x)\,.$$

Proof. Proposition 3 tells us that there is a pair of left Clifford holomorphic functions f_1 and f_2 such that

$$h(x) = (x - a) f_1(x) + f_2(x)$$

on $B(a, R)$. So $h(a) = f_2(a)$, and we have previously shown that

$$\frac{1}{\omega_n} \int_{\partial B(a,r)} f_2(x)\, d\sigma(x) = f_2(a).$$

Now

$$\int_{\partial B(a,r)} (x - a) f_1(x)\, d\sigma(x) = r \int_{\partial B(a,r)} n(x) f_1(x)\, d\sigma(x) = 0.$$

The following is an immediate consequence of Proposition 3.

Proposition 4. *If $h_l(x)$ is a harmonic polynomial homogeneous of degree l, then*

$$h_l(x) = p_l(x) + x p_{l-1}(x)$$

where p_l is a left Clifford holomorphic polynomial homogeneous of degree l while p_{l-1} is a left monogenic polynomial which is homogeneous of degree $l - 1$.

It is well known that pairs of homogeneous harmonic polynomials of differing degrees of homogeneity are orthogonal with respect to the usual inner product over the unit sphere. Proposition 4 offers a further refinement to this. Suppose that f and g are $C\ell_n$-valued functions defined on S^{n-1}, and each component of f and g is square integrable. If we define the $C\ell_n$ inner product of f and g to be

$$< f, g >= \frac{1}{\omega_n} \int_{S^{n-1}} \overline{f(x)} g(x)\, d\sigma(x),$$

then if f and g are both real valued, this inner product is equal to

$$\frac{1}{\omega_n}\int_{S^{n-1}} f(x)g(x)\,d\sigma(x),$$

which is the usual inner product for real-valued square integrable functions defined on S^{n-1}. Now

$$\begin{aligned} < xp_{l-1}(x), p_l(x) > &= -\frac{1}{\omega_n}\int_{S^{n-1}} \overline{p}_{l-1}(x)xp_l(x)\,d\sigma(x) \\ &= -\frac{1}{\omega_n}\int_{S^{n-1}} \overline{p}_{l-1}(x)n(x)p_l(x)\,d\sigma(x) = 0. \end{aligned}$$

The evaluation of the last integral is an application of Cauchy's theorem.

Let us denote the space of $C\ell_n$-valued functions defined on S^{n-1}, and such that each component is square integrable, by $L^2(S^{n-1}, C\ell_n)$. Clearly, the space of real valued square integrable functions defined on S^{n-1} is a subset of $L^2(S^{n-1}, C\ell_{n-1})$. The space $L^2(S^{n-1}, C\ell_n)$ is a $C\ell_n$-module.

We have shown that by introducing the module $L^2(S^{n-1}, C\ell_n)$, Proposition 4 provides a further orthogonal decomposition of harmonic polynomials, using left Clifford holomorphic polynomials. We shall return to this theme later. This decomposition was introduced for the case $n = 4$ by Sudbery [47], and independently extended for all n by Sommen [43].

Let us now consider higher order iterates of the Dirac operator D. In the same way that $DH(x) = G(x)$, there is a function $G_3(x)$ defined on $\mathbb{R}^n\backslash\{0\}$ such that $DG_3(x) = H(x)$. Specifically,

$$G_3(x) = C(n,3)\frac{x}{\|x\|^{n-2}},$$

for some dimensional constant $C(n,3)$. Continuing inductively, we may find a function $G_k(x)$ on $\mathbb{R}^n\backslash\{0\}$ such that $DG_k(x) = G_{k-1}(x)$. Specifically,

$$G_k(x) = C(n,k)\frac{x}{\|x\|^{n-k+1}}$$

where when n is odd, so is k.

$$G_k(x) = C(n,k)\frac{1}{\|x\|^{n-k}}$$

where when n is odd, k is even.

$$G_k(x) = C(n,k)\frac{x}{\|x\|^{n-k+1}}$$

where when n is even, k is odd and $k < n$.

$$G_k(x) = C(n,k)\frac{1}{\|x\|^{n-k}}$$

where when n is even, k is even and $k < n$.

$$G_k(x) = C(n,k)(x^{k-n}\log\|x\| + A(n,k)x^{k-n})$$

where when n is even, $k \geq n$. In the last expression, $A(n,k)$ is a real constant dependent on n and k, and $C(n,k)$ is a constant dependent on n and k throughout.

It should be noted that $G_1(x) = G(x)$ and $G_2(x) = H(x)$. It should also be noted that $D^k G_k(x) = 0$.

Here is a simple technique for constructing solutions to the equation $D^k g = 0$ from left Clifford holomorphic functions. The special case $k = 2$ was illustrated in Proposition 2.

Proposition 5. *Suppose that f is a left Clifford holomorphic function on U. Then $D^k x^{k-1} f(x) = 0$.*

Proof. The proof is by induction. We have already seen the result to be true in the case $k = 2$ in Proposition 2. If k is odd, then

$$Dx^{k-1}f(x) = (k-1)x^{k-2}f(x).$$

If k is even, then

$$Dx^{k-1}f(x) = -n(k-1)x^{k-2}f(x) + x^{k-2}\sum_{j=1}^{n} e_j x \frac{\partial f(x)}{\partial x_j}.$$

By arguments presented in Proposition 5, this expression is equal to

$$-n(k-1)x^{k-2}f(x) + x^{k-2}\sum_{j=1}^{n} x_j \frac{\partial f(x)}{\partial x_j}.$$

The induction hypothesis tells us that the only term we need consider is

$$x^{k-2}\sum_{j=1}^{n} x_j \frac{\partial f(x)}{\partial x_j}.$$

However, $\sum_{j=1}^{n} x_j \frac{\partial f(x)}{\partial x_j}$ is a left Clifford holomorphic function. So the proof by induction is now complete. □

We shall refer to a function $g : U \to C\ell_n$, which satisfies the equation $D^k g = 0$, as a left k-monogenic function. Similarly, if $h : U \to C\ell_n$ satisfies the equation $hD^k = 0$, then h is a right k-monogenic function. In the case where $k = 1$, we return to the setting of left, or right, Clifford holomorphic functions, and when $k = 2$ we return to the setting of harmonic functions. When $k = 4$, the equations $D^4 g = 0$ and $gD^4 = 0$ correspond to the equations $\Delta_n^2 g = 0$ and $\Delta_n^2 h = 0$. So left or right 4-monogenic functions are in fact biharmonic functions.

More generally, if k is even, then a left or right k-monogenic function f automatically satisfies the equation $\Delta_n^{\frac{k}{2}} f = 0$.

Proposition 6. *Suppose that p is a left k-monogenic polynomial homogeneous of degree q. Then there are left Clifford holomorphic polynomials $f_0, \ldots, f_{k-1}$ such that*

$$p(x) = f_0(x) + \ldots + x^{k-1} f_{k-1}(x),$$

and each polynomial f_j is homogeneous of degree $q - j$, whenever $q - j \geq 0$, and identically zero otherwise.

Proof. The proof is via induction on k. The case $k = 2$ is established immediately after the proof of Proposition 2. Let us now consider $Dp(x)$. This is a left $(k-1)$-monogenic polynomial homogeneous of degree $q-1$. So by the induction hypothesis,

$$Dp(x) = g_1(x) + \ldots + x^{k-2} g_{k-1}(x),$$

where each g_j is a left Clifford holomorphic polynomial homogeneous of degree $q - j$, whenever $q - j \geq 0$, and is equal to zero otherwise. Using Euler's lemma, and the observations made after the proof of Proposition 5, one may now find left Clifford holomorphic polynomials $f_1(x), \ldots, f_{k-1}(x)$ such that

$$D(x f_1(x) + \ldots + x^{k-1} f_{k-1}(x)) = Dp(x),$$

and

$$f_j(x) = c_j g_j(x)$$

for some $c_j \in \mathbb{R}$, and where $1 \leq j \leq k-1$. It follows that

$$p(x) - \sum_{j=1}^{k-1} x^j f_j(x)$$

is a left Clifford holomorphic polynomial f_0, homogeneous of degree q. □

One may now use Proposition 6, and the arguments used to establish Proposition 3, to deduce:

Theorem 6. *Suppose that f is a left k-monogenic function defined in a neighborhood of the ball $B(0, R)$. Then there are left monogenic functions $f_0, \ldots, f_{k-1}$ defined on $B(0, R)$ such that $f(x) = f_0(x) + \ldots + x^{k-1} f_{k-1}(x)$ on $B(0, R)$.*

Theorem 6 establishes an Almansi decomposition for left k-monogenic functions in terms of left Clifford holomorphic functions over any open ball. It also follows from this theorem that each left k-monogenic function is a real analytic function. It is also reasonably well known that if h is a biharmonic function defined in a neighborhood of $B(0, R)$, then there are harmonic functions h_1 and h_2 defined on $B(0, R)$ such that

$$h(x) = h_1(x) + \|x\|^2 h_2(x).$$

In the special case where $k = 4$, Theorem 6 both establishes this result and refines it.

Since each left k-monogenic function is a real analytic function, we can immediately use Stokes' theorem to deduce the following Cauchy–Green type formula.

Theorem 7. *Suppose that f is a left k-monogenic function defined on some domain U, and suppose that S is a piecewise C^1 compact surface lying in U and bounding a bounded subdomain V of U. Then for each $y \in V$,*

$$f(y) = \frac{1}{\omega_n} \int_S (\sum_{j=1}^{k} (-1)^{j-1} G_j(x-y) n(x) D^{j-1} f(x))\, d\sigma(x).$$

3.5 Conformal groups and Clifford analysis

Here we examine the role played by the conformal group within parts of Clifford analysis. Our starting point is to ask what type of diffeomorphisms acting on subdomains of $\mathbb{R}^n$ preserve Clifford holomorphic functions. If a diffeomorphism ϕ can transform the class of left Clifford holomorphic functions on one domain U to a class of left Clifford holomorphic functions on the domain $\phi(U)$ and do the same for the class of right Clifford holomorphic functions on U, then it must preserve Cauchy's theorem. If f and g are left and right Clifford holomorphic on U, respectively, and these functions are transformed to f' and g', left and right Clifford holomorphic functions on $\phi(U)$, then

$$\int_S g(x) n(x) f(x)\, d\sigma(x) = 0 = \int_{\phi(S)} g'(y) n(y) f'(y)\, d\sigma(y)$$

where S is a piecewise C^1-compact surface lying in U and $y = \phi(x)$. An important point to note here is that we need to assume that ϕ preserves vectors orthogonal to the tangent spaces at x and $\phi(x)$. As the choice of x and S is arbitrary, it follows that the diffeomorphism ϕ is angle preserving. In other words, ϕ is a conformal transformation. A theorem of Liouville [24] tells us that for dimensions 3 and greater the only conformal transformations on domains are Möbius transformations.

In order to deal with Möbius transformations using Clifford algebras, we have seen in a previous chapter that one can use Vahlen matrices. We now proceed to show that each Möbius transformation preserves monogenicity. Sudbery [47], and also Bojarski [3], have established this fact. We will need the following two lemmas.

Lemma 1. *Suppose that $\phi(x) = (ax+b)(cx+d)^{-1}$ is a Möbius transformation; then*

$$G(u-v) = J(\phi, x)^{-1} G(x-y) \tilde{J}(\phi, y)^{-1},$$

where $u = \phi(x)$, $v = \phi(y)$ and

$$J(\phi, x) = \frac{\widetilde{(cx+d)}}{\|cx+d\|^n}.$$

Proof. The proof essentially follows from the fact that

$$(x^{-1} - y^{-1}) = x^{-1}(y - x)y^{-1}.$$

Consequently,

$$\|x^{-1} - y^{-1}\| = \|x\|^{-1}\|x - y\|\|y\|^{-1}, \quad \text{and} \quad ax\tilde{a} - ay\tilde{a} = a(x - y)\tilde{a}.$$

If one breaks the transformation down into terms arising from the generators of the Möbius group, using the previous set of equations, then one readily arrives at the result. □

Lemma 2. *Suppose that $y = \phi(x) = (ax + b)(cx + d)^{-1}$ is a Möbius transformation, and for domains U and V we have $\phi(U) = V$. Then*

$$\int_S f(u)n(u)g(u)\, d\sigma(u) = \int_{\psi^{-1}(S)} f(\psi(x))\tilde{J}(\psi, x)n(x)J(\psi, x)g(\psi(x))\, d\sigma(x)$$

where $u = \psi(x)$, S is a orientable hypersurface lying in U, and

$$J(\psi, x) = \frac{\widetilde{cx + d}}{\|cx + d\|^n}.$$

Proof Outline. On breaking ψ up into the generators of the Möbius group, the result follows by noting that

$$\frac{\partial x^{-1}}{\partial x_j} = -x^{-1}e_j x^{-1}.$$

It follows from Cauchy's Theorem that if $g(u)$ is a left Clifford holomorphic function in the variable u, then $J(\psi, x)f(\psi(x))$ is left Clifford holomorphic in the variable x.

When $\phi(x)$ is the Cayley transformation

$$y = (e_n x + 1)(x + e_n)^{-1},$$

we can use this transformation to establish a Cauchy–Kowalewska extension in a neighborhood of the sphere. If $f(x)$ is a real analytic function defined on an open subset U of $S^{n-1}\backslash\{e_n\}$, then

$$l(y) = J(\phi^{-1}, y)^{-1} f(\phi(y))$$

is a real analytic function on the open set $V = \phi^{-1}(U)$. This function has a Cauchy–Kowalewska extension to a left Clifford holomorphic function $L(y)$ defined on an open neighborhood $V(g) \subset \mathbb{R}^n$ of V. Consequently,

$$F(x) = J(\phi^{-1}, x)L(\phi^{-1}(x))$$

is a left Clifford holomorphic defined on an open neighborhood

$$U(f)=\phi^{-1}(V(g))$$

of U. Moreover $F_{|U} = f$. Combining with similar arguments for the other Cayley transformation

$$y = (-e_n x + 1)(x - e_n)^{-1},$$

one can deduce:

Theorem 8 (Cauchy–Kowalewska Theorem). *Suppose that f is a $C\ell_n$-valued real analytic function defined on S^{n-1}. Then there is a unique left Clifford holomorphic function F defined on an open neighborhood $U(f)$ of S^{n-1} such that $F_{|S^{n-1}} = f$.*

In fact, if $f(u)$ is defined on some domain and satisfies the equation $D^k f = 0$, then the function $J_k(\psi, x) f(\psi(x))$ satisfies the same equation, where

$$J_k(\psi, x) = \frac{\widetilde{cx + d}}{\|cx + d\|^{n-k+1}}.$$

Theorem 9 (Fueter–Sce Theorem). *Suppose that $f = u + iv$ is a holomorphic function on a domain $\Omega \subset \mathbb{C}$ and that $\Omega = \overline{\Omega}$ and $f(\overline{z}) = \overline{f(z)}$. Then the function*

$$F(\underline{x}) = u(x_1, \|x'\|) + e_1^{-1} \frac{x'}{\|x'\|} v(x_1, \|x'\|)$$

is a unital left$(n-1)$-monogenic function on the domain $\{\underline{x} : x_1 + i\|x'\| \in \Omega\}$ whenever n is even. Here $x' = x_2 e_2 + \ldots + x_n e_n$.

Proof. First let us note that $x^{-1}e_1$ is left $n-1$ monogenic whenever n is even. It follows that

$$\frac{\partial^k}{\partial x_1^k} x^{-1} e_1 = c_k x^{-k-1} e_1$$

is $n-1$ left monogenic for each positive integer k. Here c_k is some nonzero real number. Using Kelvin inversion, it follows that $x^k e_1$ is left $n-1$ monogenic for each positive integer k. By taking translations and Taylor series expansions for the function f, the result follows. □

This result was first established for the case $n = 4$ by Fueter [12], see also Sudbery [47]. It was extended to all even dimensions by Sce [40], though the methods used do not make use of the conformal group. This result has been applied in [32, 33] to study various types of singular integral operators acting on L^p spaces of Lipschitz perturbations of the sphere.

3.6 Conformally flat spin manifolds

The invariance of monogenic functions under Möbius transformations, described in the previous section, makes use of a conformal weight factor $J(\psi, x)$. This invariance can be seen as an automorphic form invariance, which leads to a natural generalization of the concept of a Riemann surface to the Euclidean setting. A manifold M is said to be conformally flat if there is an atlas $\mathcal{A}$ of M whose transition functions are Möbius transformations. For instance, via the Cayley transformations

$$(e_{n+1}x+1)(x+e_{n+1})^{-1} \quad \text{and} \quad (-e_{n+1}x+1)(x-e_{n+1})^{-1},$$

one can see that the sphere $S^n \subset \mathbb{R}^{n+1}$ is an example of a conformally flat manifold. Another way of constructing conformally flat manifolds is to take a simply connected domain U of $\mathbb{R}^n$, and consider a Kleinian subgroup Γ of the Möbius group at acts discontinuously on U. Then the factorization $U\backslash\Gamma$ is a conformally flat manifold. For instance, let $U = \mathbb{R}^n$ and let Γ be the integer lattice

$$Z^k = Ze_1 + \ldots + Ze_k$$

for some positive integer $k \leq n$. In this case, $\mathbb{R}^n \backslash Z^k$ gives the cylinder C_k, and when $k = n$ we get the n-torus. Also, if we let

$$U = \mathbb{R}^n \backslash \{0\} \quad \text{and} \quad \Gamma = \{2^k : k \in Z\},$$

the resulting manifold is $S^1 \times S^{n-1}$.

We locally construct a spinor bundle over M by making the identification (u, X) with either $(x, \pm J(\psi, x)X)$, where

$$u = \psi(x) = (ax+b)(cx+d)^{-1} = (-ax-b)(-cx-d)^{-1}.$$

If we can compatibly choose the signs, then we have created a spinor bundle over the conformally flat manifold. Note, it might be possible to create more than one spinor bundle over M. For instance, consider the cylinder C_k. If we make the identification (x, X) with $(x + \underline{m}, (-1)^{m_1+\ldots+m_l}X)$, where l is a fixed integer with $l \leq k$, and $\underline{m} = m_1e_1 + \ldots + m_le_l + \ldots + m_ke_k$, then we have created k different spinor bundles $E^1, \ldots, E^k$ over C_k.

We have used the conformal weight function $J(\psi, x)$ to construct the spinor bundle E. It is easy to see that a section $f : M \to E$ could be called a left monogenic section if it is locally a left monogenic function. It is now natural to ask if one can construct Cauchy integral formulas for such sections. To do this, we need to construct a kernel over the Euclidean domain U that is periodic with respect to Γ, and then use the projection map $p : U \to M$ to construct from this kernel a Cauchy kernel for U. In [19], we show that the Cauchy kernel for C_k, with spinor bundle E^l, is constructed from the kernel

$$\cot_{k,l}(x,y) = \sum_{\underline{m}\in Z^l, \underline{n}\in Z^{k-l}} (-1)^{m_1+\ldots+m_l} G(x-y+\underline{m}+\underline{n}),$$

where $\underline{n} = n_{l+1}e_{l+1} + \ldots + n_n e_n$. While for the conformally flat spin manifold $S^1 \times S^{n-1}$ with trivial bundle $C\ell_n$, the Cauchy kernel is constructed from the kernel

$$\sum_{k=0}^{\infty} G(2^k x - 2^k y) + 2^{2-2n} G(x) (\sum_{k=-1}^{-\infty} G(2^{-k}x^{-1} - 2^{-k}y^{-1})) G(y).$$

See [17–19] for more details and related results.

It should be noted that one may set up a Dirac operator over arbitrary Riemannian manifolds, see for instance [4], and one may set up Cauchy integral formulas for functions annihilated by these Dirac operators [6, 28].

3.7 Boundary behavior and Hardy spaces

Possibly the main topic that unites all that has been previously discussed here on Clifford analysis is its applications to boundary value problems. This, in turn, leads to a study of boundary behavior of classes of Clifford holomorphic functions and Hardy spaces. Let us look first at one of the simplest cases. Previously, we noted that if θ is a square integrable function defined on the sphere S^{n-1}, then there is a harmonic function h defined on the unit ball in $\mathbb{R}^n$ with boundary value θ almost everywhere. Also, we have seen that $h(x) = f_1(x) + xf_2(x)$ where f_1 and f_2 are left Clifford holomorphic. However, on S^{n-1} the function $G(x) = x$. One can see that on S^{n-1} we have $\theta(x) = f_1(x) + g(x)$ almost everywhere. Here, f_1 is left monogenic on the unit ball $B(0,1)$ and g is left Clifford holomorphic on $\mathbb{R}^n \backslash \overline{B(0,1)}$, where $\overline{B(0,1)}$ is the closure of the open unit ball. Let $H^2(B(0,1))$ denote the space of Clifford holomorphic functions defined on $B(0,1)$, with extensions to a square integrable functions on S^{n-1}, and let $H^2(\mathbb{R}^n \backslash \overline{B(0,1)}$ denote the class of left Clifford holomorphic functions defined on $\mathbb{R}^n \backslash \overline{B(0,1)}$, with square integrable extensions to S^{n-1}. What we have so far outlined is that

$$L^2(S^{n-1}) = H^2(B(0,1)) \oplus H^2(\mathbb{R}^n \backslash \overline{B(0,1)}),$$

where $L^2(S^{n-1})$ is the space of $C\ell_n$ valued Lebesgue square integrable functions defined on S^{n-1}. This is the Hardy 2-space decomposition of $L^2(S^{n-1})$. It is also true if we replace 2 by p where $1 < p < \infty$. We will not go into more details here, since it is beyond the scope of the material presented here.

Let us now take an alternative look at a way of obtaining this decomposition. This method will generalize to all reasonable surfaces. We will clarify what we mean by a reasonable surface later. Instead of considering an arbitrary square integrable function on S^{n-1}, let us instead assume that θ is a continuously differentiable function. Let us now consider the integral

$$\frac{1}{\omega_n} \int_{S^{n-1}} G(x-y) n(x) \theta(x) \, d\sigma(x)$$

where $y \in B(0,1)$. This defines a left Clifford holomorphic function on $B(0,1)$. Now let the point y approach a point $z \in S^{n-1}$ along a differentiable path $y(t)$. Let us also assume that $\frac{dy(t)}{dt}$ is evaluated at $t = 1$, so that $y(t) = z$ is not tangential to S^{n-1} at z. We can essentially ignore this last point at a first read. We want to evaluate

$$\lim_{t\to 1} \frac{1}{\omega_n} \int_{S^{n-1}} G(x-y(t))n(x)\theta(x)\,d\sigma(x).$$

We do this by removing a small ball on $B(0,1)$ from S^{n-1}. The ball is centered at z and is of radius ϵ. We denote this ball by $b(z,\epsilon)$. The previous integral now splits into an integral over $b(z,\epsilon)$ and an integral over $S^{n-1}\backslash b(z,\epsilon)$. On $b(z,\epsilon)$, we can express $\theta(x)$ as $(\theta(x)-\theta(z))+\theta(z)$. As θ is continuously differentiable,

$$\|\theta(x)-\theta(z)\| < C\|x-z\|$$

for some $C \in \mathbb{R}^+$. It follows that

$$\lim_{\epsilon\to 0}\lim_{t\to 1} \int_{b(z,\epsilon)} \|G(x-y(t))n(x)(\theta(x)-\theta(z))\|\,d\sigma(x) = 0.$$

Moreover, the term

$$\lim_{\epsilon\to 0}\lim_{t\to 1} \frac{1}{\omega_n} \int_{b(z,\epsilon} G(x-y(t))n(x)\theta(z)\,d\sigma(x)$$

can be replaced by the term

$$\lim_{\epsilon\to 0}\lim_{t\to 1} \int_{B(0,1)\cap\partial B(z,\epsilon)} G(x-y(t))n(x)\theta(z)\,d\sigma(x),$$

since $\theta(z)$ is a Clifford holomorphic function. By the residue theorem the limit of this integral evaluates to $\frac{1}{2}\theta(z)$.

We leave it to the interested reader to note that the singular integral or principal valued integral

$$\lim_{\epsilon\to 0}\lim_{t\to 1} \frac{1}{\omega_n} \int_{S^{n-1}\backslash b(z,\epsilon)} G(x-y(t))n(x)\theta(x)\,d\sigma(x)$$
$$= P.V.\frac{1}{\omega_n} \int_{S^{n-1}} G(x-z)n(x)\theta(x)\,d\sigma(x)$$

is bounded.

We have established that

$$\lim_{t\to 1} \int_{S^{n-1}} G(x-y(t))n(x)\theta(x)\,d\sigma(x)$$
$$= \tfrac{1}{2}\theta(z) + P.V.\frac{1}{\omega_n} \int_{S^{n-1}} G(x-z)n(x)\theta(x)\,d\sigma(x).$$

If we now assumed that $y(t)$ is a path tending to z on the complement of $\overline{B(0,1)}$, then similar arguments give

$$\lim_{t\to 1}\int_{S^{n-1}} G(x-y(t))n(x)\theta(x)\,d\sigma(x)$$
$$= -\tfrac{1}{2}\theta(z) + P.V.\frac{1}{\omega_n}\int_{S^{n-1}} G(x-z)n(x)\theta(x)\,d\sigma(x).$$

We will write these expressions as

$$(\pm\tfrac{1}{2}I + C_{S^{n-1}})\theta.$$

If we consider the limit

$$\lim_{t\to 1}\frac{1}{\omega_n}\int_{S^{n-1}} G(x-y(t))n(x)(\tfrac{1}{2}I + C_{S^{n-1}})\theta(x)\,d\sigma(x),$$

we may determine that

$$(\tfrac{1}{2}I + C_{S^{n-1}})^2 = \tfrac{1}{2}I + C_{S^{n-1}}.$$

Furthermore,

$$(\tfrac{1}{2}I + C_{S^{n-1}})(-\tfrac{1}{2}I + C_{S^{n-1}}) = 0 \quad\text{and}\quad (-\tfrac{1}{2}I + C_{S^{n-1}})^2 = -\tfrac{1}{2}I + C_{S^{n-1}}.$$

It is known that each function $\psi \in L^2(S^{n-1})$ can be approximated by a sequence of functions, each with the same properties as θ. This tells us that the previous formulas can be repeated, but this time simply for $\theta \in L^2(S^{n-1})$. It follows that for such a θ we have

$$\theta = (\tfrac{1}{2}I + C_{S^{n-1}})\theta + (\tfrac{1}{2}I - C_{S^{n-1}})\theta.$$

This formula gives the Hardy space decomposition of $L^2(S^{n-1})$. In fact, if one looks more carefully at the previous calculations used to obtain these formulas we see that it is not so significant that the surface used is a sphere, and we can redo the calculations for any "reasonable" hypersurface S. In this case, we get

$$\theta = (\tfrac{1}{2}I + C_S)\theta + (\tfrac{1}{2}I - C_S)\theta$$

where θ now belongs to $L^2(S)$, and

$$C_S\theta = P.V.\frac{1}{\omega_n}\int_S G(x-y)n(x)\theta(x)\,d\sigma(x).$$

This gives rise to the Hardy space decomposition

$$L^2(S) = H^2(S^+) \oplus H^2(S^-),$$

where $S^{\pm}$ are the two domains that complement the surface S (we are assuming that S divides $\mathbb{R}^n$ into two complementary domains).

Lastly, one should address the smoothness of S. In some parts of the literature, one simply assumes that S is compact and C^2. More recently, one assumes that S has rougher conditions, usually that the surface is Lipschitz continuous, see for instance [21, 22, 27]. The formulas given above, involving the singular integral operator C_S, are called Plemelj formulas. It is a simple exercise to show that these formulas are conformally invariant. Using Kelvin inversion, or even a Cayley transformation, one can show that these formulas and the Hardy space decompositions are also valid on unbounded surfaces and domains. A great deal of recent Clifford analysis has been devoted to the study of such Hardy spaces and singular integral operators. This is due to an idea of R. Coifman, that various hard problems in classical harmonic analysis, studied in Euclidean space, might be more readily handled using tools from Clifford analysis, particularly the singular Cauchy transform and associated Hardy spaces.

In particular, Coifman speculated that a more direct proof of the celebrated Coifman–McIntosh–Meyer Theorem [7], could be derived using Clifford analysis. The Coifman–McIntosh–Meyer Theorem establishes the L^2 boundedness of the double layer potential operator for Lipschitz graphs in $\mathbb{R}^n$. Coifman's observation was that the double layer potential operator is the real or scalar part of the singular Cauchy transform arising in Clifford analysis and discussed earlier. If one can establish the L^2 boundedness of the singular Cauchy transform for a Lipschitz graph in $\mathbb{R}^n$, then one automatically has the L^2 boundedness for the double layer potential operator for the same graph. The L^2 boundedness of the singular Cauchy transform was first established for Lipschitz graphs with small constant by Murray [30], and extended to the general case by McIntosh, see [26, 27]. One very important reason for needing to know that the double layer potential operator is L^2 bounded for Lipschitz graphs is to be able to solve boundary value problems for domains with Lipschitz graphs as boundaries. Such boundary value problems would include the Dirichlet problem and Neuman problem for the Laplacian. See [26, 27] for more details. In [49] Clifford analysis, and more precisely the Hardy space decomposition mentioned here, is specifically used to solve the water wave problem in three dimensions.

3.8 More on Clifford analysis on the sphere

In the previous section, we saw that $L^2(S^{n-1})$ splits into a direct sum of Hardy spaces for the corresponding complementary domains $B(0,1)$ and $\mathbb{R}^n \backslash \overline{B(0,1)}$. In an earlier section, we saw that any left Clifford holomorphic function $f(x)$ can be expressed as a locally uniformly convergent series $\sum_{j=0}^{\infty} f_j(x)$, where each $f_j(x)$ is left Clifford holomorphic and homogeneous of degree j. Now following [47], consider the operator

$$D = x^{-1}xD = x^{-1}(\sum_{i<k} e_i e_k (x_i \frac{\partial}{\partial x_k} - x_k \frac{\partial}{\partial x_i} - \sum_{j=1}^{n} x_j \frac{\partial}{\partial x_j})).$$

By letting the last term in this expression act on homogeneous polynomials, one may determine from Euler's lemma that

$$\sum_{j=1}^{n} x_j \frac{\partial}{\partial x_j}$$

is the radial operator $r\frac{\partial}{\partial r}$. So $r\frac{\partial}{\partial r} f_j(x) = j f_j(x)$. Since each polynomial f_k is Clifford holomorphic, it follows that each f_j is an eigenvector of the spherical Dirac operator

$$x\Lambda_{n-1} = x \sum_{i<k} e_i e_k \big(x_i \frac{\partial}{\partial x_k} - x_k \frac{\partial}{\partial x_i}\big)$$

with eigenvalue k. Now using Kelvin inversion, we know that f_k is homogeneous of degree k and left Clifford holomorphic if and only if $G(x) f_k(x^{-1})$ is homogeneous of degree $-n+1-k$ and is left Clifford holomorphic. On restricting $G(x) f_k(x^{-1})$ to the unit sphere, this function becomes $x f_k(x^{-1})$ and this function is an eigenvector for the spherical Dirac operator $x\Lambda_{n-1}$. Since each $f \in H^2(\mathbb{R}^n \backslash \overline{B(0,1)})$ can be written as

$$\sum_{k=0}^{\infty} G(x) f_k(x^{-1}),$$

where each f_k is homogeneous of degree k and is left Clifford holomorphic, it follows that if $h \in L^2(S^{n-1})$, then

$$\Lambda_{n-1} x h(x) = (1-n) x h(x) - x \Lambda_{n-1} h(x).$$

Similarly, if we replace S^{n-1} by the n-sphere S^n embedded in $\mathbb{R}^{n+1}$, then we have the identity

$$\Lambda_n x h(x) = -n x h(x) - x \Lambda_n h(x)$$

for each $h \in L^2(S^n)$. As all C^∞ functions defined on S^n belong to $L^2(S^n)$, this identity holds for all such functions too.

It should be noted that for each $x \in S^n$, if we restrict the operator $x\Lambda_n$ to the tangent bundle TS^n_x, then we obtain the Euclidean Dirac operator acting on this tangent space.

By using the Cayley transformation

$$x = \psi(y)(e_{n+1} y + 1)(y + e_{n+1})^{-1}$$

from $\mathbb{R}^n$ to $S^n \backslash \{e_{n+1}\}$, one can transform left Clifford holomorphic functions from domains in $\mathbb{R}^n$ to functions defined on domains lying on the sphere. If $f(y)$ is left Clifford holomorphic on the domain U lying in $\mathbb{R}^n$, then we obtain a function $f'(x) = J(\psi^{-1}, x) f(\psi^{-1}(x)$ defined on the domain $U' = \psi(U)$ lying on S^n. Here

$$J(\psi^{-1}, x) = \frac{x+1}{\|x+1\|^n}.$$

Similarly, if $g(y)$ is right Clifford holomorphic on U, then

$$g'(x) = g(\psi^{-1}(x)J(\psi^{-1}, x)$$

is a well-defined function on U'. Moreover for any smooth, compact hypersurface S bounding a subdomain V of U, we have from the conformal invariance of Cauchy's Theorem,

$$\int_{S'} g'(x)n(x)f'(x)\, d\sigma'(x) = 0$$

where $S' = \psi(S)$, and $n(x)$ is the unit vector lying in the tangent space TS^n_x of S^n at x and outer normal to S' at x. Furthermore σ' is the Lebesgue measure on S'.

From Lemma 1, it now follows that for each point $y' \in V' = \psi(V)$, we have the following version of Cauchy's Integral Formula:

$$f'(y') = \frac{1}{\omega_n}\int_{S'} G(x - y')n(x)f'(x)\, d\sigma(x),$$

where, as before,

$$G(x - y') = \frac{x - y'}{\|x - y'\|^n},$$

but now x and $y' \in S^n$. It would now appear that the functions f' and g' are solutions to some spherical Dirac equations. We need to isolate this Dirac operator. We shall achieve this by applying the operator $x\Lambda_n$ to the kernel $G_s(x, y') = G(x - y')$. Since x and $y' \in S^n$,

$$\|x - y'\|^2 = 2 - 2 < x, y' >,$$

where $< x, y' >$ is the inner product of x and y'. So

$$G_s(x, y') = 2^{\frac{-n}{2}} \frac{x - y'}{(1- < x, y' >)^{\frac{n}{2}}}.$$

In calculating $x\Lambda_n G_s(x, y')$, we need to know what $\Lambda_n < x, y' >$ evaluates to. It is a simple exercise to determine that

$$\Lambda_n < x, y' >= xy' + < x, y' >,$$

which is the wedge product $x \wedge y'$ of x with y'.

Now let us calculate $x\Lambda_n G_s(x, y')$:

$$x\Lambda_n G_s(x, y') = 2^{\frac{-n}{2}}(x\Lambda_n \frac{x}{(1- < x, y' >)^{\frac{n}{2}}} - x\Lambda_n \frac{y'}{(1- < x, y' >)^{\frac{n}{2}}}) =$$

$$2^{\frac{n}{2}}(-x\frac{nx}{(1- < x, y' >)^{\frac{n}{2}}} + \Lambda_n \frac{1}{(1- < x, y' >)^{\frac{-n}{2}}} - x\Lambda_n \frac{1}{(1- < x, y' >)^{\frac{n}{2}}} y').$$

First,

$$\Lambda_n \frac{1}{(1-<x,y'>)^{\frac{n}{2}}} = \frac{n}{2}\frac{x\wedge y'}{(1-<x,y'>)^{\frac{n+2}{2}}},$$

so

$$x\Lambda_n G_s(x,y')$$
$$= 2^{\frac{-n}{2}} \frac{n}{2(1-<x,y>)^{\frac{n+2}{2}}}(2(1-<x,y'>) + x\wedge y' - x(x\wedge y')y').$$

The expression

$$2(1-<x,y'>) + x\wedge y' - x(x\wedge y')y'$$

is equal to

$$2-2<x,y'> + xy' + <x,y'> - x(xy + <x,y'>)y'.$$

This expression simplifies to

$$1-<x,y'> + xy' - xy'<x,y'>,$$

which in turn simplifies to

$$(1-<x,y'>)(1+xy') = -x(1-<x,y'>)(x-y').$$

So

$$x\Lambda_n G_s(x,y') = \frac{-n}{2}xG_s(x,y').$$

Hence

$$x(\Lambda_n + \frac{n}{2})G_s(x,y') = 0,$$

so the Dirac operator, D_s, over the sphere is $x(\Lambda_n + \frac{n}{2})$. It follows from our Cauchy integral formula for the sphere that $D_s f'(x) = 0$. For more details on this operator, related operators and their properties see [8, 25, 38, 39, 48].

Besides the operator D_s, we also need a Laplacian $\triangle_s$ acting on functions defined on domains on S^n. To do this, we will work backwards and look for a fundamental solution to $\triangle_n$. A strong candidate for such a fundamental solution is the kernel

$$H_s(x,y') = \frac{1}{n-2}\frac{1}{\|x-y'\|^{n-2}}.$$

By similar considerations to those made in the previous calculation, we find that

$$D_s H_s(x,y') = -xH_s(x,y') + G_s(x,y').$$

So

$$(D_s + x)H_s(x,y') = G_s(x,y').$$

Therefore we may define our Laplacian $\triangle_s$ to be $D_s(D_s + x)$.

Definition 2. *Suppose h is a $C\ell_n$ valued function defined on a domain U' of S^n. Then h is called a harmonic function on U' if $\Delta_s h = 0$.*

In much the same way as one would derive Green's Theorem in $\mathbb{R}^n$, one now has

Theorem 10. *Suppose U' is a domain on S^n and $h : U' \to C\ell_n$ is a harmonic function on U'. Suppose also that S' is a smooth hypersurface lying in U' and that S' bounds a subdomain V' of U' and that $y' \in V'$. Then*

$$h(y') = \frac{1}{\omega_n}\int_{S'}(G_s(x,y')n(x)h(x) + H_s(x,y')n(x)D_s h(x))\,d\sigma'(x).$$

See [25] for more details.

3.9 The Fourier transform and Clifford analysis

Closely related to Hardy spaces is the Fourier transform. Here we will consider $\mathbb{R}^n$ as divided into the upper and lower half spaces $\mathbb{R}^{n+}$ and $\mathbb{R}^{n-}$, where

$$\mathbb{R}^{n+} = \{x = x_1e_1 + \ldots + x_ne_n : x_n > 0\}$$

and

$$\mathbb{R}^{n-} = \{x = x_1e_1 + \ldots + x_ne_n : x_n < 0\}.$$

These two domains have $\mathbb{R}^{n-1} = \text{span} < e_1, \ldots, e_{n-1} >$ as a common boundary. As before,

$$L^2(\mathbb{R}^{n-1}) = H^2(\mathbb{R}^{n+}) \oplus H^2(\mathbb{R}^{n-}).$$

Let us now consider a function $\psi \in L^2(\mathbb{R}^{n-1})$. Then

$$\psi(y) = \left(\tfrac{1}{2}\psi(y) + \frac{1}{\omega_n}P.V.\int_{\mathbb{R}^{n-1}} G(x'-y)e_n\psi(x')\,dx^{n-1}\right) + \left(\tfrac{1}{2}\psi(y) - \frac{1}{\omega_n}P.V.\int_{\mathbb{R}^{n-1}} G(x'-y)e_n\psi(x')\,dx^{n-1}\right)$$

almost everywhere. Here

$$\tfrac{1}{2}\psi(y) + \frac{1}{\omega_n}P.V.\int_{\mathbb{R}^{n-1}} G(x'-y)e_n\psi(x')\,dx^{n-1}$$

is the nontangential limit of

$$\frac{1}{\omega_n}\int_{\mathbb{R}^{n-1}} G(x'-y(t))e_n\psi(x')\,dx^{n-1},$$

as $y(t)$ tends to y nontangentially through a smooth path in the upper half space, and

$$\tfrac{1}{2}\psi(y) - \frac{1}{\omega_n}P.V.\int_{\mathbb{R}^{n-1}} G(x'-y)e_n\psi(x')\,dx^{n-1}$$

is the nontangential limit of

$$\frac{-1}{\omega_n}\int_{\mathbb{R}^{n-1}} G(x'-y(t))e_n\psi(x')\,dx^{n-1}$$

as $y(t)$ tends nontangentially to y through a smooth path in lower half space.

Consider now the Fourier transform, $\mathcal{F}(\psi)=\hat{\psi}$, of ψ. In order to proceed, we need to calculate

$$\mathcal{F}(\tfrac{1}{2}\psi \pm \frac{1}{\omega_n} P.V.\int_{\mathbb{R}^{n-1}} G(x'-y)e_n\psi(x')\,dx^{n-1}).$$

In particular, we need to determine

$$\mathcal{F}(\frac{1}{\omega_n} P.V.\int_{\mathbb{R}^{n-1}} G(x'-y)e_n\psi(x')\,dx^{n-1}).$$

Following [26], it may be determined that this is

$$\tfrac{1}{2}i\frac{\zeta}{\|\zeta\|}e_n\hat{\psi}(\zeta),$$

where $\zeta=\zeta_1 e_1+\ldots+\zeta_{n-1}e_{n-1}$. So

$$\mathcal{F}(\tfrac{1}{2}\psi \pm \frac{1}{\omega_n}\int_{\mathbb{R}^{n-1}} G(x'-y)e_n\psi(x')\,d\sigma(x'))=\tfrac{1}{2}(1\pm i\frac{\zeta}{\|\zeta\|}e_n).$$

Now, as observed in [26],

$$(\tfrac{1}{2}(1\pm i\frac{\zeta}{\|\zeta\|}e_n))^2=\tfrac{1}{2}(1\pm i\frac{\zeta}{\|\zeta\|}e_n),\quad \tfrac{1}{2}(1\pm i\frac{\zeta}{\|\zeta\|}e_n)\tfrac{1}{2}(1\mp i\frac{\zeta}{\|\zeta\|}e_n)=0$$

and

$$i\zeta e_n\tfrac{1}{2}(1\pm i\frac{\zeta}{\|\zeta\|}e_n)=\|\zeta\|\tfrac{1}{2}(1\pm i\frac{\zeta}{\|\zeta\|}e_n).$$

Taking the Cauchy–Kowalewska extension of $e^{i<x',\zeta>}$, we get

$$\exp(i<x',\zeta>-ix_n e_n\zeta)$$

defined on some neighborhood in $\mathbb{R}^n$ of $\mathbb{R}^{n-1}$.

Now consider

$$\exp(i<x',\zeta>-ix_n e_n\zeta)\tfrac{1}{2}(1\pm i\frac{\zeta}{\|\zeta\|}e_n),$$

which simplifies to

$$e^{i<x',\zeta>-x_n\|\zeta\|}\tfrac{1}{2}(1+i\frac{\zeta}{\|\zeta\|}e_n)$$

if $x_n > 0$, and to

$$e^{i<x',\zeta>+x_n\|\zeta\|}\frac{1}{2}(1 - i\frac{\zeta}{\|\zeta\|}e_n)$$

if $x_n < 0$. The first of these series converges locally uniformly on the upper half space while the second series converges locally uniformly on the lower half space. We denote these two functions by $e_{\pm}(x,\zeta)$, respectively. The integrals

$$\frac{1}{\omega_n}\int_{\mathbb{R}^{n-1}} e_{\pm}(x,\zeta)\hat{\psi}(\zeta)\,d\sigma(\zeta)$$

define the left monogenic functions $\Psi_{\pm}(x)$ on the upper and lower half spaces, respectively. Moreover,

$$\Psi_{\pm} \in H^2(\mathbb{R}^{n\pm})$$

explicitly gives the Hardy space decomposition of $\psi \in L^2(\mathbb{R}^{n-1})$. The links between the Fourier transform and Clifford analysis were first found in [44], and later independently rediscovered and applied in [22].

The integrals

$$\frac{1}{\omega_n}\int_{\mathbb{R}^{n-1}} e_{\pm}(x,\zeta)\,d\zeta^{n-1}$$

can be expressed in polar coordinates as

$$\frac{1}{\omega_n}\int_{S^{n-2}}\int_0^\infty e^{ir<x',\zeta'>\pm x_n r}r^{n-2}\frac{1}{2}(1 \pm \frac{\zeta}{\|\zeta\|}e_n)\,dr dS^{n-2},$$

where $\zeta' = \frac{\zeta}{\|\zeta\|}$. In [22], Chun Li observed that the integrals

$$\int_0^\infty e^{ir<x',\zeta'>\pm x_n r}\frac{1}{2}(1 \pm \zeta' e_n)r^{n-2}\,dr$$

are Laplace transforms of the function $f(R) = R^{n-2}$. So the integral

$$\int_0^\infty e^{ir<x',\zeta'>-x_n r}\frac{1}{2}(1 - i\zeta' e_n)\,dr$$

evaluates to

$$(-i)^n(n-2)!(< x',\zeta' > +ix_n)^{-n+1}.$$

Hence the integral

$$\frac{1}{\omega_n}\int_{\mathbb{R}^{n-1}} e_{+}(x,\zeta)\,d\zeta^{n-1}$$

becomes

$$\frac{1}{\omega_n}\int_{S^{n-2}}(-i)^n(n-2)!(< x',\zeta' > +ix_n)^{-n+1}\frac{1}{2}(1 + ie_n\zeta')\,dS^{n-2}.$$

For any complex number $a + ib$, the product $(a + ib)\frac{1}{2}(1 + ie_n\zeta')$ is equal to $(a - be_n)\frac{1}{2}(1 + ie_n\zeta')$. Thus, the previous integral becomes

$$\frac{1}{\omega_n}\int_{S^{n-2}}(-i)^n(<x',\zeta'> -x_ne_n\zeta')^{-n+1}\tfrac{1}{2}(1+ie_n\zeta')\,dS^{n-2}.$$

The imaginary, or $iC\ell_n$ part of this integral is the integral of an odd function, so when n is even, the integral becomes

$$\frac{1}{\omega_n}\int_{S^{n-2}}\frac{(n-2)!e_n\zeta'}{2(<x',\zeta'> -x_ne_n\zeta')^{n-1}}\,dS^{n-2},$$

and when n is odd, the integral becomes

$$\frac{-1}{\omega_n}\int_{S^{n-2}}\frac{(n-2)!}{(<x',\zeta'> -x_ne_n\zeta')^{n-1}}\,dS^{n-2}.$$

These integrals are the plane wave decompositions of the Cauchy kernel for the upper half space, described by Sommen in [45]. It should be noted that while introducing the Fourier transform and exploring some of its links with Clifford analysis, we have also been forced to complexify the Clifford algebra $C\ell_n$ to the complex Clifford algebra $C\ell_n(\mathbb{C})$. Furthermore, the functions $\frac{1}{2}(1 \pm i\frac{\zeta}{\|\zeta\|e_n})$ are defined on spheres lying in the null cone

$$\{x_ne_n + iw' : w' \in \mathbb{R}^{n-1},\ x_n^2 - \|w'\|^2 = 0\}.$$

This leads naturally to the question: What domains in $\mathbb{C}^n$ do the functions $e_\pm(x,\zeta)$ extend to?

Here we are replacing the real vector variable $x \in \mathbb{R}^n$ by a complex vector variable $\underline{z} = z_1e_1 + \ldots + z_ne_n \in \mathbb{C}^n$, where $z_1, \ldots, z_n \in \mathbb{C}$. Letting $\underline{z} = x + iy$ where x and y are real vector variables, the term

$$e^{-<\zeta,y'>-x_n\|\zeta\|}$$

is well defined for $x_n\|\zeta\| > | <\zeta, y'> |$. In this case, $iy' + x_ne_n$ varies over the interior of the forward null cone

$$\{iy' + x_ne_n : x_n > 0 \quad \text{and} \quad x_n > \|y'\|\},$$

so $e_+(\underline{z},\zeta)$ is well defined for each

$$\underline{z} = x + iy = x' + iy' + (x_n + iy_n)e_n \in \mathbb{C}^n,$$

where $x' \in \mathbb{R}^{n-1}$, $y_n \in \mathbb{R}$, $x_n > 0$ and $\|y'\| < x_n$. Similarly, $e_-(x,\zeta)$ holomorphically extends to

$$\{\underline{z} = x' + iy' + (x_n + iy_n)e_n : x' \in \mathbb{R}^{n-1},\ x_n < 0,\ y_n \in \mathbb{R},\ \|y'\| < |x_n|\}.$$

We denote these domains by $C^\pm$, respectively. It should be noted that $\Psi^\pm$ holomorphically continue to $C^\pm$, respectively. We denote these holomorphic continuations of $\Psi^\pm$ by $\Psi'^\pm$. The domains $C^\pm$ are examples of tube domains.

3.10 Complex Clifford analysis

In the previous section, we showed that any left Clifford holomorphic function $f \in H^2(\mathbb{R}^{n,+})$ can be holomorphically continued to a function $f^\dagger$ defined on a tube domain C^+ in $\mathbb{C}^n$. In this section, we will briefly show how this type of holomorphic continuation happens for all Clifford holomorphic functions defined on a domain $U \subset \mathbb{R}^n$.

Let S be a compact smooth hypersurface lying in U, and suppose that S bounds a subdomain V of U. Cauchy's integral formula gives

$$f(y) = \frac{1}{\omega_n} \int_S G(x-y)n(x)f(x)\,d\sigma(x)$$

for each $y \in V$. Let us now complexify the Cauchy kernel. The function $G(x)$ holomorphically continues to

$$\frac{\underline{z}}{(\underline{z}^2)^{\frac{n}{2}}}.$$

In even dimensions this is a well-defined function on $\mathbb{C}^n \backslash N(0)$, where $N(0) = \{\underline{z} \in \mathbb{C}^n : \underline{z}^2 = 0\}$. In odd dimensions this lifts to a well-defined function on a complex n-dimensional Riemann surface double covering $\mathbb{C}^n \backslash N(0)$. Though things work out well in odd dimensions, for simplicity we will work with the cases where n is even. In holomorphically extending $G(x-y)$ in the variable y, we obtain a function

$$G^\dagger(x-\underline{z}) = \frac{x-\underline{z}}{((x-\underline{z})^2)^{\frac{n}{2}}}.$$

This function is well defined on $\mathbb{C}^n \backslash N(x)$, where

$$N(x) = \{\underline{z} \in \mathbb{C}^n : (x-\underline{z})^2 = 0\}.$$

It follows that the integral

$$\frac{1}{\omega_n} \int_S G^\dagger(x-\underline{z})n(x)f(x)\,d\sigma(x)$$

is well defined provided $\underline{z}$ is not in $N(x)$ for any $x \in S$. The set

$$\mathbb{C}^n \backslash \cup_{x \in S} N(x)$$

is an open set in $\mathbb{C}^n$. We shall take the component of this open set which contains V and denote it by $V^\dagger$. It follows that the left Clifford holomorphic function $f(y)$ has a holomorphic extension $f^\dagger(\underline{z})$ to $V^\dagger$. Furthermore, this function is given by the integral formula

$$f^\dagger(\underline{z}) = \frac{1}{\omega_n} \int_S G^\dagger(x-\underline{z})n(x)f(x)\,d\sigma(x).$$

The holomorphic function $f^\dagger$ is a solution to the complex Dirac equation $D^\dagger f^\dagger = 0$, where

$$D^\dagger = \sum_{j=1}^{n} e_j \frac{\partial}{\partial z_j}.$$

By allowing the hypersurface to deform and move out to include more of U in its interior, we see that $f^\dagger$ is a well-defined holomorphic function on $U^\dagger$, where $U^\dagger$ is the component of

$$\mathbb{C}^n \backslash \cup_{x \in \overline{U} \backslash U} N(x)$$

which contains U. In the special cases where U is either of $\mathbb{R}^{n,\pm}$, then $U^\dagger = C^\pm$. See [34–36] for more details.

3.11 References

[1] L. V. Ahlfors, Möbius transformations in $\mathbb{R}^n$ expressed through 2×2 matrices of Clifford numbers, *Complex Variables* **5** (1986), 215–224.

[2] M. F. Atiyah, R. Bott and A. Shapiro, Clifford modules, *Topology* **3** (1965), 3–38.

[3] B. Bojarski, Conformally covariant differential operators, *Proceedings, XXth Iranian Math. Congress*, Tehran, 1989.

[4] B. Booss-Bavnbek and K. Wojciechowski, *Elliptic Boundary Problems for Dirac Operators*, Birkhäuser, Boston, 1993.

[5] F. Brackx, R. Delanghe and F. Sommen, *Clifford Analysis*, Pitman, London, 1982.

[6] D. Calderbank, *Clifford analysis for Dirac operators on manifolds with boundary*, Max-Plank-Institut für Mathematik preprint, Bonn, 1996.

[7] R. Coifman, A. McIntosh and Y. Meyer, L'intégrale de Cauchy définit un opérateur borné sur L^2 pour les courbes lipschitziennes, *Annals of Mathematics* **116** (1982), 361–387.

[8] J. Cnops and H. Malonek, *An introduction to Clifford analysis*, Univ. Coimbra, Coimbra, 1995.

[9] R. Delanghe, On regular-analytic functions with values in a Clifford algebra, *Math. Ann.* **185** (1970), 91–111.

[10] R. Delanghe, F. Sommen and V. Soucek, *Clifford Algebra and Spinor-Valued Functions*, Kluwer, Dordrecht, 1992.

[11] A. C. Dixon, On the Newtonian potential, *Quarterly Journal of Mathematics*, 35, 1904, 283–296.

[12] R. Fueter, Die Funktionentheorie der Differentialgleichungen $\Delta u = 0$ und $\Delta\Delta u = 0$ mit Vier Reallen Variablen, *Commentarii Mathematici Helvetici* **7** (1934–1935), 307–330.

[13] K. Guerlebeck and W. Sproessig, *Quaternionic and Clifford Calculus for Physicists and Engineers*, Wiley, Chichester, 1998.

[14] J. Gilbert and M. A. M. Murray, *Clifford Algebras and Dirac Operators in Harmonic Analysis*, CUP, Cambridge, 1991.

[15] D. Hestenes, Multivector functions, *J. Math. Anal. Appl.* **24** (1968), 467–473.

[16] V. Iftimie, Fonctions hypercomplexes, *Bull. Math. de la Soc. Sci. Math. de la R. S. de Roumanie* **9** (1965), 279–332.

[17] R. S. Kraußhar, Automorphic forms in Clifford analysis, *Complex Variables* **47** (2002), 417–440.

[18] R. S. Kraußhar and J. Ryan, Clifford and harmonic analysis on cylinders and tori, to appear.

[19] R. S. Kraußhar and J. Ryan, Some conformally flat spin manifolds, Dirac operators and automorphic forms, to appear.

[20] V. Kravchenko and M. Shapiro, *Integral Representations for Spatial Models of Mathematical Physics*, Pitman Research Notes in Mathematics, London, No. 351, 1996.

[21] C. Li, A. McIntosh and S. Semmes, Convolution singular integrals on Lipschitz surfaces, *JAMS* **5** (1992), 455–481.

[22] C. Li, A. McIntosh and T. Qian, Clifford algebras, Fourier transforms, and singular convolution operators on Lipschitz surfaces, *Revista Mathematica Iberoamericana* **10** (1994), 665–721.

[23] T. E. Littlewood and C. D. Gay, Analytic spinor fields, *Proc. Roy. Soc., A313* (1969), 491–507.

[24] J. Liouville, *Extension au cas des trois dimensions de la question du trace géographique. Applications de l'analyse à la géométrie*, G. Monger, Paris, 1850, pp. 609–616.

[25] H. Liu and J. Ryan, Clifford analysis techniques for spherical pde, *Journal of Fourier Analysis and its Applications* **8** (2002), 535–564.

[26] A. McIntosh, Clifford algebras, Fourier theory, singular integrals, and harmonic functions on Lipschitz domains, *Clifford Algebras in Analysis and Related Topics*, ed. J. Ryan, CRC Press, Boca Raton, 1996, pp. 33–88.

[27] M. Mitrea, *Clifford Wavelets, Singular Integrals and Hardy Spaces*, Lecture Notes in Mathematics, Springer-Verlag, Heidelberg, No. 1575, 1994.

[28] M. Mitrea, Generalized Dirac operators on non-smooth manifolds and Maxwell's equations, *Journal of Fourier Analysis and its Applications* **7** (2001), 207–256.

[29] Gr. C. Moisil and N. Theodorescu, Fonctions holomorphes dans l'espace, *Mathematica (Cluj)* **5** (1931), 142–159.

[30] M. Murray, The Cauchy integral, Calderon commutation, and conjugation of singular integrals in $\mathbb{R}^n$, *Trans. of the AMS* **298** (1985), 497–518.

[31] E. Obolashvili, *Partial Differential Equations in Clifford Analysis*, Pitman Monographs and Surveys in Pure and Applied Mathematics No. 96, Harlow, 1998.

[32] T. Qian, Singular integrals on star shaped Lipschitz surfaces in the quaternionic space, *Math Ann.* **310** (1998), 601–630.

[33] T. Qian, Fourier analysis on starlike Lipschitz surfaces, *Journal of Functional Analysis* **183** (2001), 370–412.

[34] J. Ryan, Cells of Harmonicity and generalized Cauchy integral formulae, *Proc. London Math. Soc* **60** (1990), 295–318.

[35] J. Ryan, Intrinsic Dirac operators in C^n, *Adv. in Math.* **118** (1996), 99–133.

[36] J. Ryan, Complex Clifford analysis and domains of holomorphy, *Journal of the Australian Mathematical Society, Series A* **48** (1990), 413–433.

[37] J. Ryan, Ed., *Clifford Algebras in Analysis and Related Topics*, Studies in Advanced Mathematics, CRC Press, Boca Raton, 1995.

[38] J. Ryan, Basic Clifford analysis, *Cubo Matematica Educacional* **2** (2000), 226–256.

[39] J. Ryan, Dirac operators on spheres and hyperbolae, *Bolletin de la Sociedad Matematica a Mexicana* **3** (1996), 255–270.

[40] M. Sce, Osservaziono sulle series di potenzi nei moduli quadratici, *Atti. Acad. Nax. Lincei Rend Sci Fis Mat. Nat*, **23** (1957), 220–225.

[41] S. Semmes, Some remarks concerning integrals of curvature on curves and surfaces, to appear.

[42] F. Sommen, Spherical monogenics and analytic functionals on the unit sphere, *Tokyo J. Math.* **4** (1981), 427–456.

[43] F. Sommen, A product and an exponential function in hypercomplex function theory, *Appl. Anal.* **12** (1981), 13–26.

[44] F. Sommen, Microfunctions with values in a Clifford algebra II, *Scientific Papers of the College of Arts and Sciences, The University of Tokyo*, 36, 1986, pp. 15–37.

[45] F. Sommen, Plane wave decomposition of monogenic functions, *Ann. Polon. Math.* **49** (1988), 101–114.

[46] E. M. Stein and G. Weiss, *Introduction to Fourier Analysis on Euclidean Space*, Princeton University Press, Princeton, 1971.

[47] A. Sudbery, Quaternionic analysis, *Math. Proc. of the Cambridge Philosophical Soc.* **85** (1979), 199–225.

[48] P. Van Lancker, Clifford analysis on the sphere, *Clifford Algebras and their Applications in Mathematical Physics*, V. Diedrich et al., editors, Kluwer, Dordrecht, 1998, pp. 201–215.

[49] S. Wu, Well-posedness in Sobolev spaces of the full water wave problem in 3-D, *JAMS* **12** (1999), 445–495.

John Ryan
Department of Mathematics
University of Arkansas
Fayetteville, AR 72701, U. S. A.
E-mail: jryan@uark.edu

Received: May 12, 2002; Revised: March 15, 2003.

4

Applications of Clifford Algebras in Physics

William E. Baylis

ABSTRACT Clifford's geometric algebra is a powerful language for physics that clearly describes the geometric symmetries of both physical space and spacetime. Some of the power of the algebra arises from its natural spinorial formulation of rotations and Lorentz transformations in classical physics. This formulation brings important quantum-like tools to classical physics and helps break down the classical/quantum interface. It also unites Newtonian mechanics, relativity, quantum theory, and other areas of physics in a single formalism and language. This lecture is an introduction and sampling of a few of the important applications in physics.

4.1 Introduction

Clifford's geometric algebra is an ideal language for physics because it maximally exploits geometric properties and symmetries. It is known to physicists mainly as the algebras of Pauli spin matrices and of Dirac gamma matrices, but its utility goes far beyond the applications to quantum theory and spin for which these matrix forms were introduced. In particular, Clifford algebra

- generalizes cross products to wedge products, which have transparent geometric meanings and work in spaces of any dimension,
- constructs the unit imaginary i as a geometric object, thereby explaining its important role in physics,
- extends complex analysis to more than two dimensions,
- simplifies calculations of duals and thereby clears potential confusion of pseudovectors and pseudoscalars,
- reduces rotations and Lorentz transformations to algebraic multiplication, and more generally allows computational geometry without matrices or tensors,

AMS Subject Classification: 15A66, 17B37, 20C30, 81R25.
Keywords: Paravectors, duals, Maxwell's equations, light polarization, Lorentz transformations, spin, gauge transformations, eigenspinors, Dirac equation, quaternions, Liénard–Wiechert potentials.

- formulates classical physics in an efficient spinorial formulation with tools that are closely related to ones familiar in quantum theory such as spinors ("rotors") and projectors, and
- thereby unites Newtonian mechanics, relativity, quantum theory, and more in a single formalism as simple as the algebra of Pauli spin matrices.

In this lecture, there is space to discuss only a few examples of the many applications of Clifford algebras in physics. For other applications, the reader will be referred to published articles and books. The mathematical background for this chapter is given by the lecture [1] of the late Professor Pertti Lounesto earlier in the volume.

4.2 Three Clifford algebras

The three most commonly employed Clifford algebras in physics are the quaternion algebra $\mathbb{H} \simeq C\ell_{0,2}$, the algebra of physical space (APS) $C\ell_3$, and the spacetime algebra (STA) $C\ell_{1,3}$. They are closely related. Hamilton's quaternion algebra $\mathbb{H}$, [2] introduced in 1843 to handle rotations, is the oldest, and provided the concept (again, by Hamilton) of vectors. The superiority of $\mathbb{H}$ for matrix-free and coordinate-free computations of rotations has been recently rediscovered by space programs, the computer-games industry, and robotics engineering. Furthermore, $\mathbb{H}$ is a division algebra and has been investigated as a replacement of the complex field in an extension of Dirac theory [3]. Quaternions were used by Maxwell and Tait to express Maxwell's equations of electromagnetism in compact form, and they motivated Clifford to find generalizations based on Grassmann theory. Hamilton's biquaternions (complex quaternions) are quaternions over the complex field $\mathbb{H} \otimes \mathbb{C}$, and with them, Conway (1911) and others were able to express Maxwell's equations as a single equation.

The complex quaternions are isomorphic to APS: $\mathbb{H} \otimes \mathbb{C} \simeq C\ell_3$, which is familiar to physicists as the algebra of the Pauli spin matrices. The even subalgebra $C\ell_3^+$ is isomorphic to $\mathbb{H}$, and the correspondences $\boldsymbol{i} \leftrightarrow \mathbf{e}_{32}$, $\boldsymbol{j} \leftrightarrow \mathbf{e}_{13}$, $\boldsymbol{k} \leftrightarrow \mathbf{e}_{21}$ identify pure quaternions with bivectors in APS, and hence with generators of rotations in physical space. APS distinguishes cleanly between vectors and bivectors, in contrast to most approaches with complex quaternions. The volume element $\mathbf{e}_{123}$ in APS squares to -1 and commutes with vectors and their products, and it can therefore be identified with the unit imaginary i. Every element of APS can then be expressed as a complex *paravector*, that is the sum of a scalar and a vector [4–6]. The identification $i \equiv \mathbf{e}_{123}$ endows the unit imaginary with geometrical significance and helps explain the widespread use of complex numbers in physics [7]. The sign of i is reversed under parity inversion, and imaginary scalars and vectors correspond to pseudoscalars and pseudovectors, respectively. The dual of any element x is simply $-ix$, and in particular, one sees that the vector cross product $\boldsymbol{x} \times \boldsymbol{y}$ of two vectors $\boldsymbol{x}, \boldsymbol{y}$ is the vector dual to the bivector

$\boldsymbol{x} \wedge \boldsymbol{y}$. It is traditional in APS to denote reversion by a dagger, $\tilde{x} = x^\dagger$, [6, 8–10] since reversal corresponds to Hermitian conjugation in any matrix representation that uses Hermitian matrices (such as the Pauli spin matrices) for the basis vectors $\mathbf{e}_1, \mathbf{e}_2, \mathbf{e}_3$.[1]

The quadratic form of vectors in APS is the traditional dot product and implies a Euclidean metric. Paravectors constitute a four-dimensional linear subspace of APS, but as shown below, the quadratic form they inherit implies the pseudo-Euclidean metric of Minkowski spacetime. Paravectors can therefore be used to model spacetime vectors in relativity [8, 11]. The algebra of paravectors and their products is no different from the algebra of vectors, and in particular the paravector volume element is i and duals are defined in the same way. APS admits interpretations in both spacetime (paravector) and spatial (vector) terms, and the formulation of relativity in APS marries covariant spacetime notation with the vector notation of spatial vectors. Matrix elements and tensor components are not needed, although they can be found by expanding multiparavectors in the basis elements of APS.

Another approach to spacetime is to assume the Minkowski spacetime metric and use the corresponding algebra $C\ell_{1,3}$ or $C\ell_{3,1}$. In STA, $C\ell_{1,3}$ is generated by products of the basis vectors γ_μ, $\mu = 0, 1, 2, 3$, satisfying

$$\tfrac{1}{2}\left(\gamma_\mu\gamma_\nu + \gamma_\nu\gamma_\mu\right) = \eta_{\mu\nu} = \begin{cases} 1, & \mu = \nu = 0 \\ -1, & \mu = \nu = 1, 2, 3 \\ 0, & \mu \neq \nu \end{cases} .$$

Dirac's electron theory (1928) was based on a matrix representation of $C\ell_{1,3}$ over the complex field, and Hestenes [10] (1966) pioneered the use of STA (real $C\ell_{1,3}$) in several areas of physics. The even subalgebra is isomorphic to APS: $C\ell^+_{1,3} \simeq C\ell_3$. The volume element in STA is $\boldsymbol{I} = \gamma_0\gamma_1\gamma_2\gamma_3$. Although it is referred to as the unit pseudoscalar and squares to -1, it anticommutes with vectors, thus behaving more like an additional spatial dimension than a scalar.

This lecture mainly uses APS, although generalizations to $C\ell_n$, the Clifford algebra of n-dimensional Euclidean space, are made where convenient, and one section is devoted to the relation of APS to STA.

4.2.1 Bivectors as plane areas

The Clifford (or geometric) product of vectors was introduced in Lecture 1 and given there as the sum of dot and wedge products. The dot product is familiar from traditional vector analysis, but the wedge product of two vectors is a new entity:

[1] The argument is easily extended to spaces of n Euclidean dimensions. There are also other reasons for using the dagger (†) for reversion: (1) it is clearest to use, whenever possible, the same notation in the algebra as in its standard matrix representation, (2) in small print, $\tilde{x}$ is easily confused with $\bar{x}$, and (3) when comparing APS and STA below, it is essential to have separate notations for reversal of elements in STA and in APS.

the bivector. The bivector $\boldsymbol{v} \wedge \boldsymbol{w}$ represents the plane containing $\boldsymbol{v}$ and $\boldsymbol{w}$, and as discussed in Lecture 1, it has an intrinsic orientation given by the circulation in the plane from $\boldsymbol{v}$ to $\boldsymbol{w}$ and a size or *magnitude* $|\boldsymbol{v} \wedge \boldsymbol{w}|$ given by the area of the parallelogram with sides $\boldsymbol{v}$ and $\boldsymbol{w}$. Bivectors enter physics in many ways: as areas, planes, and the generators of reflections and rotations. In the old vector analysis of Gibbs and Heaviside, bivectors are replaced by cross products, which are vectors perpendicular to the plane. However, this ploy is useful only in three dimensions, and it hides the intrinsic properties of bivectors. As discussed below, cross products are actually examples of algebraic duals.

The *reversion*, as defined in Lecture 1, reverses the order of vector factors in the product. Because noncollinear vectors do not commute, reversion gives us a way of distinguishing collinear and orthogonal components. Any element invariant under reversal is said to be *real* whereas elements that change sign are *imaginary*. Every element x can be split into real and imaginary parts:

$$x = \frac{x + x^\dagger}{2} + \frac{x - x^\dagger}{2} \equiv \langle x \rangle_{\text{Re}} + \langle x \rangle_{\text{Im}} .$$

Scalars and vectors (in a Euclidean space) are thus real, whereas bivectors are imaginary.

Exercise 1. *Verify that the dot and wedge product of any vectors $\boldsymbol{u}, \boldsymbol{v} \in C\ell_n$ can be identified as the real and imaginary parts of the geometric product $\boldsymbol{uv}$.*

Exercise 2. *Let $\boldsymbol{r}$ be the position vector of a point that moves with velocity $\boldsymbol{v} = \dot{\boldsymbol{r}}$. Show that the magnitude of the bivector $\boldsymbol{r} \wedge \boldsymbol{v}$ is twice the rate at which the time-dependent $\boldsymbol{r}$ sweeps out area. Explain how this relates the conservation of angular momentum $\boldsymbol{r} \wedge \boldsymbol{p}$, with $\boldsymbol{p} = m\boldsymbol{v}$, to Kepler's second law for planetary orbits, namely that equal areas are swept out in equal times.*

4.2.2 *Bivectors as operators*

The fact that the bivector of a plane commutes with vectors orthogonal to the plane and anticommutes with ones in the plane means that we can easily use unit bivectors to represent reflections. In particular, in an n-dimensional Euclidean space, $n > 2$, the two-sided transformation

$$\boldsymbol{v} \longrightarrow \mathbf{e}_{12} \boldsymbol{v} \mathbf{e}_{12}$$

reflects any vector $\boldsymbol{v}$ in the $\mathbf{e}_{12} = \mathbf{e}_1\mathbf{e}_2$ plane.

As shown in Lecture 1, bivectors generate rotations. The demonstration there was for rotations in $C\ell_2$ introduced in Lecture 1, but it is easily extended to three or more dimensions. The bivector $\mathbf{e}_{12}$ *generates a rotation* in the $\mathbf{e}_{12}$ plane: any vector $\boldsymbol{v}$ in the $\mathbf{e}_{12}$ plane is rotated by θ in the plane in the sense that takes $\mathbf{e}_1 \rightarrow \mathbf{e}_2$ by

$$\boldsymbol{v} \longrightarrow \boldsymbol{v} \exp(\mathbf{e}_{12}\theta) = \exp(\mathbf{e}_{21}\theta)\, \boldsymbol{v}. \tag{2.1}$$

The rotation element $\exp(e_{12}\theta)$ is the product of two unit vectors in the rotation plane:

$$\exp(e_{12}\theta) = e_1 e_1 \exp(e_{12}\theta) \equiv e_1 n,$$

where $n = e_1 \exp(e_{12}\theta)$ is the unit vector obtained from e_1 by a rotation of θ in the e_{12} plane. In general, every product mn of unit vectors m and n can be interpreted as a rotation operator in the plane containing m and n. The product mn does not depend on the actual directions of m and n, but only on the plane in which they lie and on the angle between them. This is the basis of a convenient representation of $SU(2)$ rotations in $C\ell_3$ as vectors on the surface of the unit sphere [12].

Relation (2.1) shows two ways to rotate any vector v in a plane. Note that the left and right rotation operators are *not* the same, but rather inverses of each other:

$$\exp(e_{21}\theta)\exp(e_{12}\theta) = \exp(e_{21}\theta)\exp(-e_{21}\theta) = 1.$$

In spaces of more than two dimensions, we may want to rotate a vector v that has components perpendicular to the rotation plane. Then the products $v\exp(e_1 e_2\theta)$ and $\exp(e_2 e_1\theta)\, v$ no longer work; they include terms like $e_3 e_1 e_2$ that are not even vectors. (They are trivectors as will be discussed in the next section.) We need a form for the rotation that leaves perpendicular components invariant. Recall that perpendicular vectors commute with the bivector for the plane. For example, $e_3 e_{12} = e_{12} e_3$. Therefore, for the rotation we use

$$v \longrightarrow R v R^{-1}, \tag{2.2}$$

where $R = \exp\left(\frac{1}{2}e_{21}\theta\right)$ is a *rotor* and $R^{-1} = \exp\left(\frac{1}{2}e_{12}\theta\right)$ is its inverse. The transformation (2.2) is linear, because $R\alpha v R^{-1} = \alpha R v R^{-1}$ for any scalar α, and $R(v+w)R^{-1} = RvR^{-1} + RwR^{-1}$ for any two vectors v and w. The two-sided form of the rotation leaves anything that commutes with the rotation plane invariant. This includes vector components perpendicular to the rotation plane as well as scalars. The form may remind you of transformations of operators in quantum theory. It is sometimes called a *spin transformation* to distinguish it from one-sided transformations common with matrices operating on column vectors.

The ability to rotate in any plane of n-dimensional space without components, tensors, or matrices is a major strength of geometric algebra in physics. A product of rotors is a rotor (rotations form a group), and in physical space, where $n = 3$, any rotor can be factored into a product of Euler-angle rotors

$$R = \exp\left(-e_{12}\frac{\phi}{2}\right)\exp\left(-e_{31}\frac{\theta}{2}\right)\exp\left(-e_{12}\frac{\psi}{2}\right). \tag{2.3}$$

However, such factorization is not necessary since there is always a simpler expression $R = \exp(\mathbf{\Theta})$, where $\mathbf{\Theta}$ is a bivector whose orientation gives the plane of rotation and whose magnitude $\Theta = |\mathbf{\Theta}|$ gives half the angle of rotation. By using the rotor $R = \exp(\mathbf{\Theta})$ instead of a product of three rotations in specified planes, one can avoid degeneracies. The simple rotor expression allows smooth rotations in a single plane and thus interpolations between arbitrary orientations [12].

Relation to rotation by matrix multiplication

Spin transformations rotate vectors, and when we expand $\boldsymbol{v} = v^j \boldsymbol{e}_j$, it is the basis vectors and not the coefficients that are directly rotated:

$$\boldsymbol{v} = v^j \boldsymbol{e}_j \to \boldsymbol{v}' = v^j \boldsymbol{e}_j', \qquad \boldsymbol{e}_j \to \boldsymbol{e}_j' = R\boldsymbol{e}_j R^{-1}.$$

It is easy to find the components of the rotated vector $\boldsymbol{v}'$ in the original (unprimed) basis:

$$v'^i = \boldsymbol{e}_i \cdot \boldsymbol{v}' = \left(\boldsymbol{e}_i \cdot \boldsymbol{e}_j'\right) v^j.$$

The matrix of scalar values $\boldsymbol{e}_i \cdot \boldsymbol{e}_j'$ is the usual rotation matrix.[2] For example, in $C\ell_3$ with $R = \exp\left(\frac{1}{2}\mathbf{e}_{21}\theta\right)$,

$$\begin{pmatrix} \mathbf{e}_1 \cdot \mathbf{e}_1' & \mathbf{e}_1 \cdot \mathbf{e}_2' & \mathbf{e}_1 \cdot \mathbf{e}_3' \\ \mathbf{e}_2 \cdot \mathbf{e}_1' & \mathbf{e}_2 \cdot \mathbf{e}_2' & \mathbf{e}_2 \cdot \mathbf{e}_3' \\ \mathbf{e}_3 \cdot \mathbf{e}_1' & \mathbf{e}_3 \cdot \mathbf{e}_2' & \mathbf{e}_3 \cdot \mathbf{e}_3' \end{pmatrix} = \begin{pmatrix} \cos\theta & -\sin\theta & 0 \\ \sin\theta & \cos\theta & 0 \\ 0 & 0 & 1 \end{pmatrix}.$$

Exercise 3. *Let* $R = \exp\left(\frac{1}{2}\mathbf{e}_{21}\theta\right)$ *and assume that*

$$\boldsymbol{n} = R\mathbf{e}_1 R^\dagger = \exp\left(\mathbf{e}_{21}\theta\right)\mathbf{e}_1.$$

Show that

$$R = \left(\boldsymbol{n}\mathbf{e}_1\right)^{1/2} = \frac{\left(\boldsymbol{n}\mathbf{e}_1 + 1\right)}{\sqrt{2\left(1 + \boldsymbol{n}\cdot\mathbf{e}_1\right)}}.$$

Hint: find the unit vector that bisects $\boldsymbol{n}$ *and* $\mathbf{e}_1$.

Relation of rotations to reflections

Evidently $R^2 = \exp\left(\mathbf{e}_{21}\theta\right) = \boldsymbol{n}\mathbf{e}_1$ is the rotor for a rotation in the plane of $\boldsymbol{n}$ and $\mathbf{e}_1$ by 2θ. Its inverse is $\mathbf{e}_1\boldsymbol{n}$, and it maps any vector $\boldsymbol{v}$ to $\boldsymbol{n}\mathbf{e}_1\boldsymbol{v}\mathbf{e}_1\boldsymbol{n}$. If $\mathbf{e}_3$ is a unit vector normal to the plane of rotation,

$$\boldsymbol{v} \to \boldsymbol{n}\mathbf{e}_3\mathbf{e}_3\mathbf{e}_1\boldsymbol{v}\mathbf{e}_3\mathbf{e}_1\boldsymbol{n}\mathbf{e}_3 = \left(\boldsymbol{n}\mathbf{e}_3\right)\mathbf{e}_{31}\boldsymbol{v}\mathbf{e}_{31}\left(\boldsymbol{n}\mathbf{e}_3\right),$$

which represents the rotation as reflections in two planes with unit bivectors $\boldsymbol{n}\mathbf{e}_3$ and $\mathbf{e}_{31}$. The planes intersect along $\mathbf{e}_3$ and have a dihedral angle of θ.

Example 4. *Mirrors in clothing stores are often arranged to give double reflections so that you can see yourself rotated rather than reflected. Two mirrors with a dihedral angle of* 90° *will rotate your image by* 180°. *This corresponds to the above transformation with* $\boldsymbol{n}$ *replaced by* $\mathbf{e}_2$.

Exercise 5. *How could you orient two mirrors so that you see yourself from the side, that is, rotated by* 270°?

---

[2] While one can readily compute the transformation matrices by writing the spin transformations explicitly for basis vectors, neither the matrices nor even the vector components are needed in the algebraic formulation.

The rotation that results from successive reflections in two nonparallel planes in physical space depends only on the line of intersection and the dihedral angle between the planes; it is independent of rotations for both planes about their common axis.

Exercise 6. *Corner cubes are used on the moon and in the rear lenses on cars to reverse the direction of the incident light. Consider a sequence of three reflections, first in the* $\mathbf{e}_{12}$ *plane, followed by one in the* $\mathbf{e}_{23}$ *plane, followed by one in the* $\mathbf{e}_{31}$ *plane. Show that when applied to any vector* $\boldsymbol{v}$*, the result is* $-\boldsymbol{v}$*.*

Time-dependent rotations

An additional infinitesimal rotation by $\boldsymbol{\Omega}'\,dt$ during the time interval dt changes a rotor R to

$$R + dR = \left(1 + \tfrac{1}{2}\boldsymbol{\Omega}'\,dt\right) R.$$

The time-rate of change of R thus has the form

$$\dot{R} = \frac{dR}{dt} = \tfrac{1}{2}\boldsymbol{\Omega}' R \equiv \tfrac{1}{2} R \boldsymbol{\Omega},$$

where the bivector $\boldsymbol{\Omega} = R^{-1}\boldsymbol{\Omega}' R$ is the *rotation rate* as viewed in the rotating frame. For the special case of a constant rotation rate, we can take the rotor to be

$$R(t) = R(0)\, e^{\boldsymbol{\Omega} t/2}.$$

If $R(0) = 1$, then $\boldsymbol{\Omega} = \boldsymbol{\Omega}'$. Any vector $\boldsymbol{r}$ is thereby rotated to $\boldsymbol{r}' = R\boldsymbol{r}R^{-1}$ giving a time derivative

$$\dot{\boldsymbol{r}}' = R\left[\frac{\boldsymbol{\Omega}\boldsymbol{r} - \boldsymbol{r}\boldsymbol{\Omega}}{2} + \dot{\boldsymbol{r}}\right] R^{-1} = R\left[\langle \boldsymbol{\Omega}\boldsymbol{r}\rangle_{\mathrm{Re}} + \dot{\boldsymbol{r}}\right] R^{-1},$$

where we noted that $\boldsymbol{r}$ is real and the bivector $\boldsymbol{\Omega}$ is imaginary. Since $\boldsymbol{r}$ can be any vector, we can replace it by $\langle \boldsymbol{\Omega}\boldsymbol{r}\rangle_{\mathrm{Re}} + \dot{\boldsymbol{r}}$ to determine the second derivative

$$\begin{aligned} \ddot{\boldsymbol{r}}' &= R\left[\langle \boldsymbol{\Omega}\left(\langle \boldsymbol{\Omega}\boldsymbol{r}\rangle_{\mathrm{Re}} + \dot{\boldsymbol{r}}\right)\rangle_{\mathrm{Re}} + \langle \boldsymbol{\Omega}\dot{\boldsymbol{r}}\rangle_{\mathrm{Re}} + \ddot{\boldsymbol{r}}\right] R^{-1} \\ &= R\left[\langle \boldsymbol{\Omega}\langle \boldsymbol{\Omega}\boldsymbol{r}\rangle_{\mathrm{Re}}\rangle_{\mathrm{Re}} + 2\langle \boldsymbol{\Omega}\dot{\boldsymbol{r}}\rangle_{\mathrm{Re}} + \ddot{\boldsymbol{r}}\right] R^{-1} \\ &= R\left[\boldsymbol{\Omega}^2\boldsymbol{r}^{\triangle} + 2\boldsymbol{\Omega}\dot{\boldsymbol{r}}^{\triangle} + \ddot{\boldsymbol{r}}\right] R^{-1}, \end{aligned} \tag{2.4}$$

where $\langle \boldsymbol{\Omega}\boldsymbol{r}\rangle_{\mathrm{Re}} = \boldsymbol{\Omega}\boldsymbol{r}^{\triangle}$, and $\boldsymbol{r}^{\triangle}$ is the part of $\boldsymbol{r}$ coplanar with $\boldsymbol{\Omega}$. A force law $\boldsymbol{f}' = m\ddot{\boldsymbol{r}}'$ in the inertial system is seen to be equivalent to an effective force

$$\boldsymbol{f} = m\ddot{\boldsymbol{r}} = R^{-1}\boldsymbol{f}'R - m\boldsymbol{\Omega}^2\boldsymbol{r}^{\triangle} - 2m\boldsymbol{\Omega}\dot{\boldsymbol{r}}^{\triangle}$$

in the rotating frame. The second and third terms on the RHS are identified as the centrifugal and Coriolis forces, respectively.

4.2.3 *Higher-grade multivectors in* $C\ell_n$

Higher-order products of vectors also play important roles in physics. Products of k orthonormal basis vectors $\mathbf{e}_j$ can be reduced if two of them are the same, but if they are all distinct, their product is a basis k-vector. In an n-dimensional space, the algebra contains $\binom{n}{k}$ such linearly independent multivectors of *grade k*, which form a basis of the *homogeneous linear subspace* $\langle C\ell_n\rangle_k$ of the algebra.[3] The highest-grade element is the volume element, proportional to

$$\mathbf{e}_T \equiv \mathbf{e}_{123\cdots n} = \mathbf{e}_1\mathbf{e}_2\mathbf{e}_3\cdots\mathbf{e}_n.$$

Exercise 7. *Find the number of linearly independent elements in the geometric algebra* $C\ell_5$ *of 5-dimensional space. How is this subdivided into vectors, bivectors, and so on?*

A general element of the algebra is a mixture of different grades. We use the notation $\langle x\rangle_k$ to isolate the part of x with grade k. Thus, $\langle x\rangle_0$ is the scalar part of x, $\langle x\rangle_1$ is the vector part, and by $\langle x\rangle_{2,1}$ we mean the sum $\langle x\rangle_2 + \langle x\rangle_1$ of the bivector and vector parts. Evidently

$$x = \langle x\rangle_{0,1,2,\ldots,n} = \sum_{k=0}^{n} \langle x\rangle_k .$$

In APS, in addition to scalars, vectors, and bivectors, there are also trivectors, elements of grade 3:

$$\begin{aligned}\mathbf{u}\wedge\mathbf{v}\wedge\mathbf{w} \equiv \langle\mathbf{uvw}\rangle_3 &= \sum_{j,k,l} u^j v^k w^l \langle\mathbf{e}_j\mathbf{e}_k\mathbf{e}_l\rangle_3 = \sum_{j,k,l} \varepsilon_{jkl} u^j v^k w^l \mathbf{e}_{123}\\ &= \mathbf{e}_T \det\begin{pmatrix} u^1 & v^1 & w^1\\ u^2 & v^2 & w^2\\ u^3 & v^3 & w^3\end{pmatrix},\end{aligned} \tag{2.5}$$

where we have used the fact that in 3-dimensional space $\langle\mathbf{e}_j\mathbf{e}_k\mathbf{e}_l\rangle_3 = \varepsilon_{jkl}\mathbf{e}_{123}$ (here ε_{jkl} is the *Levi-Civita symbol*). The determinant in (2.5) ensures that the wedge product vanishes if the vector factors are linearly dependent.

While the component expressions can be useful for comparing results with other work, the component-free versions $\mathbf{u}\wedge\mathbf{v}\wedge\mathbf{w} \equiv \langle\mathbf{uvw}\rangle_3$ are simpler and more efficient to work with. In the trivector $\langle\mathbf{uvw}\rangle_3$, the factor $\mathbf{uv}$ can be split into scalar (grade-0) and bivector (grade-2) parts $\mathbf{uv} = \langle\mathbf{uv}\rangle_0 + \langle\mathbf{uv}\rangle_2$, but $\langle\mathbf{uv}\rangle_0\,\mathbf{w}$ is a (grade-1) vector, so that only the bivector piece contributes: $\langle\mathbf{uvw}\rangle_3 = \langle\langle\mathbf{uv}\rangle_2\,\mathbf{w}\rangle_3$. Now split $\mathbf{w}$ into components coplanar with $\langle\mathbf{uv}\rangle_2$ and orthogonal to it:

$$\mathbf{w} = \mathbf{w}^{\triangle} + \mathbf{w}^{\perp},$$

[3] In the notation of Lecture 1, $\langle C\ell_n\rangle_k$ is the same as the subspace $\wedge^k V$ of the exterior algebra $\wedge V$, where V is n-dimensional Euclidean space.

and recall that $w^{\triangle}$ and $w^{\perp}$ anticommute and commute with $\langle uv\rangle_2$, respectively. The coplanar part $w^{\triangle}$ is linearly dependent on u and v and therefore does not contribute to the trivector. We are left with

$$\begin{aligned}\langle uvw\rangle_3 &= \langle\langle uv\rangle_2\, w\rangle_3 = \langle\langle uv\rangle_2\, w^{\perp}\rangle_3 \\ &= \tfrac{1}{2}\left(\langle uv\rangle_2\, w + w\,\langle uv\rangle_2\right) = \langle\langle uv\rangle_2\, w\rangle_{\mathrm{Im}}\,.\end{aligned}$$

Similarly, the vector part of the product $\langle uv\rangle_2\, w$ is

$$\langle\langle uv\rangle_2\, w\rangle_1 = \langle\langle uv\rangle_2\, w^{\triangle}\rangle_1 = \tfrac{1}{2}\left(\langle uv\rangle_2\, w - w\,\langle uv\rangle_2\right) = \langle\langle uv\rangle_2\, w\rangle_{\mathrm{Re}}\,.$$

Exercise 8. *Show that while* $\langle uvw\rangle_3 = \langle\langle uv\rangle_2\, w\rangle_3$, *the difference*

$$\langle uvw\rangle_1 - \langle\langle uv\rangle_2\, w\rangle_1 = \langle\langle uv\rangle_0\, w\rangle_1$$

does not generally vanish.

A couple of important results follow easily.

Theorem 9. $\langle\langle uv\rangle_2\, w\rangle_{\mathrm{Re}} = u\,(v\cdot w) - v\,(u\cdot w)\,.$

Proof. Expand $\langle\langle uv\rangle_2\, w\rangle_{\mathrm{Re}} = \frac{1}{2}\left(\langle uv\rangle_2\, w - w\,\langle uv\rangle_2\right)$, add and subtract term uwv and vwu, and collect:

$$\begin{aligned}&\langle\langle uv\rangle_2\, w\rangle_{\mathrm{Re}} \\ &= \tfrac{1}{4}\left[(uv - vu)\,w - w\,(uv - vu)\right] \\ &= \tfrac{1}{4}\left[uvw + uwv - vuw - vwu - wuv - uwv + wvu + vwu\right] \\ &= \tfrac{1}{2}\left[u\,(v\cdot w) - v\,(u\cdot w) - (w\cdot u)\,v + (w\cdot v)\,u\right] \\ &= u\,(v\cdot w) - v\,(u\cdot w)\,.\end{aligned}$$

Let B be the bivector $\langle uv\rangle_2$. If we expand $u = u^j e_j$ and $v = v^k e_k$, we find

$$B = u^j v^k \langle e_j e_k\rangle_2 = \tfrac{1}{2} B^{jk} \langle e_j e_k\rangle_2$$

where $B^{jk} = u^j v^k - u^k v^j$ and we used the antisymmetry of

$$\langle e_j e_k\rangle_2 = -\langle e_k e_j\rangle_2\,.$$

From the last theorem, the vector $\langle Bw\rangle_1$ lies in the plane of B and is orthogonal to w. In terms of components,

$$\langle Bw\rangle_1 = \tfrac{1}{2} B^{jk} w^l \langle\langle e_j e_k\rangle_2 e_l\rangle_1 = \tfrac{1}{2} B^{jk} w^l \left(e_j \delta_{kl} - e_k \delta_{jl}\right) = B^{jl} w_l e_j$$

is the *contraction* $B\cdot w = -w\cdot B$ of the bivector B by the vector w.[4] It lies in the *intersection* of the plane of B with the hypersurface orthogonal to w. Similar relations for wedge and dot products among higher-grade elements can be formulated [4, 9] but are not needed here.

---

[4]The dot is commonly used [10] to indicate contractions of mixed-grade products, as used below. When restricted to the case as here that B contains only grades equal to or higher than those of w, we can use the notation of Lecture 1: $\langle Bw\rangle_1 = B \llcorner w = -w \lrcorner B$.

Duals

Note that in $C\ell_n$, the number of independent k-grade multivectors is the same as the number of independent $(n-k)$-grade elements. For example, both the vectors (grade 1 elements) and the pseudovectors (grade $n-1$ elements) occupy linear spaces of n dimensions. We can therefore establish a one-to-one mapping between such elements. We define the Clifford-Hodge dual[5] $^*x$ of an element x by

$$^*x \equiv x\mathbf{e}_T^{-1}.$$

The dual of a dual is $\mathbf{e}_T^{-2} = \pm 1$ times the original element. If x is a k-vector, each term in a k-vector basis expansion of x will cancel k of the basis vector factors[6] in $\mathbf{e}_T$, leaving $^*x = x\mathbf{e}_T^{-1}$ as an $(n-k)$-vector.

A *simple* k-vector is a single product of k orthogonal vectors that span a k-dimensional subspace. Every vector in that subspace is orthogonal to vectors whose products comprise the $(n-k)$-vector $^*x$. In this sense, any simple element and its dual are fully orthogonal. The dual of a scalar is a volume element, known as a *pseudoscalar*; the dual of a vector is the hypersurface orthogonal to that vector, known as a *pseudovector*; and so on.[7]

In physical space ($n = 3$), the dual to a bivector is the vector normal to the plane of the bivector. Thus, $\mathbf{e}_T = \mathbf{e}_{123}$ and $\mathbf{e}_T^{-1} = -\mathbf{e}_{123} = -\mathbf{e}_T$, and for example

$$^*(\mathbf{e}_{12}) = \mathbf{e}_{12}(-\mathbf{e}_{123}) = \mathbf{e}_3.$$

We recognize this as the *cross product* and we can write more generally

$$^*(\boldsymbol{u} \wedge \boldsymbol{v}) = \boldsymbol{u} \times \boldsymbol{v}\,,$$

which when taken between polar vectors emphasizes its relation to the plane of $\boldsymbol{u}$ and $\boldsymbol{v}$. The volume element $\mathbf{e}_T$ squares to -1 and commutes with all vectors and hence all elements. Because of its association with the *unit imaginary*,[8] it is convenient to denote $\mathbf{e}_T$ by i, i.e., we let $i = \mathbf{e}_T = \mathbf{e}_{123}$. Thus, $^*(\boldsymbol{u} \wedge \boldsymbol{v}) = (\boldsymbol{u} \wedge \boldsymbol{v})\, i^{-1}$ and

$$\boldsymbol{u} \wedge \boldsymbol{v} = i(\boldsymbol{u} \times \boldsymbol{v}).$$

---

[5]This definition is the same as used in my text [6] and gives the commonly used sign for the Hodge dual of the electromagnetic field. However, Lounesto [4] uses $\mathbf{e}_T$ instead of $\mathbf{e}_T^{-1}$ in his definition of the dual. The difference is the sign $\mathbf{e}_T^2 = \pm 1$.

[6]Within an overall sign arising from the anticommutativity of the basis vectors. Formally, the cancellation is a left grade contraction of $\mathbf{e}_T^{-1}$ by x.

[7]These names are most suitable in APS where the pseudoscalar belongs to the center of the algebra, but they are also applied in cases with an even number n of dimensions, where $\mathbf{e}_T$ anticommutes with the vectors.

[8]The association is given formally by the isomorphism between the center (commuting part) of the algebra of physical space (APS) and the field of complex numbers $\mathbb{C}$ [1]. It is helpful to remember the geometric meaning of i thus introduced into the algebra: it is the unit volume element, which defines the dual relationship. Its appearance helps make sense of some of the many complex numbers that appear in *real* physics, and the dual relationship helps avoid confusion associated with pseudoscalars, pseudovectors, and their behavior under inversion. The bivector in APS, for example, is the pseudovector whose dual is the vector normal to the plane of the bivector: $\mathbf{e}_{12} = \mathbf{e}_{123}\mathbf{e}_3 = i\mathbf{e}_3$.

However, whereas the cross product of two vectors is not generally defined in $n > 3$ dimensions and is nonassociative as well as noncommutative, the exterior wedge product is defined in spaces of any dimension and is associative.[9] It also emphasizes the essential properties of the plane and is an operator on vectors that generates rotations.

Exercise 10. *Verify by calculation of some explicit values that the Levi-Civita symbol is the dual to the volume element $\langle \mathbf{e}_j \mathbf{e}_k \mathbf{e}_l \rangle_3$ in APS:*

$$\varepsilon_{jkl} = {}^* \langle \mathbf{e}_j \mathbf{e}_k \mathbf{e}_l \rangle_3 = \langle \mathbf{e}_j \mathbf{e}_k \mathbf{e}_l \rangle_3 \, \mathbf{e}_T^{-1}.$$

This definition is easily extended to higher dimensions.

We can use duals to express a rotor in physical space in terms of the *axis of rotation*, which is dual to the rotation plane. For example

$$R = \exp\left(\tfrac{1}{2}\mathbf{e}_2\mathbf{e}_1\theta\right) = \exp\left(-\tfrac{1}{2}i\mathbf{e}_3\theta\right)$$

is the rotor for a rotation $\boldsymbol{v} \to R\boldsymbol{v}R^{-1}$ by θ about the $\mathbf{e}_3$ axis in physical space. Furthermore, in APS the volume of the parallelepiped with sides $\boldsymbol{a}, \boldsymbol{b}, \boldsymbol{c}$, is dual to a pseudoscalar, namely the trivector

$$\begin{aligned} \boldsymbol{a} \wedge \boldsymbol{b} \wedge \boldsymbol{c} = \langle \boldsymbol{abc} \rangle_3 &= \langle \boldsymbol{abc} \rangle_{\text{Im}} = \langle \langle \boldsymbol{ab} \rangle_2 \, \boldsymbol{c} \rangle_{\text{Im}} \\ &= \langle i \, {}^* \langle \boldsymbol{ab} \rangle_2 \, \boldsymbol{c} \rangle_{\text{Im}} = i \, \langle (\boldsymbol{a} \times \boldsymbol{b}) \, \boldsymbol{c} \rangle_{\text{Re}} = i \, (\boldsymbol{a} \times \boldsymbol{b}) \cdot \boldsymbol{c}. \end{aligned}$$

Exercise 11. *The bivector rotation rate $\boldsymbol{\Omega} = -i\boldsymbol{\omega}$ can be expressed as the dual of a vector $\boldsymbol{\omega}$ in physical space. Show that $\langle \boldsymbol{\Omega r} \rangle_{\text{Re}} = \boldsymbol{\omega} \times \boldsymbol{r}$.*

Exercise 12. *Show that in physical space $\langle \langle \boldsymbol{uv} \rangle_2 \, \boldsymbol{w} \rangle_{\text{Re}} = \boldsymbol{u}\,(\boldsymbol{v} \cdot \boldsymbol{w}) - \boldsymbol{v}\,(\boldsymbol{u} \cdot \boldsymbol{w})$ becomes $(\boldsymbol{u} \times \boldsymbol{v}) \times \boldsymbol{w} = (\boldsymbol{u} \cdot \boldsymbol{w})\boldsymbol{v} - (\boldsymbol{v} \cdot \boldsymbol{w})\boldsymbol{u}$.*

Exercise 13. *Rewrite the effective force (4.2.2) in the rotating frame in terms of $\boldsymbol{\omega} = i\boldsymbol{\Omega}$.*

Except for their normalization, reciprocal basis vectors are duals to hyperplanes formed by wedging all but one of the basis vectors. The reciprocal basis is important when the basis is not orthogonal and not necessarily normalized, as in the study of crystalline solids. We can think of the reciprocal vectors as basis 1-forms, that is linear operators $\boldsymbol{a}^k$ on vectors $\boldsymbol{a}_j$ whose operation is defined by

$$\boldsymbol{a}^k \left(\boldsymbol{a}_j\right) = \boldsymbol{a}_j \cdot \boldsymbol{a}^k = \delta_j^k .$$

[9] As Lounesto discusses [4], one can define a cross product of $n - 1$ vectors as the dual of a pseudovector, and in $n = 7$ dimensions, a trivector can be defined that gives a cross product of any two vectors as the pure part of an octonion product.

4.3 Paravectors and relativity

The space of scalars, the space of vectors, and the space of bivectors, are all linear subspaces of the full 2^n-dimensional space of the algebra $C\ell_n$. Direct sums of the subspaces are also linear subspaces of the algebra. The most important is the direct sum of the scalar and the vector subspaces. It is an $(n+1)$-dimensional linear space known as *paravector space.* In APS $(n=3)$, every element reduces to a complex paravector.[10]

Elements of real paravector space have the form $p = p^0 + \mathbf{p} = \langle p \rangle_0 + \langle p \rangle_1$, and the algebra $C\ell_n$ also includes homogeneous subspaces of multiparavectors. For example,

$$\begin{aligned}
\text{paravector space} &= \langle C\ell_n \rangle_{1,0}, \quad (n+1)\text{-dim} \\
\text{biparavector space} &= \langle C\ell_n \rangle_{2,1}, \quad \binom{n+1}{2}\text{-dim} \\
\text{k-paravector space} &= \langle C\ell_n \rangle_{k,k-1}, \quad \binom{n+1}{k}\text{-dim.}
\end{aligned}$$

In general, 0-grade paravectors are scalars in $\langle C\ell_n \rangle_0$, $(n+1)$-grade paravectors are volume elements (pseudoscalars) in $\langle C\ell_n \rangle_n$, and the linear space of k-grade multiparavectors is $\langle C\ell_n \rangle_{k,k-1} \equiv \langle C\ell_n \rangle_k \oplus \langle C\ell_n \rangle_{k-1}$, $k = 1, 2, \ldots, n$.

4.3.1 Clifford conjugation

In addition to reversion (dagger-conjugation), introduced above, we need Clifford conjugation, the anti-automorphism that changes the sign of all vector factors as well as reversing their order, and we need their combination. For any paravector p, its Clifford conjugate is

$$\bar{p} = p^0 - \mathbf{p}, \ \overline{pq} = \bar{q}\bar{p}.$$

Clifford conjugation can be used to split paravectors into scalar and vector parts, and it is convenient to extend such a split to a general element p of $C\ell_n$ by defining the *scalarlike* and *vectorlike* parts so that[11]

$$p = \frac{p+\bar{p}}{2} + \frac{p-\bar{p}}{2} = \langle p \rangle_S + \langle p \rangle_V = \text{scalarlike} + \text{vectorlike}.$$

Clifford conjugation is combined with reversion to give the *grade involution*[12]

$$\bar{p}^\dagger = \bar{p}, \ (\overline{pq})^\dagger = \overline{(pq)^\dagger} = \bar{p}^\dagger \bar{q}^\dagger$$

[10] Paravectors have other uses as well. Paravectors in $C\ell_{0,7}$ can be identified with octonions. [4]

[11] The scalarlike part of an element comprises all parts of grade k with $k \mod 4 = 0, 3$, and the vectorlike part comprises the rest.

[12] The symbol in Lecture 1 for the grade involution is the hat: $\bar{p}^\dagger \equiv \hat{p}$. However, physicists commonly use the hat decoration to indicate a unit element, and in this lecture it is preserved for that purpose.

an automorphism with which elements can be split into *even* and *odd* parts:

$$p = \frac{p+\bar{p}^{\dagger}}{2} + \frac{p-\bar{p}^{\dagger}}{2} = \langle p\rangle_{+} + \langle p\rangle_{-} = \text{even} + \text{odd}.$$

These relations offer simple ways to isolate different vector and paravector grades. In particular, for $n = 3$, (here $\cdots$ stands for any expression)

$$\begin{aligned} \langle\cdots\rangle_S &= \langle\cdots\rangle_{0,3} & \langle\cdots\rangle_V &= \langle\cdots\rangle_{1,2} \\ \langle\cdots\rangle_{\mathrm{Re}} &= \langle\cdots\rangle_{0,1} & \langle\cdots\rangle_{\mathrm{Im}} &= \langle\cdots\rangle_{2,3} \\ \langle\cdots\rangle_{+} &= \langle\cdots\rangle_{0,2} & \langle\cdots\rangle_{-} &= \langle\cdots\rangle_{1,3}\,. \end{aligned}$$

Exercise 14. *Verify that the splits can be combined to extract individual vector grades as follows:*

$$\begin{aligned} \langle\cdots\rangle_0 &= \langle\cdots\rangle_{\mathrm{Re}\,S} = \langle\cdots\rangle_{\mathrm{Re}\,+} = \langle\cdots\rangle_{S+}\,, & \langle\cdots\rangle_1 &= \langle\cdots\rangle_{\mathrm{Re}\,V} \\ \langle\cdots\rangle_2 &= \langle\cdots\rangle_{\mathrm{Im}\,V}\,, & \langle\cdots\rangle_3 &= \langle\cdots\rangle_{\mathrm{Im}\,S}\,. \end{aligned}$$

Example 15. *Let $\mathbf{B}$ be any bivector. Then $\mathbf{B}$ is even, imaginary, and vectorlike, whereas $\mathbf{B}^2$ is even, real, and scalarlike. Any analytic function $f(\mathbf{B})$ is even and $\overline{f(\mathbf{B})} = f(-\mathbf{B}) = f(\bar{\mathbf{B}}) = f\left(\mathbf{B}^{\dagger}\right)$. Spatial rotors $R(\mathbf{B}) = \exp\left(\frac{1}{2}\mathbf{B}\right)$ are even and unitary: $R^{\dagger}(\mathbf{B}) = R^{-1}(\mathbf{B}) = R(-\mathbf{B})$.*[13]

4.3.2 Paravector metric

If $\{\mathbf{e}_1, \mathbf{e}_2, \cdots, \mathbf{e}_n\}$ is an orthonormal basis of the original Euclidean space, so that

$$\langle \mathbf{e}_j \mathbf{e}_k\rangle_0 = \tfrac{1}{2}(\mathbf{e}_j\mathbf{e}_k + \mathbf{e}_k\mathbf{e}_j) = \delta_{jk},$$

the proper basis of paravector space is $\{\mathbf{e}_0, \mathbf{e}_1, \mathbf{e}_2, \cdots, \mathbf{e}_n\}$, where we identify $\mathbf{e}_0 \equiv 1$ for convenience in expanding paravectors $p = p^{\mu}\mathbf{e}_{\mu}$, $\mu = 0, 1, \cdots, n$ in the basis. The metric of paravector space is determined by a quadratic form. We need a product of a paravector p with itself or a conjugate that is scalar valued. It is easy to see that p^2 generally has vector parts, but $p\bar{p} = \langle p\bar{p}\rangle_0 = \bar{p}p$ is a scalar. Therefore it is adopted as the *quadratic form* ("square length"):

$$Q(p) = p\bar{p}.$$

By "polarization" $p \to p + q$ we find the inner product

$$\langle p, q\rangle = \langle p\bar{q}\rangle_S = \tfrac{1}{2}(p\bar{q} + q\bar{p}) = p^{\mu}q^{\nu}\langle \mathbf{e}_{\mu}\bar{\mathbf{e}}_{\nu}\rangle_S \equiv p^{\mu}q^{\nu}\eta_{\mu\nu}.$$

---

[13]We can use either grade numbers or conjugation symmetries to split an element into parts. The grade numbers emphasize the algebraic structure whereas the conjugation symmetries indicate an operational procedure to compute the part.

Exercise 16. *Show that* $\langle p\bar{q}\rangle_S = \frac{1}{2}\left[Q(p+q) - Q(p) - Q(q)\right]$.

Exercise 17. *Find the values of* $\eta_{\mu\nu} = \langle \mathbf{e}_\mu \bar{\mathbf{e}}_\nu\rangle_S$.

We recognize the matrix

$$(\eta_{\mu\nu}) = \mathrm{diag}(1, -1, -1, \ldots, -1)$$

as a natural metric of the paravector space.[14] It has the form of the *Minkowski metric*. If $\langle p\bar{q}\rangle_S = 0$, then the paravectors p and q are *orthogonal*. For any element $x \in C\ell_n$, $x\bar{x} = \overline{x\bar{x}} = \langle x\bar{x}\rangle_S$. If $n = 3$, then $x\bar{x} = \det \underline{x}$, where $\underline{x}$ is the standard 2×2 matrix representation of $x \in C\ell_3$, in which $\mathbf{e}_k \simeq \underline{\sigma}_k$. [1] If $x\bar{x} = 1$, x is *unimodular.*

The inverse of an element x can be written

$$x^{-1} = \frac{\bar{x}}{x\bar{x}}$$

but this does not exist if $x\bar{x} = 0$. The existence of nonzero elements of zero length ("nonzero zero divisors") means that $C\ell_n$, $n > 0$, and most other Clifford algebras, unlike the (Clifford) algebras of reals, complexes, and quaternions, are *not* division algebras. This may seem an annoyance at first, but it is the basis for powerful projector techniques, as we demonstrate below.

Exercise 18. *Show that the paravector* $1 + \mathbf{e}_1$ *has no inverse and is orthogonal to itself.*

Spacetime as paravector space

The paravectors of physical space provide a covariant model of spacetime. (Extensions to curved spacetimes are possible, but for simplicity we restrict ourselves here to flat spacetimes.) We use SI units with $c = 1$ and, unless specified otherwise, take $n = 3$. Spacetime vectors are represented by paravectors whose frame-dependent split into scalar and vector parts reflects the observer's ability to distinguish time and space components. In particular, any timelike spacetime displacement $dx = dt + d\mathbf{x}$ has a Lorentz-invariant length defined as the proper time interval $d\tau$:

$$d\tau^2 = dx\, d\bar{x} = \eta_{\mu\nu} dx^\mu dx^\nu.$$

The proper velocity is

$$u = \frac{dx}{d\tau} = \gamma(1 + \mathbf{v}) = u^\mu \mathbf{e}_\mu$$

where $\mathbf{v} = d\mathbf{x}/dt$ is the usual coordinate velocity and we use the summation convention for repeated indices. Other spacetime vectors are similarly represented,

[14] If we instead adopt the quadratic form $Q(p) = -p\bar{p}$, the paravector metric has the opposite signature.

for example,

$$\begin{aligned}
p &= mu = E + \mathbf{p} : \text{ paramomentum} \\
j &= j^{\mu}\mathbf{e}_{\mu} = \rho + \mathbf{j} : \text{current density} \\
A &= A^{\mu}\mathbf{e}_{\mu} = \phi + \mathbf{A} : \text{paravector potential} \\
\partial &= \partial_{\mu}\mathbf{e}_{\nu}\eta^{\mu\nu} = \partial_t - \nabla : \text{gradient operator.}
\end{aligned}$$

Exercise 19. *Consider a function $f(s)$, where s is the scalar $s = \langle k\bar{x}\rangle_S = k^{\mu}x_{\mu} = k_{\mu}x^{\mu}$ and k is a constant paravector. Use the chain rule for differentiation to prove that $\partial f(s) = kf'(s)$, where $f'(s) = df/ds$. Note that $\partial_{\mu} \equiv \partial/\partial x^{\mu}$.*

Biparavectors represent *oriented planes* in *spacetime*, for example the electromagnetic field

$$\mathbf{F} = \langle\partial\bar{A}\rangle_V = \tfrac{1}{2}F^{\mu\nu}\langle\mathbf{e}_{\mu}\bar{\mathbf{e}}_{\nu}\rangle_V = \mathbf{E} + i\mathbf{B}.$$

The basis biparavectors $\langle\mathbf{e}_{\mu}\bar{\mathbf{e}}_{\nu}\rangle_V = -\langle\mathbf{e}_{\nu}\bar{\mathbf{e}}_{\mu}\rangle_V$ are also the generators of Lorentz rotations, and the expansion of $\mathbf{F}$ in this basis gives directly the usual tensor components $F^{\mu\nu}$. However, no tensor elements are needed in APS, and the algebra offers several ways to interpret $\mathbf{F}$ geometrically. The field at any point is a covariant plane in spacetime, and for any observer it splits naturally into frame-dependent parts: a timelike plane $\mathbf{E} = \mathbf{E}\mathbf{e}_0 = \langle\mathbf{F}\rangle_1$ and a spatial plane $i\mathbf{B} = \langle\mathbf{F}\rangle_2$. Since $\mathbf{E} = \langle\mathbf{F}\mathbf{e}_0\rangle_{\text{Re}}$, the electric field $\mathbf{E}$ lies in the intersection of $\mathbf{F}$ with the spatial hyperplane dual (and thus orthogonal) to $\mathbf{e}_0$. From $i\mathbf{B} = \langle\mathbf{F}\mathbf{e}_0\rangle_{\text{Im}}$, we see that the usual magnetic field $\mathbf{B}$ is the vector dual to the spacetime hypersurface $\langle\mathbf{F}\mathbf{e}_0\rangle_{\text{Im}}$.

We need to be careful about stating that $\mathbf{F}$ is *a* plane in spacetime. The sum of electromagnetic fields is another electromagnetic field, but the sum of planes in four dimensions is not necessarily a single plane. We may get two orthogonal planes. (Of course this cannot happen in three dimensions, where the sum of any two planes is also a plane, but spacetime has four dimensions.) Thus, we distinguish simple fields, which are single spacetime planes from compound ones, which occupy two orthogonal (and hence commuting) planes. How do we distinguish a simple field from a compound one? All we need to do is square it. The square of any simple field is a real scalar, whereas the square of a compound field contains a spacetime volume element, that is a pseudoscalar, given in APS as an imaginary scalar.

Example 20. *The biparavector $\langle p\bar{q}\rangle_V = \frac{1}{2}(p\bar{q} - q\bar{p})$ is simple and represents a spacetime plane containing paravectors p and q. Its square*

$$\langle p\bar{q}\rangle_V^2 = \tfrac{1}{4}(p\bar{q} - q\bar{p})^2 = \tfrac{1}{4}(p\bar{q} + q\bar{p})^2 - \tfrac{1}{2}(p\bar{q}q\bar{p} + q\bar{p}p\bar{q}) = \langle p\bar{q}\rangle_0^2 - p\bar{p}q\bar{q}$$

is seen to be a real scalar.

If paravectors p and q are orthogonal to both r and s, then $\langle p\bar{q}+r\bar{s}\rangle_V$ is a compound biparavector with orthogonal spacetime planes $\langle p\bar{q}\rangle_V$ and $\langle r\bar{s}\rangle_V$. However, the planes in a compound biparavector may not be unique: if p, q, r, s are mutually orthogonal paravectors and if $p\bar{p} = r\bar{r}$ and $q\bar{q} = s\bar{s}$, then

$$\langle p\bar{q}+r\bar{s}\rangle_V = \tfrac{1}{2}\left\langle (p+r)\,\overline{(q+s)} + (p-r)\,\overline{(q-s)}\right\rangle_V$$

expresses the compound biparavector as two sets of orthogonal planes. Orthogonal planes in spacetime are proportional to each other's dual. As a result, any compound biparavector can be expressed as a simple biparavector times a complex number.

Exercise 21. *Show that* $\mathbf{e}_1\bar{\mathbf{e}}_0 + \mathbf{e}_3\bar{\mathbf{e}}_2 = \mathbf{e}_1\bar{\mathbf{e}}_0\,(1+i)$ *for any compound biparavector* $\mathbf{e}_1\bar{\mathbf{e}}_0 + \mathbf{e}_3\bar{\mathbf{e}}_2$.

The square of $\boldsymbol{F} = \boldsymbol{E} + i\boldsymbol{B}$ is

$$\boldsymbol{F}^2 = \boldsymbol{E}^2 - \boldsymbol{B}^2 + 2i\boldsymbol{E}\cdot\boldsymbol{B}$$

and $\boldsymbol{F}$ is evidently simple if and only if $\boldsymbol{E}\cdot\boldsymbol{B} = 0$. A null field, with $\boldsymbol{F}^2 = 0$, is thus simple. It can be written $\boldsymbol{F} = \left(1+\hat{\boldsymbol{k}}\right)\boldsymbol{E}$, where $\hat{\boldsymbol{k}}\boldsymbol{E} = i\boldsymbol{B}$.

Exercise 22. *Show that* $\hat{\boldsymbol{k}}\boldsymbol{F} = \boldsymbol{F} = -\boldsymbol{F}\hat{\boldsymbol{k}}$ *for any null field* $\boldsymbol{F} = \left(1+\hat{\boldsymbol{k}}\right)\boldsymbol{E}$.

Lorentz transformations

Much of the power of Clifford's geometric algebra in relativistic applications arises from the form of Lorentz transformations. In APS, we identify paravector space with spacetime, and physical (restricted) Lorentz transformations are rotations in paravector space. They take the form of spin transformations

$$\begin{aligned} p &\to LpL^\dagger, && \text{odd multiparavector grade} \\ \boldsymbol{F} &\to L\boldsymbol{F}\bar{L}, && \text{even multiparavector grade,} \end{aligned}$$

where the Lorentz rotors L are unimodular, $L\bar{L} = 1$, with matrix representations in $SL(2,\mathbb{C})$, and they have the form

$$L = \pm\exp\left(\tfrac{1}{2}\boldsymbol{W}\right), \quad \boldsymbol{W} = \tfrac{1}{2}W^{\mu\nu}\left\langle \mathbf{e}_\mu\bar{\mathbf{e}}_\nu\right\rangle_{1,2}.$$

Every L can be factored into a boost $B = B^\dagger$ (a real factor) and a spatial rotation $R = \bar{R}^\dagger$ (a unitary factor): $L = BR$.

For any paravectors p, q, the scalar product

$$\langle p\bar{q}\rangle_S \to \left\langle LpL^\dagger\bar{L}^\dagger\bar{q}\bar{L}\right\rangle_S = \langle p\bar{q}\rangle_S$$

is Lorentz invariant. In particular, the square length $p\bar{p} = \left(p^0\right)^2 - \boldsymbol{p}^2$ is invariant and can be timelike (> 0), spacelike (< 0), or lightlike (null, $= 0$). Null paravectors are orthogonal to themselves. Similarly $\boldsymbol{F}^2$ is Lorentz invariant, and simple fields can be classified as predominantly electric ($\boldsymbol{F}^2 > 0$), predominantly magnetic ($\boldsymbol{F}^2 < 0$), or null ($\boldsymbol{F}^2 = 0$).

With Lorentz transformations, we can easily transform properties between inertial frames. The *position coordinate* x of a particle instantaneously *at rest* changes only by its time, the proper time $\tau : dx_{\text{rest}} = d\tau$. We transform this to the lab, in which the particle moves with proper velocity $u = dx/d\tau$, by

$$dx = L dx_{\text{rest}} L^\dagger = LL^\dagger d\tau = u d\tau = dt + d\boldsymbol{x} = dt\,(1 + \boldsymbol{v}) .$$

With $L = BR$ it follows that

$$LL^\dagger = B^2 = u = \frac{dt}{d\tau}\,(1 + \boldsymbol{v}) .$$

Now $LL^\dagger = L\mathbf{e}_0 L^\dagger = u$ is just the Lorentz rotation of the unit basis paravector $\mathbf{e}_0$, and its square length is invariant:

$$u\bar{u} = 1 = \gamma^2\left(1 - \boldsymbol{v}^2\right),$$

where $\gamma = dt/d\tau$ is the *time-dilation factor.*

Example 23. *Consider the transformation of a paravector $p = p^\mu \mathbf{e}_\mu$ in a system that is boosted from rest to a velocity $\boldsymbol{v} = v\mathbf{e}_3$:*

$$p \to LpL^\dagger = BpB = p^\mu \boldsymbol{u}_\mu$$

where $B = \exp\left(\frac{1}{2} w \mathbf{e}_3 \bar{\mathbf{e}}_0\right) = u^{1/2}$ represents a rotation in the $\mathbf{e}_3\bar{\mathbf{e}}_0$ paravector plane and $\boldsymbol{u}_\mu = B\mathbf{e}_\mu B$ is the boosted proper basis paravector. Evidently

$$\begin{aligned}
\boldsymbol{u}_0 &= B\mathbf{e}_0 B = B^2 \mathbf{e}_0 = u\mathbf{e}_0 = \gamma\left(\mathbf{e}_0 + v\mathbf{e}_3\right) \\
\boldsymbol{u}_1 &= \mathbf{e}_1 \\
\boldsymbol{u}_2 &= \mathbf{e}_2 \\
\boldsymbol{u}_3 &= B\mathbf{e}_3 B = u\mathbf{e}_3 = \gamma\left(\mathbf{e}_3 + v\mathbf{e}_0\right)
\end{aligned}$$

with $\gamma = \left(1 - \frac{v^2}{c^2}\right)^{-\frac{1}{2}}$.

As in the case of spatial rotations, if we put $LpL^\dagger = p' = p'^\nu \mathbf{e}_\nu$, we can easily find

$$p'^\nu = p^\mu \left\langle \boldsymbol{u}_\mu \bar{\mathbf{e}}^\nu \right\rangle_S$$

and hence the usual 4×4 matrix relating the components of p before and after the boost, but we have no need of it. The relations for $\boldsymbol{u}_\mu$ are useful for drawing spacetime diagrams. Thus, if $v = 0.6$, then $\gamma = 1.25$ and

$$\boldsymbol{u}_0 = \tfrac{1}{4}\left(5\mathbf{e}_0 + 3\mathbf{e}_3\right), \qquad \boldsymbol{u}_3 = \tfrac{1}{4}\left(5\mathbf{e}_3 + 3\mathbf{e}_0\right).$$

We can take this further and look at planes in spacetime, as shown in Fig. 4.1.

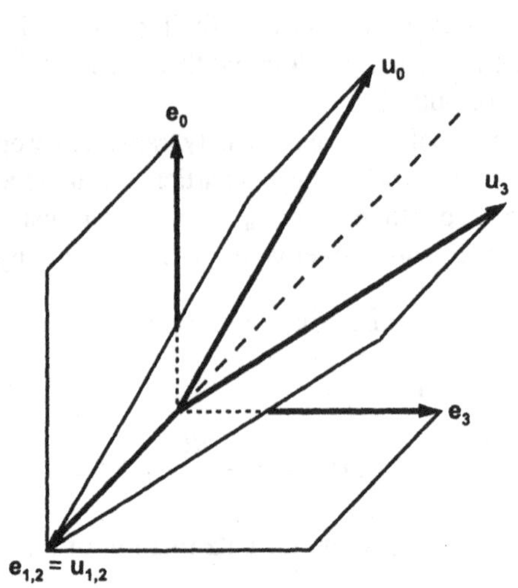

FIGURE 4.1. Spacetime diagram showing a boost to $\boldsymbol{v} = 0.6\,\mathbf{e}_3$.

Exercise 24. *Show that the biparavectors* $\mathbf{e}_3\bar{\mathbf{e}}_0$ *and* $\mathbf{e}_1\bar{\mathbf{e}}_2$ *are invariant under any boost B along* $\mathbf{e}_3$.

Exercise 25. *Let system B have proper velocity* u_{AB} *with respect to A, and let system C have proper velocity* u_{BC} *as seen by an observer in B. Show that the proper velocity of C as viewed by A is*

$$u_{AC} = u_{AB}^{1/2} u_{BC} u_{AB}^{1/2}$$

and that this reduces to the product $u_{AC} = u_{AB} u_{BC}$ *when the spatial velocities are collinear. Writing each proper velocity in the form* $u = \gamma(1 + \boldsymbol{v})$, *show that in the collinear case*

$$\boldsymbol{v}_{AC} = \frac{\langle u_{AC}\rangle_V}{\langle u_{AC}\rangle_S} = \frac{\boldsymbol{v}_{AB} + \boldsymbol{v}_{BC}}{1 + \boldsymbol{v}_{AB}\cdot\boldsymbol{v}_{BC}}.$$

Example 26. *Consider a boost of the photon wave paravector*

$$k = \omega\left(1 + \hat{\boldsymbol{k}}\right) \to k' = BkB = u\left(\omega + \boldsymbol{k}^{\parallel}\right) + \boldsymbol{k}^{\perp}$$

with $\boldsymbol{k}^{\parallel} = \boldsymbol{k}\cdot\hat{\boldsymbol{v}}\,\hat{\boldsymbol{v}} = \boldsymbol{k} - \boldsymbol{k}^{\perp}$ *and* $u = \gamma(1 + \boldsymbol{v})$. *This describes what happens to the photon momentum when the light source is boosted. Evidently* $\boldsymbol{k}^{\perp}$ *is unchanged, but there is a Doppler shift and a change in* $\boldsymbol{k}^{\parallel}$:

$$\begin{aligned}
\omega' &= \left\langle u\left(\omega + \boldsymbol{k}^{\parallel}\right)\right\rangle_S = \gamma\omega\left(1 + \hat{\boldsymbol{k}}\cdot\boldsymbol{v}\right)\\
\boldsymbol{k}'\cdot\hat{\boldsymbol{v}} &= \left\langle u\hat{\boldsymbol{v}}\left(\omega + \boldsymbol{k}^{\parallel}\right)\right\rangle_S = \gamma\omega\left(v + \hat{\boldsymbol{k}}\cdot\hat{\boldsymbol{v}}\right) = \omega'\cos\theta'.
\end{aligned}$$

Thus the photons are thrown forward

$$\cos\theta' = \frac{v + \cos\theta}{1 + v\cos\theta}. \qquad (3.6)$$

in what is called the "headlight" effect (see Fig. 4.2).

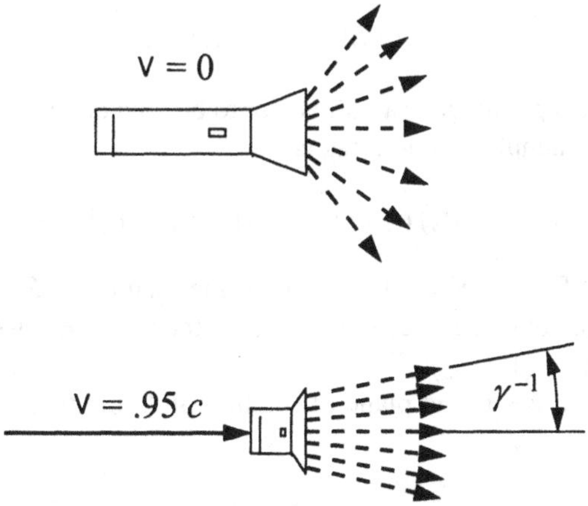

FIGURE 4.2. Headlight effect in boosted light source.

Exercise 27. *Solve (3.6) for* $\cos\theta$ *and show that the result is the same as in (3.6) except that* v *is replaced by* $-v$ *and* θ *and* θ' *are interchanged.*

Exercise 28. *Show that at high velocities, the radiation from the boosted source is concentrated in the cone of angle* γ^{-1} *about the forward direction.*

Simple rotors of Lorentz transformations can be expressed as a product of paravectors in the spacetime plane of rotation. Consider a biparavector $\langle p\bar{q}\rangle_V$ and a paravector r with a coplanar component $r^{\triangle}$ and an orthogonal component $r^{\perp}$. The coplanar component is a linear combination of p and q : $r^{\triangle} = \alpha p + \beta q$, where α and β are scalars. Now the product of $\langle p\bar{q}\rangle_V$ with p satisfies

$$\langle p\bar{q}\rangle_V\, p = \tfrac{1}{2}\left(p\bar{q}p - q\bar{p}p\right) = p\left\langle \bar{q}p\right\rangle_V = p\left\langle p\bar{q}\right\rangle_V^{\dagger},$$

Similarly with q and thus with any linear combination of p and q. It follows that if L is a simple Lorentz rotor in the plane $\langle p\bar{q}\rangle_V$, that the coplanar component obeys

$$Lr^{\triangle} = r^{\triangle}L^{\dagger}.$$

On the other hand $r^{\perp}$ is orthogonal to the plane and thus coplanar with its dual: $\langle p\bar{r}^{\perp}\rangle_S = 0 = \langle q\bar{r}^{\perp}\rangle_S$, so that

$$\langle p\bar{q}\rangle_V r^{\perp} = \tfrac{1}{2}\left(p\bar{q}r^{\perp} - q\bar{p}r^{\perp}\right) = \tfrac{1}{2}r^{\perp}\left(\bar{p}q - \bar{q}p\right) = -r^{\perp}\langle p\bar{q}\rangle_V^{\dagger}$$

and consequently for any Lorentz rotor L that rotates in the plane of $\langle p\bar{q}\rangle_V$,

$$\bar{L}r^{\perp} = r^{\perp}L^{\dagger}.$$

The Lorentz transformation of r thus gives

$$LrL^{\dagger} = r^{\perp} + L^2 r^{\triangle}.$$

In spacetime, it is possible for a null vector to be both coplanar and orthogonal to a null flag. An example of a null flag is

$$\boldsymbol{F} = (1 + \mathbf{e}_3)\,\mathbf{e}_1 = (1 + \mathbf{e}_3)\,\mathbf{e}_3\mathbf{e}_1 = i\,(1 + \mathbf{e}_3)\,\mathbf{e}_2.$$

The dual flag $^{*}\boldsymbol{F} = -i\boldsymbol{F} = (1 + \mathbf{e}_3)\,\mathbf{e}_2$ is the rotation of $\boldsymbol{F}$ about $\mathbf{e}_3$ by $\frac{1}{2}\pi$. A Lorentz transformation generated by $\boldsymbol{F}$ leaves the flagpole $1 + \mathbf{e}_3$ invariant, since $\boldsymbol{F}(1 + \mathbf{e}_3) = 0 = (1 + \mathbf{e}_3)\,\boldsymbol{F}^{\dagger}$.

Suppose that r lies in the plane of rotation of L and that $s = LrL^{\dagger} = L^2 r$. Then, as long as $r\bar{r} \neq 0$,

$$L = \left(sr^{-1}\right)^{1/2} = \frac{(s + r)\,r^{-1}}{\sqrt{2\,\langle sr^{-1} + 1\rangle_S}} = \frac{sr^{-1} + 1}{\langle 2\,(sr^{-1} + 1)\rangle_S^{1/2}}.$$

Exercise 29. *Verify that* $L\bar{L} = 1$ *and*

$$L^2 r = \frac{\left(sr^{-1} + 1\right)(s + r)}{\langle 2\,(sr^{-1} + 1)\rangle_S} = \frac{2 + sr^{-1} + rs^{-1}}{2 + sr^{-1} + \bar{r}^{-1}\bar{s}}\,s = s.$$

(Hint: note that $s\bar{s} = r\bar{r} \neq 0$ *so that* $\bar{r}^{-1}\bar{s} = r\bar{s}/(r\bar{r}) = r\bar{s}/(s\bar{s}) = rs^{-1}$.*)*

If we rotate a null paravector $k = \omega\left(1 + \hat{\boldsymbol{k}}\right)$ in a spacetime plane that contains k, then $k \to k' = LkL^{\dagger} = L^2 k$. In the case of a boost, $L = B = \exp\left(\frac{1}{2}w\hat{\boldsymbol{k}}\right)$, we find

$$k' = e^{w}k$$

with $e^{w} = \langle k'\rangle_S / \langle k\rangle_S = \omega'/\omega = \gamma\,(1 + v) = \sqrt{(1 + v)\,/\,(1 - v)}$.

Lorentz rotations in the same or in dual planes commute, but otherwise they generally do not. Furthermore, whereas any product of spatial rotations is another spatial rotation, the product of noncommuting boosts generally does not give a pure boost, but rather the product of a boost and a rotation. Lorentz rotations can also be expressed as the product of spacetime reflections. Up to four reflections may be needed.

4.3.3 Relation of APS to STA

An alternative to the paravector model of spacetime in APS is the spacetime algebra (STA) introduced by David Hestenes [10]. They are closely related, and it is the purpose of this section to show how.

STA is the geometric algebra $C\ell_{1,3}$ of Minkowski spacetime. It starts with a 4-dimensional basis $\{\gamma_0, \gamma_1, \gamma_2, \gamma_3\} \equiv \{\gamma_\mu\}$ satisfying

$$\gamma_\mu \gamma_\nu + \gamma_\nu \gamma_\mu = 2\eta_{\mu\nu}$$

in each frame. The chosen frame can be independent of the observer and her frame $\{\hat{\gamma}_\mu\}$.[15] Any spacetime vector $p = p^\mu \hat{\gamma}_\mu$ can be multiplied by $\hat{\gamma}_0$ to give the *spacetime split*

$$p\hat{\gamma}_0 = p \cdot \hat{\gamma}_0 + p \wedge \hat{\gamma}_0 = p^0 + p^k \mathbf{e}_k \,,$$

where $\mathbf{e}_0 = 1$ and $\mathbf{e}_k \equiv \hat{\gamma}_k \hat{\gamma}_0$ are the proper basis paravectors of the system in APS. (This association establishes the previously mentioned isomorphism between the even subalgebra of STA and APS.) More general paravector basis elements $\boldsymbol{u}_\mu$ in APS arise when the basis $\{\gamma_\mu\}$ used for the expansion $p = p^\mu \gamma_\mu$ is for a frame in motion with respect to the observer:[16]

$$\boldsymbol{u}_\mu = \gamma_\mu \hat{\gamma}_0 \,.$$

In particular, $\boldsymbol{u}_0 = \gamma_0 \hat{\gamma}_0$ is the proper velocity of the frame $\{\gamma_\mu\}$ with respect to the observer frame $\{\hat{\gamma}_\mu\}$. The basis vectors in APS are relative; they always relate two frames, but those in STA can be considered absolute.

Clifford conjugation in APS corresponds to reversion in STA, indicated by a tilde:

$$\bar{\boldsymbol{u}}_\mu = (\gamma_\mu \hat{\gamma}_0)^{\sim} = \hat{\gamma}_0 \gamma_\mu .$$

For example, if the proper velocity of frame $\{\gamma_\mu\}$ with respect to $\hat{\gamma}_0$ is $\boldsymbol{u}_0 = \gamma_0 \hat{\gamma}_0$, then the proper velocity of frame $\{\hat{\gamma}_\mu\}$ with respect to γ_0 is $\bar{\boldsymbol{u}}_0 = \hat{\gamma}_0 \gamma_0$. It is not possible to make all of the basis vectors in any STA frame Hermitian, but one usually takes $\hat{\gamma}_0^\dagger = \hat{\gamma}_0$ and $\hat{\gamma}_k^\dagger = -\hat{\gamma}_k$ in the observer's frame $\{\hat{\gamma}_\mu\}$. Hermitian conjugation in STA then combines reversion with the reflection in the observer's time axis $\hat{\gamma}_0$: $\Gamma^\dagger = \hat{\gamma}_0 \tilde{\Gamma} \hat{\gamma}_0$, for example

$$\boldsymbol{u}_\mu^\dagger = \hat{\gamma}_0 (\gamma_\mu \hat{\gamma}_0)^{\sim} \hat{\gamma}_0 = \gamma_\mu \hat{\gamma}_0 = \boldsymbol{u}_\mu \,,$$

---

[15] The hat here is not related to the grade involution. It is used to designate the observer frame and to emphasize that the basis vectors $\hat{\gamma}_\mu$ have unit magnitudes: $(\hat{\gamma}_\mu)^2 = \pm 1$.

[16] A double arrow might be thought more appropriate than an equality here, because $\boldsymbol{u}_\mu$ and $\gamma_\mu, \hat{\gamma}_0$ act in different algebras. However, we are identifying $C\ell_3$ with the even subalgebra of $C\ell_{1,3}$, so that the one algebra is embedded in the other. Caution is still needed to avoid statements such as

$$i = \mathbf{e}_1 \mathbf{e}_2 \mathbf{e}_3 = \mathbf{e}_1 \wedge \mathbf{e}_2 \wedge \mathbf{e}_3 \overset{\text{wrong!}}{=} \hat{\gamma}_1 \wedge \hat{\gamma}_0 \wedge \hat{\gamma}_2 \wedge \hat{\gamma}_0 \wedge \hat{\gamma}_3 \wedge \hat{\gamma}_0 = 0.$$

This is not valid because the wedge products on either side of the third equality refer to different algebras and are not equivalent.

which shows that all the basis paravectors $\boldsymbol{u}_\mu$ are Hermitian. It is important to note that Hermitian conjugation is frame dependent in STA just as Clifford conjugation of paravectors is in APS.

Example 30. *The Lorentz-invariant scalar part of the paravector product $p\bar{q}$ in APS thus becomes*

$$\langle p\bar{q}\rangle_S = \tfrac{1}{2}p^\mu q^\nu \left(\boldsymbol{e}_\mu \bar{\boldsymbol{e}}_\nu + \boldsymbol{e}_\nu \bar{\boldsymbol{e}}_\mu\right) = \tfrac{1}{2}p^\mu q^\nu \left(\gamma_\mu \hat{\gamma}_0 \hat{\gamma}_0 \gamma_\nu + \gamma_\nu \hat{\gamma}_0 \hat{\gamma}_0 \gamma_\mu\right) = p^\mu q^\nu \eta_{\mu\nu}\,.$$

Biparavector basis elements in APS become basis bivectors in STA:

$$\tfrac{1}{2}\left(\boldsymbol{e}_\mu \bar{\boldsymbol{e}}_\nu - \boldsymbol{e}_\nu \boldsymbol{e}_\mu\right) = \tfrac{1}{2}\left(\gamma_\mu \hat{\gamma}_0 \hat{\gamma}_0 \gamma_\nu - \gamma_\nu \hat{\gamma}_0 \hat{\gamma}_0 \gamma_\mu\right) = \tfrac{1}{2}\left(\gamma_\mu \gamma_\nu - \gamma_\nu \gamma_\mu\right).$$

Lorentz transformations in STA are effected by $\gamma_\mu \to L\gamma_\mu L^{\sim}$, with $LL^{\sim} = 1$. Every product of basis vectors transforms the same way. An *active transformation* keeps the observer frame fixed and transforms only the system frame:

$$\boldsymbol{e}_\mu = \hat{\gamma}_\mu \hat{\gamma}_0 \to \boldsymbol{u}_\mu = \gamma_\mu \hat{\gamma}_0 = L\hat{\gamma}_\mu L^{\sim} \hat{\gamma}_0 = L\hat{\gamma}_\mu \hat{\gamma}_0 \left(\hat{\gamma}_0 L^{\sim} \hat{\gamma}_0\right) = L\boldsymbol{e}_\mu L^\dagger.$$

We noted that the $\hat{\gamma}_0$ in the definition of $\boldsymbol{e}_\mu$ is always the observer's time axis. In a *passive transformation*, it is the system frame that stays the same and the observer's frame that changes. Let us suppose that the observer moves from frame $\{\gamma_\mu\}$ to frame $\{\hat{\gamma}_\mu\}$ where $\gamma_\mu = L\hat{\gamma}_\mu L^{\sim}$. Then

$$\boldsymbol{e}_\mu = \gamma_\mu \gamma_0 \to \boldsymbol{u}_\mu = \gamma_\mu \hat{\gamma}_0\,.$$

To re-express the transformed relative coordinates $\boldsymbol{u}_\mu$ in terms of the original $\boldsymbol{e}_\mu$, we must expand the system frame vectors γ_μ in terms of the observer's transformed basis vectors. Thus

$$\boldsymbol{u}_\mu = L\hat{\gamma}_\mu L^{\sim} \hat{\gamma}_0 = L\boldsymbol{e}_\mu L^\dagger.$$

The mathematics is identical to that for the active transformation, but the interpretation is different. Since the transformations can be realized by changing the observer frame and keeping the system frame constant, the physical objects can be taken to be fixed in STA, giving what is referred to as an *invariant* treatment of relativity.

The Lorentz rotation is the same whether we rotate the object forward or the observer backward or some combination. This is trivially seen in APS where only the object frame relative to the observer enters. Furthermore, as seen above, the space/time split of a property in APS is simply a result of expanding into vector grades in the observer's proper basis $\{\boldsymbol{e}_\mu\}$:

$$p = \langle p\rangle_0 + \langle p\rangle_1 = p^0 + \boldsymbol{p}, \qquad \boldsymbol{F} = \langle \boldsymbol{F}\rangle_1 + \langle \boldsymbol{F}\rangle_2 = \boldsymbol{E} + i\boldsymbol{B}.$$

To get the split for a different observer, you can expand p in his paravector basis $\{\boldsymbol{u}_\mu\}$ and $\boldsymbol{F}$ in his biparavector basis $\left\{\langle \boldsymbol{u}_\mu \bar{\boldsymbol{u}}_\nu\rangle_{1,2}\right\}$, where $u = \boldsymbol{u}_0$ is his proper velocity relative to the original observer. Then with the transformation $u_\mu = Le_\mu L^\dagger$ you re-express the result in its proper basis before splitting vector grades. The physical fields, momenta, etc. are transformed and are not invariant in APS, but *covariant*, that is the form of the equations remains the same but not the vectors and multivectors themselves.[17]

4.4 Eigenspinors

A Lorentz rotor of particular interest is the ***eigenspinor*** Λ that relates the particle reference frame to the observer. It transforms distinctly from paravectors and their products: $\Lambda \to L\Lambda$ and is a generally reducible element of the spinor carrier space of Lorentz rotations in $SL(2,\mathbb{C})$. This property makes Λ a *spinor*. The ***eigen*** part refers to its association with the particle. Indeed any property of the particle in the reference frame is easily transformed by Λ to the lab (= observer's frame). For example, the proper velocity of a massive particle can be taken to be $u = 1$ in the reference frame. In the lab it is then

$$u = \Lambda e_0 \Lambda^\dagger = \Lambda\Lambda^\dagger,$$

which is seen to be the timelike basis vector of a frame moving with proper velocity u (with respect to the observer).

If we write $\Lambda = BR$, then u is independent of R. Traditional particle dynamics gives only u and by integration the world line. The eigenspinor gives more, namely the orientation and the full moving frame $\{u_\mu = \Lambda e_\mu \Lambda^\dagger\}$. While u_0 is the proper velocity, u_3 is essentially the Pauli–Lubański spin paravector [13].

4.4.1 *Time evolution*

The eigenspinor Λ changes in time, with $\Lambda(\tau)$ giving the Lorentz rotation at time τ. For boosts (rotations in timelike planes) this means that Λ relates the observer frame to the commoving inertial frame of the object at τ. Eigenspinors at different times can be related by

$$\Lambda(\tau_2) = L(\tau_2, \tau_1)\,\Lambda(\tau_1),$$

where the time-evolution operator is

$$L(\tau_2, \tau_1) = \Lambda(\tau_2)\,\bar{\Lambda}(\tau_1)$$

and is also seen to be a Lorentz rotation.

The time evolution is in principle found by solving the equation of motion

$$\dot{\Lambda} = \tfrac{1}{2}\boldsymbol{\Omega}\Lambda = \tfrac{1}{2}\Lambda\boldsymbol{\Omega}_r \tag{4.7}$$

---

[17] You can have *absolute frames* in APS, if you want them for use in passive transformations, by introducing an *absolute observer*.

with the spacetime rotation rate $\boldsymbol{\Omega} = \dot{\Lambda}\bar{\Lambda} - \Lambda\bar{\dot{\Lambda}} = \Lambda\boldsymbol{\Omega}_r\bar{\Lambda}$, where $\boldsymbol{\Omega}_r$ is its biparavector value in the reference frame. This relation allows us to compute time-rates of change of any property that is known in the reference frame. We take the reference frame of a massive particle to be the commoving inertial frame of the particle, in which $u = 1$. For example, the acceleration in the lab is

$$\dot{u} = \langle \boldsymbol{\Omega} u \rangle_{\mathrm{Re}} = \langle \Lambda \boldsymbol{\Omega}_r \Lambda^\dagger \rangle_{\mathrm{Re}} = \Lambda \langle \boldsymbol{\Omega}_r \rangle_{\mathrm{Re}} \Lambda^\dagger .$$

The proper velocity u of a particle can always be obtained from an eigenspinor that is a pure boost:

$$\Lambda = u^{1/2} = (1+u) / \sqrt{2\langle 1+u\rangle_S}.$$

The spacetime rotation rate is then

$$\begin{aligned} \boldsymbol{\Omega} = 2\dot{\Lambda}\bar{\Lambda} &= \left\langle \left(\frac{d}{d\tau} \frac{1+u}{\langle 1+u\rangle_S^{\frac{1}{2}}} \right) \left(\frac{1+\bar{u}}{\langle 1+u\rangle_S^{\frac{1}{2}}} \right) \right\rangle_V \\ &= \frac{\langle \dot{u}(1+\bar{u})\rangle_V}{1+\gamma} = \dot{\boldsymbol{u}} - \frac{\dot{\gamma}\boldsymbol{u}}{1+\gamma} - i\frac{\dot{\boldsymbol{u}} \times \boldsymbol{u}}{1+\gamma}, \end{aligned}$$

where we noted that $(1+u)(1+\bar{u})$ is a scalar. The negative imaginary part $\dot{\boldsymbol{u}} \times \boldsymbol{u}/(1+\gamma)$ is the *spatial* rotation rate, known as the proper *Thomas precession rate.*[18]

4.4.2 Charge dynamics in uniform fields

A standard problem in particle dynamics is to find the motion of a charge in constant, uniform electric and magnetic fields. We saw above an interpretation of the field $\boldsymbol{F}$ as a spacetime plane. Its *definition* is given in this *operational* or dynamic sense: it is the spacetime rotation rate of a test charge with a unit charge-to-mass ratio. The Lorentz-force equation follows from the eigenspinor evolution (4.7) with the spacetime rotation rate $\boldsymbol{\Omega} = e\boldsymbol{F}/m$:

$$\dot{\Lambda} = \frac{e}{2m}\boldsymbol{F}\Lambda. \tag{4.8}$$

Exercise 31. *From the relation $p = mu = m\Lambda\Lambda^\dagger$ for the momentum p of the charge, prove that the identification above of the spacetime rotation rate leads to the covariant Lorentz-force equation*[19]

$$\dot{p} = e\langle \boldsymbol{F} u\rangle_{\mathrm{Re}} \equiv \frac{e}{2}\left(\boldsymbol{F}u + u\boldsymbol{F}^\dagger\right).$$

---

[18] This one-line derivation is not only much neater but considerably clearer than the usual cumbersome one based on differentials!

[19] One of the advantages of treating EM relativistically is that, provided we know how quantities transform, we can determine general laws from behavior in the rest frame. Thus the Lorentz force equation is the covariant extension of the *definition of the electric field,* viz. the force per unit charge

For any constant $\boldsymbol{F}$, the eigenspinor satisfying (4.8) has the form

$$\Lambda(\tau) = L(\tau)\Lambda(0), \; L(\tau) = \exp\left(\frac{e}{2m}\boldsymbol{F}\tau\right)$$

and this implies a spacetime rotation of the proper velocity

$$u(\tau) = \Lambda(\tau)\Lambda^{\dagger}(\tau) = L(\tau)u(0)L^{\dagger}(\tau).$$

This is a trivial solution that works for all constant, uniform fields, whether or not they are spacelike, timelike, or null, simple or compound. Traditional texts usually treat the simple spacelike case (or occasionally the simple timelike case) by finding a drift frame in which the electromagnetic field is purely magnetic (or electric). We see that a more general solution is much easier. Furthermore, it is readily extended. If $\boldsymbol{F}$ varies in time but commutes with itself at all different times, the solution has the form above with $\boldsymbol{F}\tau$ replaced by $\int_0^{\tau}\boldsymbol{F}(\tau')\,d\tau'$. Also, if you have a nonnull simple field, it can always be factored into

$$\boldsymbol{F} = u_d\boldsymbol{F}_d = u_d^{1/2}\boldsymbol{F}_d\bar{u}_d^{1/2}$$

where $\boldsymbol{F}_d$ is the field in the *drift frame* and u_d is the proper velocity of the drift frame with respect to the lab. Its vector part is orthogonal to $\boldsymbol{F}_d$.

Example 32. *Consider the field*

$$\boldsymbol{F} = (5\mathbf{e}_1 + 4i\mathbf{e}_3)E_0 = (5\mathbf{e}_1 + 4\mathbf{e}_1\mathbf{e}_2)E_0.$$

We note that $\boldsymbol{F}^2 > 0$, *so we expect the field in the drift frame to be purely electric. We therefore factor out the electric field* $E_0\mathbf{e}_1$, *leaving*

$$\boldsymbol{F} = 5\left(1 - \tfrac{4}{5}\mathbf{e}_2\right)E_0\mathbf{e}_1 = u_d\boldsymbol{F}_d.$$

In the last step, we normalize the velocity factor so that u_d *is a unit paravector:*

$$u_d = \frac{\left(1 - \frac{4}{5}\mathbf{e}_2\right)}{\sqrt{1 - \frac{16}{25}}} = \tfrac{5}{3}\left(1 - \tfrac{4}{5}\mathbf{e}_2\right) \equiv \gamma(1 + \boldsymbol{v})$$

which leaves $\boldsymbol{F}_d = 3E_0\mathbf{e}_1$.

---

in the rest frame of the charge: $\dot{p}_{\text{rest}} = e\boldsymbol{E}_{\text{rest}}$. This rest-frame relation is NOT covariant. The LHS is the rest-frame value of the covariant *paravector* $\dot{p}$, where the dot indicates differentiation with respect to proper time, whereas the RHS is the *real part* of the covariant *biparavector* field $\boldsymbol{F}$, which transforms distinctly. The covariant Lorentz-force equation follows when we boost $\dot{p}$ from rest to the lab:

$$\dot{p} = \Lambda\dot{p}_{\text{rest}}\Lambda^{\dagger} = e\Lambda\langle\boldsymbol{F}_{\text{rest}}\rangle_{\text{Re}}\Lambda^{\dagger} = e\langle\Lambda\boldsymbol{F}_{\text{rest}}\Lambda^{\dagger}\rangle_{\text{Re}} = e\langle\Lambda\left(\bar{\Lambda}\boldsymbol{F}\Lambda\right)\Lambda^{\dagger}\rangle_{\text{Re}} = e\langle\boldsymbol{F}u\rangle_{\text{Re}}.$$

Exercise 33. *Factor the electromagnetic field*

$$\mathbf{F} = (3\mathbf{e}_1 - 5i\mathbf{e}_3)\, E_0$$

into a drift velocity and electric or magnetic drift field.
Solution: $\mathbf{F} = u_d \mathbf{F}_d = \frac{5}{4}\left(1 + \frac{3}{5}\mathbf{e}_2\right)\left(-i4\mathbf{e}_3 E_0\right)$.

For *compound fields*, the drift field is a combination of collinear magnetic and electric fields, and the solution easily gives a *rifle transformation*: a commuting boost and rotation. Traditional electromagnetic theory texts rarely treat this case. For *null fields*, $\mathbf{F}$ with $\mathbf{F}^2 = 0$, the drift frame idea is not useful, since the drift velocity is at the speed of light. Our simple algebraic solution above still works, however, and indeed is then especially easy to evaluate since

$$\exp\left(\tfrac{1}{2}\boldsymbol{\Omega}\tau\right) = 1 + \tfrac{1}{2}\boldsymbol{\Omega}\tau$$

when $\boldsymbol{\Omega}^2 = 0$.

4.5 Maxwell's equation

Maxwell's famous equations were written as a single quaternionic equation by Conway (1911) [14, 15], Silberstein (1912, 1914) [16, 17], and others. In APS we can write

$$\bar{\partial}\mathbf{F} = \mu_0 \bar{j}\,, \tag{5.9}$$

where $\mu_0 = \varepsilon_0^{-1} = 4\pi \times \dot{3}0$ Ohm is the impedance of the vacuum, with $\dot{3} \equiv 2.99792458$. The usual four equations are simply the four vector grades of this relation, extracted as the real and imaginary, scalarlike and vectorlike parts. The relation is also seen as the *necessary covariant extension* of Coulomb's law $\nabla \cdot \mathbf{E} = \rho/\varepsilon_0$. The covariant field is not $\mathbf{E}$ but $\mathbf{F} = \mathbf{E} + i\mathbf{B}$, the divergence is part of the covariant gradient $\bar{\partial}$, and ρ must be part of $\bar{j} = \rho - \mathbf{j}$. The combination is Maxwell's equation.[20]

Exercise 34. *Derive the continuity equation* $\langle \partial \bar{j} \rangle_{\underline{S}} = 0$ *from Maxwell's equation in one step. (Hint: note that the D'Alembertian* $\partial\bar{\partial}$ *is a scalar operator and that* $\langle \mathbf{F} \rangle_S = 0$.*)*

4.5.1 *Directed plane waves*

In source-free space ($\bar{j} = 0$), there are solutions $\mathbf{F}(s)$ that depend on spacetime position only through the Lorentz invariant $s = \langle k\bar{x} \rangle_0 = \omega t - \mathbf{k} \cdot \mathbf{x}$, where $k =$

[20] We have assumed that the source is a real paravector current and that there are no contributing pseudoparavector currents. Known currents are of the real paravector type, and a pseudoparavector current would behave counter-intuitively under parity inversion. Our assumption is supported experimentally by the apparent lack of magnetic monopoles.

$\omega + \boldsymbol{k} \neq 0$ is a constant propagation paravector. Since $\partial \langle k\bar{x}\rangle_0 = k$, Maxwell's equation gives

$$\bar{\partial}\boldsymbol{F} = \bar{k}\boldsymbol{F}'(s) = 0. \tag{5.10}$$

In a division algebra, we could divide by $\bar{k}$ and conclude that $\boldsymbol{F}'(s) = 0$, a rather uninteresting solution. There is another possibility here because APS is not a division algebra: $\bar{k}$ may have no inverse. Then k has the form $k = \omega\left(1 + \hat{\boldsymbol{k}}\right)$, and after integrating (5.10) from some s_0 at which $\boldsymbol{F}$ is presumed to vanish, we get $\left(1 - \hat{\boldsymbol{k}}\right)\boldsymbol{F}(s) = 0$, which means

$$\boldsymbol{F}(s) = \hat{\boldsymbol{k}}\boldsymbol{F}(s).$$

The scalar part of $\boldsymbol{F}$ vanishes and consequently $\langle\hat{\boldsymbol{k}}\boldsymbol{F}(s)\rangle_S = \hat{\boldsymbol{k}} \cdot \boldsymbol{F}(s) = 0$ so that the fields $\boldsymbol{E}$ and $\boldsymbol{B}$ are perpendicular to $\hat{\boldsymbol{k}}$ and thus anticommute with it. Furthermore, equating imaginary parts gives $i\boldsymbol{B} = \hat{\boldsymbol{k}}\boldsymbol{E}$ and it follows that

$$\boldsymbol{F} = \boldsymbol{E} + i\boldsymbol{B} = \left(1 + \hat{\boldsymbol{k}}\right)\boldsymbol{E}(s)$$

with $\boldsymbol{E} = \langle\boldsymbol{F}\rangle_1$ real. This is a plane-wave solution with $\boldsymbol{F}$ constant on all spatial planes perpendicular to $\hat{\boldsymbol{k}}$. Such planes propagate at the speed of light along $\hat{\boldsymbol{k}}$. In spacetime, $\boldsymbol{F}$ is constant on the light cones $\hat{\boldsymbol{k}} \cdot \boldsymbol{x} - t =$constant.

However, $\boldsymbol{F}$ is not necessarily monochromatic, since $\boldsymbol{E}(s)$ can have any functional form, including a pulse, and the scale factor ω, although it has dimensions of frequency, may have nothing to do with any physical oscillation. Note further that $\boldsymbol{F}$ is *null*:[21]

$$\boldsymbol{F}^2 = \left(1 + \hat{\boldsymbol{k}}\right)\boldsymbol{E}\left(1 + \hat{\boldsymbol{k}}\right)\boldsymbol{E} = \left(1 + \hat{\boldsymbol{k}}\right)\left(1 - \hat{\boldsymbol{k}}\right)\boldsymbol{E}^2 = 0.$$

The energy density $\mathcal{E} = \frac{1}{2}\left(\varepsilon_0\boldsymbol{E}^2 + \boldsymbol{B}^2/\mu_0\right)$ and Poynting vector $\boldsymbol{S} = \boldsymbol{E} \times \boldsymbol{B}/\mu_0$ are given by

$$\tfrac{1}{2}\varepsilon_0\boldsymbol{F}\boldsymbol{F}^\dagger = \mathcal{E} + \boldsymbol{S} = \varepsilon_0\boldsymbol{E}^2\left(1 + \hat{\boldsymbol{k}}\right).$$

Example 35. *Monochromatic plane wave of frequency ω linearly polarized along $\boldsymbol{E}_0$:*

$$\boldsymbol{F} = \left(1 + \hat{\boldsymbol{k}}\right)\boldsymbol{E}_0\cos s\,.$$

Example 36. *Monochromatic plane wave of frequency 5ω linearly polarized along $\boldsymbol{E}_0$:*

$$\boldsymbol{F} = \left(1 + \hat{\boldsymbol{k}}\right)\boldsymbol{E}_0\cos 5s\,.$$

[21] In fact, $\boldsymbol{F}$ is what Penrose calls a *null flag*. The *flagpole* $\left(1 + \hat{\boldsymbol{k}}\right)$ lies in the plane of the flag but is orthogonal to it. This becomes important below when we discuss charge dynamics. The null flag structure is beautiful and powerful, but you miss it entirely if you write only separate electric and magnetic fields!

Example 37. *Monochromatic plane wave circularly polarized with helicity* κ :

$$F = \left(1 + \hat{k}\right) E_0 \exp\left(i\kappa s\hat{k}\right) = \left(1 + \hat{k}\right) E_0 \exp\left(-i\kappa s\right).$$

Note that the rotation factor has become a phase factor (a "duality rotation") in the last expression. This is a result of the "Pacwoman property" in which $1 + \hat{k}$ *gobbles neighboring factors of* $\hat{k}$: $\left(1 + \hat{k}\right)\hat{k} = \left(1 + \hat{k}\right)$:

$$\begin{aligned} \left(1 + \hat{k}\right) E_0 \exp\left(i\kappa s\hat{k}\right) &= \left(1 + \hat{k}\right) \exp\left(-i\kappa s\hat{k}\right) E_0 \\ &= \left(1 + \hat{k}\right)\left(\cos\kappa s - i\hat{k}\sin\kappa s\right) E_0 \\ \text{[gobble!]} \quad &= \left(1 + \hat{k}\right)\left(\cos\kappa s - i\sin\kappa s\right) E_0 \\ &= \left(1 + \hat{k}\right) E_0 \exp\left(-i\kappa s\right). \end{aligned}$$

Example 38. *Linearly polarized Gaussian pulse of width* Δ/ω :

$$F = \left(1 + \hat{k}\right) E_0 \exp\left(-\tfrac{1}{2}s^2/\Delta^2\right).$$

Example 39. *A circularly polarized Gaussian pulse with center frequency* ω :

$$F = \left(1 + \hat{k}\right) E_0 \exp\left(-\tfrac{1}{2}s^2/\Delta^2 + is\right).$$

These all have the common form

$$F = \left(1 + \hat{k}\right) E = \left(1 + \hat{k}\right) E_0 f(s),$$

where $f(s)$ is a scalar function, possibly complex valued. We will use this below.[22]

4.5.2 Polarization basis

A beam of monochromatic radiation can be elliptically polarized as well as linearly or circularly polarized. There are two degrees of freedom, so that arbitrary polarization can be expressed as a linear combination of two independent polarization types. There is a close analogy to the 2-D oscillations of a pendulum

---

[22] Warning: Don't assume from the last relation that $E(s)$ is $E_0 f(s)$. It doesn't follow when f is complex. Remember that $\left(1 + \hat{k}\right)$ has no inverse, so we can't simply drop it. Instead, since E_0 is real and $\hat{k}E_0$ is imaginary,

$$E = \left\langle \left(1 + \hat{k}\right) E_0 f(s) \right\rangle_{\text{Re}} = E_0 \langle f \rangle_{\text{Re}} + \hat{k} E_0 \langle f \rangle_{\text{Im}}.$$

formed by hanging a mass on a string. Both linear and circular polarization bases are common, but we find the circular basis most convenient, partially because of the relation noted above between spatial and duality rotations. Circularly polarized waves also have the simple form used popularly by R. P. Feynman [19] in terms of the paravector potential as a rotating real vector:

$$A_\kappa = \boldsymbol{a} \exp\left(i\kappa s\hat{\boldsymbol{k}}\right),$$

with $s = \langle k\bar{x}\rangle_S = \omega t - \boldsymbol{k}\cdot\boldsymbol{x}$ and $\boldsymbol{a}\cdot\boldsymbol{k} = 0$, where $\kappa = \pm 1$ is the helicity. The corresponding field is

$$\boldsymbol{F}_\kappa = \langle\partial\bar{A}\rangle_V = i\kappa\boldsymbol{k}\boldsymbol{a}\exp\left(i\kappa s\hat{\boldsymbol{k}}\right) = \left(1+\hat{\boldsymbol{k}}\right)\boldsymbol{E}_0\exp\left(i\kappa s\hat{\boldsymbol{k}}\right),$$

with $\boldsymbol{E}_0 = i\kappa\boldsymbol{k}\boldsymbol{a} = \kappa\boldsymbol{a}\times\boldsymbol{k}$.

A linear combination of both helicities of such directed waves is given by

$$\begin{aligned}\boldsymbol{F} &= \left(1+\hat{\boldsymbol{k}}\right)\hat{\boldsymbol{E}}_0 e^{i\delta\hat{\boldsymbol{k}}}\left(E_+e^{is\hat{\boldsymbol{k}}} + E_-e^{-is\hat{\boldsymbol{k}}}\right)\\ &= \left(1+\hat{\boldsymbol{k}}\right)\hat{\boldsymbol{E}}_0 e^{-i\delta}\left(E_+e^{-is} + E_-e^{is}\right)\end{aligned}$$

where $E_\pm$ are the real field amplitudes, δ gives the rotation of $\boldsymbol{E}$ about $\hat{\boldsymbol{k}}$ at $s = 0$, and in the second line, we let Pacwoman gobble the $\hat{\boldsymbol{k}}$'s. Because every directed plane wave can be expressed in the form $\boldsymbol{F} = \left(1+\hat{\boldsymbol{k}}\right)\boldsymbol{E}(s)$, it is sufficient to determine $\boldsymbol{E}(s) = \langle\boldsymbol{F}\rangle_{\text{Re}}$:

$$\begin{aligned}\boldsymbol{E} &= \left\langle\left(1+\hat{\boldsymbol{k}}\right)\hat{\boldsymbol{E}}_0 E_+e^{-i\delta}e^{-is} + \left(1+\hat{\boldsymbol{k}}\right)\hat{\boldsymbol{E}}_0 E_-e^{-i\delta}e^{is}\right\rangle_{\text{Re}}\\ &= \left\langle\left[\left(1+\hat{\boldsymbol{k}}\right)\hat{\boldsymbol{E}}_0 E_+e^{-i\delta} + \hat{\boldsymbol{E}}_0\left(1+\hat{\boldsymbol{k}}\right)E_-e^{i\delta}\right]e^{-is}\right\rangle_{\text{Re}}\\ &= \left\langle\left(\boldsymbol{\epsilon}_+,\boldsymbol{\epsilon}_-\right)\Phi e^{-is}\right\rangle_{\text{Re}},\end{aligned}$$

where the complex polarization basis vectors $\boldsymbol{\epsilon}_\pm = 2^{-\frac{1}{2}}\left(1\pm\hat{\boldsymbol{k}}\right)\hat{\boldsymbol{E}}_0$ are basis null flags satisfying $\boldsymbol{\epsilon}_- = \boldsymbol{\epsilon}_+^\dagger, \boldsymbol{\epsilon}_+\cdot\boldsymbol{\epsilon}_+^\dagger = 1 = \boldsymbol{\epsilon}_-\cdot\boldsymbol{\epsilon}_-^\dagger$, and the *Poincaré spinor*

$$\Phi = \sqrt{2}\begin{pmatrix}E_+e^{-i\delta}\\ E_-e^{i\delta}\end{pmatrix}$$

gives the (real) electric-field amplitudes and their phases, and thus contains all the information needed to determine the polarization and intensity of the wave.[23]

---

[23]The direction of the magnetic field at $s = 0$ is $\hat{\boldsymbol{B}}_0 = \hat{\boldsymbol{k}}\times\hat{\boldsymbol{E}}_0$. In terms of this

$$\boldsymbol{\epsilon}_\pm = \frac{1}{\sqrt{2}}\left(\hat{\boldsymbol{E}}_0 \pm i\hat{\boldsymbol{B}}_0\right).$$

Stokes parameters

Physical beams of radiation are not fully monochromatic and not necessarily fully polarized. To describe partially polarized light, we can use the *coherency density*, which in the case of a single Poincaré spinor is defined by

$$\rho = \varepsilon_0 \Phi\Phi^\dagger = \rho^\mu \boldsymbol{\sigma}_\mu ,$$

where the $\boldsymbol{\sigma}_\mu$ are the usual Pauli spin matrices and the normalization factor ε_0 has been chosen to make ρ^0 the time-averaged energy density:

$$\begin{aligned}\langle \mathcal{E} + \boldsymbol{S}\rangle_{\text{t-av}} &= \left\langle \tfrac{1}{2}\varepsilon_0 \boldsymbol{F}\boldsymbol{F}^\dagger \right\rangle_{\text{t-av}} = \tfrac{1}{2}\varepsilon_0 \left\langle \left(1+\hat{\boldsymbol{k}}\right) \boldsymbol{E}^2 \left(1+\hat{\boldsymbol{k}}\right)\right\rangle_{\text{t-av}} \\ &= \varepsilon_0 \left(1+\hat{\boldsymbol{k}}\right) \left\langle \boldsymbol{E}^2\right\rangle_{\text{t-av}} = \rho^0 \left(1+\hat{\boldsymbol{k}}\right).\end{aligned}$$

The coefficients ρ^μ are the *Stokes parameters*. The coherency density can be treated algebraically in $C\ell_3$ to study all polarization and intensity properties of the beam.[24]

The Stokes parameters are given by

$$\rho^\mu = \langle \rho \boldsymbol{\sigma}_\mu \rangle_S = \tfrac{1}{2} tr\left(\rho \boldsymbol{\sigma}_\mu\right).$$

Explicitly

$$\begin{aligned}\rho^0 &= \varepsilon_0 \left(E_+^2 + E_-^2\right), & \rho^1 &= 2\varepsilon_0 E_+ E_- \cos\phi, \\ \rho^2 &= 2\varepsilon_0 E_+ E_- \sin\phi, & \rho^3 &= \varepsilon_0 \left(E_+^2 - E_-^2\right),\end{aligned}$$

where $\phi = 2\delta$ is the azimuthal angle of $\boldsymbol{\rho} = \rho^1\boldsymbol{\sigma}_k + \rho^2\boldsymbol{\sigma}_2 + \rho^3\boldsymbol{\sigma}_3$. The coherency density is a paravector in the space spanned by the basis $\{\boldsymbol{\sigma}_1, \boldsymbol{\sigma}_2, \boldsymbol{\sigma}_3\}$, namely $\rho = \rho^0 + \boldsymbol{\rho}$.

This space, called *Stokes subspace*, is a 3-D Euclidean space analogous to physical space. It is not physical space, but its geometric algebra has exactly the same form as (is isomorphic to) APS, and it illustrates how Clifford algebras can arise in physics for spaces other than physical space. As in APS, it is the algebra and not the explicit matrix representation that is significant.

The electric field $\boldsymbol{E}$ can be transformed to the familiar *Jones-vector* basis by a unitary matrix:

$$\boldsymbol{E} = \left\langle \left(\hat{\boldsymbol{E}}_0, \hat{\boldsymbol{B}}_0\right) \Phi_J\, e^{-is} \right\rangle_{\text{Re}}, \qquad \left(\hat{\boldsymbol{E}}_0, \hat{\boldsymbol{B}}_0\right) = \left(\epsilon_+, \epsilon_-\right) U_J^\dagger,$$

$$\Phi_J = U_J \Phi = \begin{pmatrix} E_+ e^{-i\delta} + E_- e^{i\delta}, \\ i\left(E_+ e^{-i\delta} - E_- e^{i\delta}\right)\end{pmatrix}, \qquad U_J = \frac{1}{\sqrt{2}}\begin{pmatrix} 1 & 1 \\ i & -i\end{pmatrix}.$$

[24]Many optics texts still use the 4 × 4 Mueller matrices for this purpose, but this strikes me as even more perverse than using 4 × 4 matrices for Lorentz transformations. The coherency density, introduced by Born and Wolf as the "coherency matrix" by the time of the third edition of their **Principles of Optics** book in 1964, is really much simpler. Transformed as here into the helicity basis, it matches the quantum formulation of the spin-$\frac{1}{2}$density matrix as well as the standard matrix representation of APS.

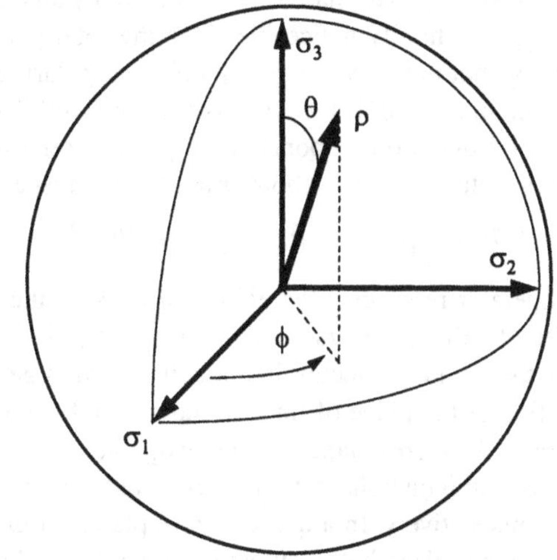

FIGURE 4.3. The direction of ρ gives the type of polarization.

As defined for a single Φ, ρ is null: $\det \rho = \rho\bar{\rho} = 0$. Thus, $\rho = \rho^0 (1 + \boldsymbol{n})$ where $\boldsymbol{n}$ is a unit vector in the direction of $\boldsymbol{\rho}$. It fully specifies the type of polarization. In particular, for positive helicity light, $\boldsymbol{n} = \boldsymbol{\sigma}_3$, for negative helicity polarization $\boldsymbol{n} = -\boldsymbol{\sigma}_3$, and for linear polarization at an angle $\delta = \frac{1}{2}\phi$ with respect to $\boldsymbol{E}_0$, $\boldsymbol{n} = \boldsymbol{\sigma}_1 \cos\phi + \boldsymbol{\sigma}_2 \sin\phi$. Other directions correspond to elliptical polarization.

Polarizers and phase shifters

The action of ideal polarizers and phase shifters on the wave is modeled mathematically by transformations on the Poincaré spinor Φ of the form $\Phi \to T\Phi$. For *polarizers* T is a *projector*

$$\mathsf{P}_{\boldsymbol{n}} = \tfrac{1}{2}(1 + \boldsymbol{n}) \,,$$

where $\boldsymbol{n}$ is a real unit vector in Stokes subspace that specifies the type of polarization. Projectors are real idempotent elements: $\mathsf{P}_{\boldsymbol{n}} = \mathsf{P}_{\boldsymbol{n}}^{\dagger} = \mathsf{P}_{\boldsymbol{n}}^2$, just as we would expect for ideal polarizers. For example, a circular polarizer allowing only waves of positive helicity corresponds to the projector $\mathsf{P}_{\boldsymbol{\sigma}_3} = \frac{1}{2}(1 + \boldsymbol{\sigma}_3)$, which when applied to Φ eliminates the contribution E_- of negative helicity without affecting the positive-helicity part:

$$\Phi := \sqrt{2}\begin{pmatrix} E_+ e^{-i\delta} \\ E_- e^{i\delta} \end{pmatrix} \to \mathsf{P}_{\boldsymbol{\sigma}_3}\Phi := \sqrt{2}\begin{pmatrix} E_+ e^{-i\delta} \\ 0 \end{pmatrix}.$$

A second application of P_{σ_3} changes nothing further. The polarizer represented by the complementary projector $\bar{\mathsf{P}}_{\boldsymbol{\sigma}_3}$ eliminates the upper component of Φ. Generally, since $\bar{\mathsf{P}}_{\boldsymbol{n}}\mathsf{P}_{\boldsymbol{n}} = \mathsf{P}_{-\boldsymbol{n}}\mathsf{P}_{\boldsymbol{n}} = 0$, opposite directions in Stokes subspace correspond to orthogonal polarizations.

Multiplication of Φ by $\exp(i\alpha)$ phase shifts the wave by an angle α. An overall phase shift in the wave is hardly noticeable since the total phase is in any case changing very rapidly, but the effect of giving different polarization components different shifts can be important. If the wave is split into orthogonal polarization components ($\pm\boldsymbol{n}$) and the two components are given a relative shift of α, the result is equivalent to rotating $\boldsymbol{\rho}$ by α about $\boldsymbol{n}$ in Stokes subspace:

$$T = \mathsf{P}_{\boldsymbol{n}} e^{i\alpha/2} + \bar{\mathsf{P}}_{\boldsymbol{n}} e^{-i\alpha/2} = e^{i\boldsymbol{n}\alpha/2}.$$

If $\boldsymbol{n} = \boldsymbol{\sigma}_3$, this operator represents the effects of passing the waves through a medium with different indices of refraction for circularly polarized light of different helicities, as in the Faraday effect or in optically active organic solutions, and the result is a rotation of the plane of linear polarization by $\frac{1}{2}\alpha$ about $\boldsymbol{k}$. On the other hand, if $\boldsymbol{n}$ lies in the $\boldsymbol{\sigma}_1\boldsymbol{\sigma}_2$ plane, the above operator T represents the effect of a birefringent medium with polarization types $\boldsymbol{n}$ and $-\boldsymbol{n}$ corresponding to the slow and fast axes, respectively. In a quarter-wave plate, for example, $\alpha = \frac{1}{2}\pi$. Incident light linearly polarized half way between the fast and slow axes will be rotated by $\frac{1}{2}\pi$ to $\pm\boldsymbol{\sigma}_3$, giving circularly polarized light.

The basic technique of splitting the light into opposite polarizations $\pm\boldsymbol{n}$, acting differently on the two polarization components, and then recombining is modeled by the operator

$$T = \mathsf{P}_{\boldsymbol{n}} A_+ + \mathsf{P}_{-\boldsymbol{n}} A_-.$$

If the actions $A_\pm$ are the same, the result is the identity operator: nothing happens. The ideal filter is the special case $A_+ = 1$, and $A_- = 0$, that is one of the polarization parts is discarded.

Coherent superpositions and incoherent mixtures

A superposition of two waves of the same frequency is *coherent* because their relative phase is fixed. Mathematically, one adds spinors in such cases:

$$\Phi = \Phi_1 + \Phi_2\,,$$

where the subscripts refer to the two waves, not to spinor components. Coherent superpositions of monochromatic waves are always fully polarized and can be represented by a single Poincaré spinor or Jones vector.

However, real beams of waves are never fully monochromatic. Two waves of different frequencies have a continually changing relative phase, and when their product $\Phi_1\Phi_2^\dagger$ is averaged over periods large relative to their beat period, such terms vanish. The waves then combine *incoherently*, and one should add their coherency densities rather than their Poincaré spinors:

$$\rho = \rho_1 + \rho_2.$$

In such an incoherent superposition, the polarization can vary from 0 to 100%.

Any transformation T of spinors, $\Phi \to T\Phi$, transforms the coherency density by

$$\rho \to T\rho T^\dagger.$$

Example 40. *Consider a sandwich of two crossed linear polarizers, with a third linear polarizer of intermediate polarization inserted in between. If ρ is initially unpolarized, the first polarizer, say of type $\boldsymbol{n}$, produces*

$$\mathsf{P}_{\boldsymbol{n}}\rho^0\mathsf{P}_{\boldsymbol{n}} = \rho^0\mathsf{P}_{\boldsymbol{n}} = \tfrac{1}{2}\rho^0\left(1+\boldsymbol{n}\right),$$

that is, fully polarized light with half the intensity. The crossed polarizer, represented by $P_{-\boldsymbol{n}}$, would annihilate the polarized beam, but if a different polarizer, say type $\boldsymbol{m}$, is applied before the $-\boldsymbol{n}$ one, we get

$$\begin{aligned}\mathsf{P}_{\boldsymbol{m}}\rho^0\mathsf{P}_{\boldsymbol{n}}\mathsf{P}_{\boldsymbol{m}} &= \rho^0\left(\mathsf{P}_{\boldsymbol{m}}\mathsf{P}_{\boldsymbol{n}} + \bar{\mathsf{P}}_{\boldsymbol{n}}\bar{\mathsf{P}}_{\boldsymbol{m}}\right)\mathsf{P}_{\boldsymbol{m}} \\ &= 2\rho^0\left\langle\mathsf{P}_{\boldsymbol{m}}\mathsf{P}_{\boldsymbol{n}}\right\rangle_S\mathsf{P}_{\boldsymbol{m}} = \tfrac{1}{2}\rho^0(1+\boldsymbol{m}\cdot\boldsymbol{n})\mathsf{P}_{\boldsymbol{m}}.\end{aligned}$$

Application of the final polarizer of type $-\boldsymbol{n}$ then yields

$$\begin{aligned}\tfrac{1}{2}\rho^0(1+\boldsymbol{m}\cdot\boldsymbol{n})P_{-\boldsymbol{n}}P_{\boldsymbol{m}}P_{-\boldsymbol{n}} &= \tfrac{1}{4}\rho^0(1+\boldsymbol{m}\cdot\boldsymbol{n})(1-\boldsymbol{m}\cdot\boldsymbol{n})P_{-\boldsymbol{n}} \\ &= \tfrac{1}{4}\rho^0\left(1-(\boldsymbol{m}\cdot\boldsymbol{n})^2\right)P_{-\boldsymbol{n}}.\end{aligned}$$

The maximum intensity is 1/8th the initial, reached when $\boldsymbol{m}\cdot\boldsymbol{n}=0$, that is when the linear polarization of the intervening polarizer is at $45°$ to each of the crossed polarizers.

In addition, many other transformations that do not preserve the polarization can be applied. For example, *depolarization* of a fraction f of the radiation is modeled by

$$\rho \to (1-f)\,\rho + f\left\langle\rho\right\rangle_S,$$

and detection itself takes the form $\rho \to \left\langle\rho D\right\rangle_S$, where the detection operator may equal $D=1$ in an ideal case, but more generally it can have different efficiencies $D_\pm$ for opposite polarization types:

$$D = D_+\mathsf{P}_{\boldsymbol{n}} + D_-\mathsf{P}_{-\boldsymbol{n}}.$$

A number of other transformations are possible.

4.5.3 *Standing waves and $\boldsymbol{E}||\boldsymbol{B}$ fields*

Standing waves are formed from the superposition of oppositely directed plane waves:

$$\begin{aligned}\boldsymbol{F} &= \left(1+\hat{\boldsymbol{k}}\right)\boldsymbol{E}_0 f\left(\omega t - \boldsymbol{k}\cdot\boldsymbol{x}\right) + \left(1-\hat{\boldsymbol{k}}\right)\boldsymbol{E}_0 f\left(\omega t + \boldsymbol{k}\cdot\boldsymbol{x}\right) \\ &= \left(f_+ - f_-\hat{\boldsymbol{k}}\right)\boldsymbol{E}_0\end{aligned}$$

where $f_\pm = f\left(\omega t + \boldsymbol{k}\cdot\boldsymbol{x}\right) \pm f\left(\omega t - \boldsymbol{k}\cdot\boldsymbol{x}\right)$. For the monochromatic circularly polarized wave $f(s) = \exp is$ of negative helicity,

$$f_+ = 2\exp\left(i\omega t\right)\cos\boldsymbol{k}\cdot\boldsymbol{x}, \qquad f_- = 2i\exp\left(i\omega t\right)\sin\boldsymbol{k}\cdot\boldsymbol{x}$$

and

$$F = 2\left(\cos \mathbf{k} \cdot \mathbf{x} - i\hat{\mathbf{k}} \sin \mathbf{k} \cdot \mathbf{x}\right) \mathbf{E}_0 \exp(i\omega t) = \underbrace{2\mathbf{E}_0 \exp(i\hat{\mathbf{k}}\, \mathbf{k} \cdot \mathbf{x})}_{\text{real: spatial rot.}} \underbrace{\exp(i\omega t)}_{\text{duality rot.}},$$

which represents electric and magnetic fields that are aligned on the radii of a spiral fixed in space. The field rotates in duality space ($\mathbf{E} \rightarrow \mathbf{B} \rightarrow -\mathbf{E}$) at every point in space. Thus at $t = 0$, there is only an electric field throughout space, and a quarter of a cycle later, there is only a magnetic field. At intermediate times, both electric and magnetic fields exist and they are aligned. To change to a wave of positive helicity plus its reflection, replace i by $-i$.

The roles of time and space are reversed if waves of opposite helicity are superimposed. In this case, the fields throughout space point in a single direction at a given instant in time, and that direction rotates about $\hat{\mathbf{k}}$ in time. How much of the field is electric and how much magnetic depends only on the spatial position along $\hat{\mathbf{k}}$. The energy density for both circular-wave superpositions are constant and their Poynting vectors vanish: $\mathcal{E} + \mathbf{S} = \frac{1}{2}\varepsilon_0 \mathbf{F}\mathbf{F}^\dagger = 2\varepsilon_0 \mathbf{E}_0^2$.[25]

4.5.4 *Charge dynamics in plane waves*

We saw above how eigenspinors made easy work of the problem of charge dynamics in constant uniform fields. What about motion in the more complicated field of a directed plane wave? A derivation of the relativistic motion of a charge in a linearly polarized monochromatic plane wave was given by A. H. Taub [20] in 1948, using 4×4 matrix Lorentz transformations of the charge. The derivation is simpler using eigenspinors (or rotors) in Clifford's geometric algebra, as shown by Hestenes in 1974 [21]. Here we review the extension [6] to arbitrary plane waves and plane-wave pulses in APS.

The fields of directed plane waves are null flags of the form

$$\mathbf{F}(s) = \left(1 + \hat{\mathbf{k}}\right) \mathbf{E}(s), \quad s = \langle k\bar{x} \rangle_S \,, \quad \mathbf{k} \cdot \hat{\mathbf{E}} = 0.$$

All null flags on a given flagpole annihilate each other and therefore commute:

$$\mathbf{F}(s_1)\, \mathbf{F}(s_2) = \left(1 + \hat{\mathbf{k}}\right) \mathbf{E}(s_1) \left(1 + \hat{\mathbf{k}}\right) \mathbf{E}(s_2) = 0 = \mathbf{F}(s_2)\, \mathbf{F}(s_1)$$

and as a result, the solution to the equation of motion (4.8) is the exponential expression

$$\Lambda(\tau) = \exp\left\{ \frac{e}{2m} \int_0^\tau \mathbf{F}\left[s\left(\tau'\right)\right] d\tau' \right\} \Lambda(0),$$

[25] There was considerable controversy about such fields when they were first proposed by Chu and Ohkawa [18] in 1982. A surprising number of physicists were convinced that the rule $\mathbf{E} \cdot \mathbf{B} = 0$ for directed plane waves and other simple fields also had to be true of superpositions in free space. The result and its interpretation are quite obvious with some geometric algebra. Now $\mathbf{E} \| \mathbf{B}$ fields are used routinely in the laser cooling of atoms to submicrokelvin temperatures. Opposite helicities, as naturally obtained by reflection, are required for the operation of the *Sisyphus effect*.

which by virtue of the nilpotency of $\boldsymbol{F}$ reduces to

$$\Lambda(\tau) = \left\{1 + \frac{e}{2m}\int_0^\tau \boldsymbol{F}\left[s\left(\tau'\right)\right] d\tau'\right\}\Lambda(0). \qquad (5.11)$$

The problem is that $\boldsymbol{F}\left[s\left(\tau\right)\right]$ is the electromagnetic field at the charge at proper time τ, and we need to know the world line $x\left(\tau\right)$ of the charge to know its value. We could get the world line by integrating $u\left(\tau\right) = \Lambda\Lambda^\dagger$, but it is Λ we're trying to find! This looks hopeless, but there's a surprising symmetry that comes to the rescue.

From $\bar{k}\boldsymbol{F} = 0$, it follows that $\bar{k}\dot{\Lambda}\left(\tau\right) = 0$, the bar conjugate of which is $\dot{\bar{\Lambda}}k = 0$, and from the definition of Λ, $\bar{\Lambda}k\bar{\Lambda}^\dagger = k_{\text{rest}}$ is the propagation paravector as seen in the instantaneous rest frame of the charge. Since k is constant in the lab,

$$\dot{k}_{\text{rest}} = \frac{d}{d\tau}\left(\bar{\Lambda}k\bar{\Lambda}^\dagger\right) = 2\left\langle\dot{\bar{\Lambda}}k\bar{\Lambda}^\dagger\right\rangle_{\text{Re}} = 0\,,$$

and it is also constant in the rest frame of the charge. Now that is unexpected because the charge, as we will see, is *accelerating*! In particular, the scalar part $\dot{s} = \left\langle k\bar{u}\right\rangle_S = \omega_{\text{rest}}$ of k_{rest} is constant. Thus, $s = s_0 + \omega_{\text{rest}}\tau$ and we can solve (5.11) above:

$$\begin{aligned}\Lambda(\tau) &= \left[1 + \frac{e}{2m\omega_{\text{rest}}}\int_{s_0}^{s} \boldsymbol{F}\left(s'\right) ds'\right]\Lambda(0) \\ &= \left[1 + \frac{e\left(1+\hat{\boldsymbol{k}}\right)\boldsymbol{E}_0}{2m\omega_{\text{rest}}}\int_{s_0}^{s} f\left(s'\right) ds'\right]\Lambda(0)\,,\end{aligned}$$

where in the second line we noted that every plane wave with flag pole $1+\hat{\boldsymbol{k}}$ can be expressed by $\left(1+\hat{\boldsymbol{k}}\right)\boldsymbol{E}_0 f\left(s\right)$. This is the solution since we know $f\left(s\right)$ and can integrate it, but there's another form, using

$$\boldsymbol{F}\left(s\right) = \partial\bar{A}\left(s\right) = k\bar{A}'\left(s\right),$$

which is valid in the Lorenz gauge $\left\langle\partial\bar{A}\right\rangle_S = 0$. The integral over $\boldsymbol{F}$ is thus trivially expressed as

$$\Lambda(\tau) - \Lambda(0) = \frac{ek}{2m\omega_{\text{rest}}}\left[\bar{A}\left(s\right) - \bar{A}\left(s_0\right)\right]\Lambda(0)\,, \qquad (5.12)$$

giving a change in the eigenspinor that is linear in the change in the paravector potential. This can lead to substantial accelerations in the plane of $\boldsymbol{k}$ and $\boldsymbol{A}\left(s\right) - \boldsymbol{A}\left(s_0\right)$, especially when the charge is injected with high velocity along $\boldsymbol{k}$ into the beam so as to produce a large Doppler shift and thus a large ratio $\omega/\omega_{\text{rest}}$. Curiously, however, the acceleration occurs always in a way that conserves $\left\langle k\bar{u}\right\rangle_S$. There can therefore be substantial first- and even second-order Doppler shifts from the acceleration caused by the field, but the *total* Doppler shift in the frame of the charge is constant.

Example 41. *For the field pulse*

$$\boldsymbol{F}(s) = k\boldsymbol{A}_0/\cosh^2 s$$

there is a net change in the paravector potential of $-2\boldsymbol{A}_0$ *so that from (5.12),*

$$\Lambda(\infty) = \left(1 + \frac{k\boldsymbol{a}_0}{\omega_{\text{rest}}}\right)\Lambda(0).$$

With injection of the charge along $\hat{\boldsymbol{k}}$, *one finds a final proper velocity*

$$u(\infty) = \Lambda(\infty)\Lambda^\dagger(\infty) = u_0 + 2\boldsymbol{a}_0 + \frac{2k\boldsymbol{a}_0^2}{\omega_{\text{rest}}}$$

where $\boldsymbol{a}_0 = e\boldsymbol{A}_0/m$ *is the dimensionless amplitude of the vector potential.*

Exercise 42. *Show that the energy gain in the last example is*

$$\frac{2\omega\boldsymbol{a}_0^2}{\omega_{\text{rest}}}m.$$

Exercise 43. *Verify that* $u\bar{u}$ *is conserved in the last example.*
(Hint: Recall $\langle k\bar{u}_0\rangle_S = \omega_{\text{rest}}$ *and note* $u_0\boldsymbol{a}_0 = \boldsymbol{a}_0\bar{u}_0$ *and* $k\boldsymbol{a}_0 = \boldsymbol{a}_0\bar{k}$.*)*

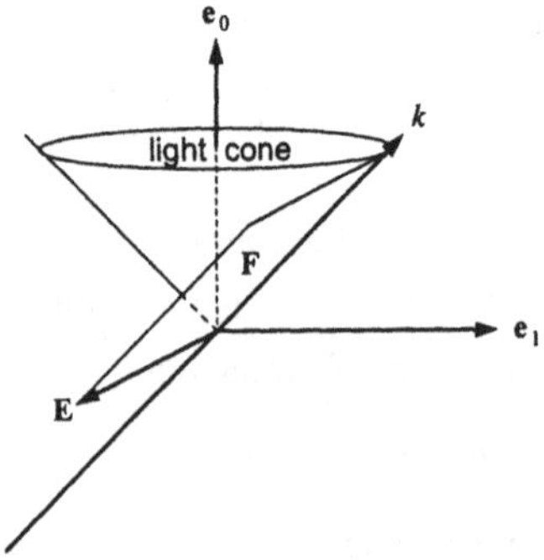

FIGURE 4.4. The electromagnetic field of a directed plane wave is a flag tangent to the light cone. Its flagpole lies along the propagation paravector k.

Insight into why k_{rest} stays constant in the frame of the accelerating charge is provided by the geometry of the null flag. The rotation occurs in the flag plane, and the propagation paravector lies along the flagpole which lies in the plane. However, the flagpole is also *orthogonal* to the plane and therefore invariant under rotations in it.

Modern lasers possess high electric fields and seem excellent candidates for particle accelerators. Our result (5.12) above shows why simply hitting a charge with a laser has rather limited effect. The net change in the eigenspinor is linear in the change in paravector potential. You can gain energy only for about a half cycle of the laser. Continue into the next half cycle and the charge starts losing energy.

Another scheme is possible that avoids these problems. It is called the autoresonant laser accelerator (ALA) and combines a constant magnetic field along the $\hat{\boldsymbol{k}}$ axis of a circularly polarized plane wave. When the cyclotron frequency of the charge is resonant with the frequency of the laser, the acceleration along $\hat{\boldsymbol{k}}$ is continuous and the frequencies remain resonant. This can lead to substantial energy gains. Eigenspinors and projectors provide a powerful tool for solving trajectories of charges in such cases, as well as for cases of superimposed electric fields [22].

The electromagnetic field in the case of a longitudinal magnetic field has the form

$$\boldsymbol{F} = k\bar{A}'(s) + iB_0\hat{\boldsymbol{k}}$$

from which the equation of motion times $\bar{k}$ follows:

$$\bar{k}\dot{\Lambda} = \frac{e}{2m}\bar{k}\boldsymbol{F}\Lambda = -\frac{i\omega_c}{2}\bar{k}\Lambda\,,$$

where $\omega_c = eB_0/m$ is the proper cyclotron frequency. The solution

$$\bar{k}\Lambda = \exp\left(-\tfrac{1}{2}i\omega_c\tau\right)\bar{k}\Lambda(0)$$

implies

$$\dot{s} = \left\langle \Lambda^\dagger \bar{k}\Lambda \right\rangle_S = \omega_{\mathbf{rest}} = \text{const.}$$

as before, so that we can again solve for $\dot{\Lambda}$. At resonance $\omega_c = \omega_{\mathbf{rest}}$ in a circularly polarized wave $A = m\exp\left(i(s-s_0)\hat{\boldsymbol{k}}\right)\boldsymbol{a}/e$, we find energy gains of $m\Delta\gamma$ with

$$\Delta\gamma = \boldsymbol{u}(0)\cdot\boldsymbol{k}\times\boldsymbol{a}\tau + \tfrac{1}{2}\omega\omega_c(\boldsymbol{a}\tau)^2\,.$$

These can be substantial, accelerating 100 MeV electrons to 1 TeV within 2 km in a 10 Tesla magnetic field with a Ti:Sapphire laser pulse. [22]

4.5.5 *Potential of moving point charge*

Accelerating charges radiate. Indeed, radiation reaction had to be calculated for the ALA, even though it turned out to be small over a wide range of parameters. The potentials and fields of point charges in general motion are derived in most electrodynamics texts. The traditional derivation of the retarded fields is tricky and the final result rather messy and opaque. APS helps clear the fog and provides insight into the origin and nature of the radiation.

The potential of a moving point charge was derived (before special relativity!) by A. Liénard (1898) and E. Wiechert (1900). It is easily obtained from the rest-frame Coulomb potential:

$$\Phi_{\mathbf{rest}} = \frac{K_0 e}{\langle R_{\mathbf{rest}}\rangle_S}$$

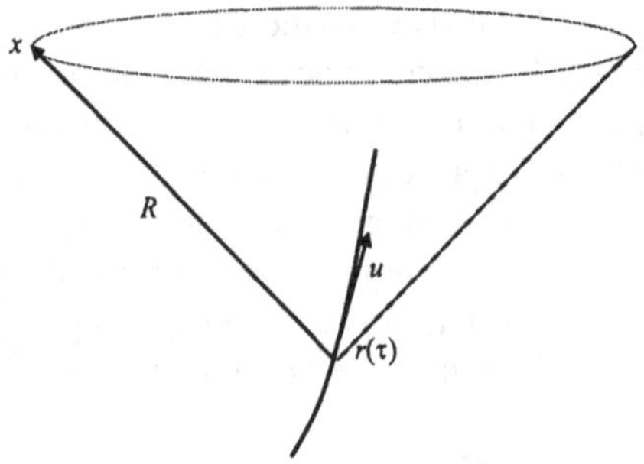

FIGURE 4.5. x is field point, $r(\tau)$ is world line of charge.

with $K_0 = (4\pi\varepsilon_0)^{-1}$ and $R(\tau) = x - r(\tau)$, the difference paravector between the field position x and the world line of the charge at the retarded proper time τ. The denominator $\langle R_{\text{rest}}\rangle_S$ is the time component of the difference in the commoving inertial frame. It is a Lorentz invariant simply because it is measured in the rest frame of the charge at the retarded time, but we make it manifestly invariant by writing

$$\langle R_{\text{rest}}\rangle_S = \langle R\bar{u}\rangle_S$$

where u is the proper velocity of the charge at τ and $R = R(\tau)$ can be in any inertial frame. We only need to boost Φ_{rest} to the proper velocity of the charge at the retarded τ to get the paravector potential for a charge in general motion:

$$A(x) = \Lambda\Phi_{\text{rest}}\Lambda^{\dagger} = \frac{K_0 e u}{\langle R\bar{u}\rangle_S}.$$

It is that simple. Note that $A(x)$ is covariant and depends only on the relative position and proper velocity of the charge at the retarded τ. It is independent of the acceleration or any other properties of the trajectory. The retarded time τ is determined by the light-cone condition $R\bar{R} = 0$, where causality demands that $R^0 > 0$.

Liénard–Wiechert field

The electromagnetic field is found from $\boldsymbol{F} = \langle \partial\bar{A}\rangle_V$ with the complication that the light-cone condition makes τ a function of x, that is a scalar field, so that there are two contributions to the gradient term:

$$\partial = (\partial)_\tau + [\partial\tau(x)]\frac{d}{d\tau}.$$

The first term is differentiation with respect to the field position x with τ held fixed. The second term arises from the dependence on x of the scalar field $\tau(x)$.

If we apply this to $R\bar{R} = 0$, we find directly

$$\partial\tau(x) = \frac{R}{\langle R\bar{u}\rangle_S}.$$

The evaluation of the field is now straightforward, and the result

$$\boldsymbol{F}(x) = \frac{K_0 e}{\langle R\bar{u}\rangle_S^3}\left(\langle R\bar{u}\rangle_V + \tfrac{1}{2}R\bar{\dot{u}}u\bar{R}\right) = \boldsymbol{F}_c + \boldsymbol{F}_r$$

has a simple interpretation: $\boldsymbol{F}_c$is the boosted Coulomb field and $\boldsymbol{F}_r$ is a directed plane wave propagating in the direction $\hat{\boldsymbol{R}}$ that is linear in the acceleration. Both $\boldsymbol{F}_c$ and $\boldsymbol{F}_r$ are simple fields, with $\boldsymbol{F}_c$ predominantly electric, $\boldsymbol{F}_c{\cdot}\boldsymbol{F}_r = 0$, and $\boldsymbol{F}_r$ a null flag. The spacetime plane of $\boldsymbol{F}_c$ is $\langle R\bar{u}\rangle_V = \gamma\left(\boldsymbol{R} - \boldsymbol{v}R^0 - \langle \boldsymbol{R}\boldsymbol{v}\rangle_V\right)$. The real part gives the electric field in the direction of $\boldsymbol{R}$ from the retarded position of the charge, minus $\boldsymbol{v}$ times the time R^0 for the radiation to get from the charge to the field point. The electric field thus points from the instantaneous "inertial image" of the charge, that is the position the charge would have at the instant the field is measured if its velocity remained the constant. If the charge moves at constant velocity, the electric field lines are straight from the instantaneous position of the charge. It is obvious that this had to be that way: in the rest frame of the charge the lines are straight from the charge, and the Lorentz transformation is linear and transforms straight lines into straight lines. The radiation field in the constant-velocity case vanishes, of course, because there is no acceleration. The magnetic field lines are normal to the spatial plane $\langle \boldsymbol{R}\boldsymbol{v}\rangle_V$ swept out by $\boldsymbol{R}$ as the charge moves along $\boldsymbol{v}$. They thus circle around the charge path. With APS, we easily decipher both the spacetime geometry and the spatial version with the same formalism.

When a charge accelerates, its inertial image can move rapidly, even at superluminal speeds, and lines need to shift laterally. The Coulomb field lines are broken, and $\boldsymbol{F}_c$ no longer satisfies Maxwell's equations by itself. The radiation field $\boldsymbol{F}_r$ is just the transverse field needed to connect the Coulomb lines. Only the *total* field $\boldsymbol{F} = \boldsymbol{F}_c + \boldsymbol{F}_r$ is generally a solution to Maxwell's equation.

The APS is sufficiently simple that calculations such as the field of a uniformly accelerated charge are easily made and provide simple analytical results that help unravel questions about the relation of such a radiating system to the nonradiating charge at rest in a uniform gravitational field [6]. One can also simplify the derivation of the Lorentz–Dirac equation for the motion of a point charge with radiation reaction and lay bare its relation to related equations such as the Landau–Lifshitz equation [23].

There are many other cases where relativistic symmetries can simplify electrodynamics, even where relativity at first seems to play no significant role. For example, the currents induced in a conductor by an incident wave are easily calculated, and the oblique incident case is simply related to the normal incident one by a boost. Similarly, wave guide modes can be generated by boosting standing waves down the guide. APS plays a crucial role in unifying relativity with vectors and simple geometry.

4.6 Quantum theory

As seen above, classical relativistic physics in Clifford algebra has a spinorial formulation that gives new geometrical insights and computational power to many problems. The formulation is closely related to standard quantum formalism. The algebraic use of spinors and projectors, together with the bilinear relations of spinors to observed currents, gives quantum-mechanical form to many classical results, and the clear geometric content of the algebra makes it an illuminating probe of the quantum/classical interface. This section summarizes how APS has provided insight into spin-$\frac{1}{2}$ systems and their measurement.

Consider an elementary particle with an extended distribution. Its current density $j(x)$ is related by an eigenspinor field $\Lambda(x)$ to the reference-frame density ρ_{ref}:

$$j(x) = \Lambda(x)\,\rho_{\text{ref}}(x)\,\Lambda^{\dagger}(x). \tag{6.13}$$

This form allows the velocity (and orientation) to be different at different spacetime positions x. The current density (6.13) can be written in terms of the density-normalized eigenspinor Ψ as

$$j = \Psi\Psi^{\dagger}, \quad \Psi = \rho_{\text{ref}}^{1/2}\Lambda\,,$$

and it is independent of *gauge rotations* $\Psi \to \Psi R$ of the reference frame. The momentum of the particle $p = mu = m\Lambda\Lambda^{\dagger}$ can be multiplied from the right by $\bar{\Lambda}^{\dagger} = \left(\Lambda^{\dagger}\right)^{-1}$ to obtain

$$p\bar{\Lambda}^{\dagger} = m\Lambda.$$

This is the *classical Dirac equation* [24, 25]. It is a real linear equation: real linear combinations of solutions are also solutions. In particular, since $\rho_{\text{ref}}^{1/2}$ is a scalar, it also holds for $\Psi = \rho_{\text{ref}}^{1/2}\Lambda$:

$$p\bar{\Psi}^{\dagger} = m\Psi. \tag{6.14}$$

The classical Dirac equation (6.14) is invariant under gauge rotations, and its real linear form suggests the possibility of wave interference.

Consider the eigenspinor Ψ_p for a free particle of well-defined constant momentum $p = mu$. In this case, we can define a proper time τ for the particle, and we look for an eigenspinor Ψ_p that depends only on τ. The continuity equation for j implies a constant ρ_{ref}:

$$\langle \partial\bar{j}\rangle_S = 0 = \langle (\partial\rho_{\text{ref}})\,\bar{u}\rangle_S = \frac{d\rho_{\text{ref}}}{d\tau}.$$

Gauge rotations $\Psi \to \Psi R$ include the possibility of a time-dependent rotation R. For the free particle of constant momentum, we assume a fixed rotation rate, and using gauge freedom to orient the reference frame, we take the rotation axis to be $\mathbf{e}_3$ in the reference frame. Including this rotation, the free eigenspinor Ψ_p has the form

$$\Psi_p(\tau) = \Psi_p(0)\,e^{-i\omega_0\tau\mathbf{e}_3}.$$

Now the proper time is given explicitly in Lorentz-invariant form by

$$\tau = \langle u\bar{x}\rangle_S = \left\langle \frac{p}{m}\bar{x}\right\rangle_S ,$$

and therefore the free eigenspinor Ψ_p has the spacetime dependence

$$\Psi_p(x) = \Psi_p(0)\, e^{-i(\omega_0/m)\langle p\bar{x}\rangle_S \mathbf{e}_3}. \tag{6.15}$$

The gauge rotation is associated with the *classical spin* of the particle, and the spacetime dependence of Ψ_p (6.15) gives the behavior of de Broglie waves provided we identify $\omega_0/m = \hbar$.

A further local gauge rotation by $\phi(x)$ about $\mathbf{e}_3$ in the reference frame can be accommodated without changing the physical momentum p in (6.15) by adding a gauge potential $A(x)$ that undergoes a compensating gauge transformation (to be determined below)

$$\Psi_p(x) = \Psi_p(0)\, e^{-i\omega_0\tau\mathbf{e}_3} = \Psi_p(0)\, e^{-i\langle(p+eA)\bar{x}\rangle_S \mathbf{e}_3/\hbar}.$$

The real linear form of the classical Dirac equation (6.14) suggests that real linear combinations

$$\Psi(x) = \int a(\mathbf{p})\, e^{-i\langle(p+eA)\bar{x}\rangle_S \mathbf{e}_3/\hbar}\, d^3\mathbf{p},$$

where $a(\mathbf{p})$ is a scalar amplitude, may form more general solutions for particles of a given mass m. Such linear combinations are indeed solutions to (6.14) if p is replaced by the momentum operator defined by

$$p\Psi = i\hbar\partial\Psi\mathbf{e}_3 - eA\Psi.$$

With this replacement, the classical Dirac equation (6.14) becomes Dirac's quantum equation. The gauge transformation in A that compensates for the local gauge transformation $\Psi \to \Psi e^{-i\phi(x)\mathbf{e}_3}$ is now seen to be

$$A \to A + \frac{\hbar}{e}\partial\phi(x). \tag{6.16}$$

The gauge potential A is identified as the electromagnetic paravector potential, and the coupling constant e is the electric charge of the particle. The gauge transformation (6.16) is easily seen leave the electromagnetic field $\boldsymbol{F} = \langle\partial\bar{A}\rangle_V$ invariant. The traditional matrix form of the Dirac equation follows by splitting (6.14) into even and odd parts and projecting both onto the minimal left ideal $C\ell_3\mathsf{P}_{\mathbf{e}_3}$, where $\mathsf{P}_{\mathbf{e}_3} = \frac{1}{2}(1+\mathbf{e}_3)$.

The Dirac theory is the basis for current understanding about the relativistic quantum theory of spin-$\frac{1}{2}$ systems. The brief discussion above suggests that its foundations lie largely in Clifford's geometric algebra of classical systems. This suggestion is explored more fully elsewhere [13], where it is shown that the basic two-valued property of spin-$\frac{1}{2}$ systems is indeed associated with a simple property of rotations in physical space. Extensions to higher dimensional spaces have succeeded in explaining the gauge symmetries of the standard model of elementary particles in terms of rotations in $C\ell_7$ [27], and extensions to multiparticle systems promise to demystify entanglement.

4.7 Conclusions

In the space of this lecture, we have only scratched the surface of the many applications of Clifford's geometric algebra to physics. There has been no attempt at a thorough review of the work, the extent of which may be gleaned from texts [4, 6, 8, 31, 32], proceedings of recent conferences and workshops [5, 28–30, 33, 34] and from several websites [35]. I have instead tried to illustrate the conceptual and computational power the algebra brings to physics. Many of its tools arise from the spinorial formulation inherent in the algebra and are familiar from quantum mechanics. Had the electrodynamics and relativity of Maxwell and Einstein been originally formulated in APS, the transition to quantum theory would have been less of a "quantum leap."

Acknowledgments: Support from the Natural Sciences and Engineering Research Council of Canada is gratefully acknowledged.

4.8 References

[1] P. Lounesto, Introduction to Clifford Algebras, Lecture 1 in *Lectures on Clifford Geometric Algebras,* ed. by R. Abłamowicz and G. Sobczyk, Birkhäuser, Boston, 2004.

[2] W. R. Hamilton, *Elements of Quaternions,* Vols. I and II, a reprint of the 1866 edition published by Longmans Green (London) with corrections by C. J. Jolly, Chelsea, New York, 1969.

[3] S. L. Adler, *Quaternionic Quantum Mechanics and Quantum Fields,* Oxford U. Press, New York, 1995.

[4] P. Lounesto, *Clifford Algebras and Spinors*, second edition, Cambridge University Press, Cambridge (UK), 2001.

[5] W. E. Baylis, editor, *Clifford (Geometric) Algebra with Applications to Physics, Mathematics, and Engineering*, Birkhäuser, Boston, 1996.

[6] W. E. Baylis, *Electrodynamics: A Modern Geometric Approach*, Birkhäuser, Boston, 1999.

[7] W. E. Baylis, J. Huschilt, and J. Wei, *Am. J. Phys.* **60** (1992), 788–797.

[8] D. Hestenes, *New Foundations for Classical Mechanics,* 2nd ed., Kluwer Academic, Dordrecht, 1999.

[9] D. Hestenes and G. Sobczyk, *Clifford Algebra to Geometric Calculus*, Kluwer Academic, Dordrecht, 1984.

[10] D. Hestenes, *Spacetime Algebra*, Gordon and Breach, New York, 1966.

[11] W. E. Baylis and G. Jones, *J. Phys. A: Math. Gen.* **22** (1989), 1–16; 17–29.

[12] W. E. Baylis and S. Hadi, Rotations in *n* dimensions as spherical vectors, in *Applications of Geometric Algebra in Computer Science and Engineering*, edited by L. Dorst, C. Doran, and J. Lasenby, Birkhäuser, Boston, 2002, pp. 79–90.

[13] W. E. Baylis, The Quantum/Classical Interface: Insights from Clifford's Geometric Algebra, in *Clifford Algebras: Applications to Mathematics, Physics, and Engineering*, R. Abłamowicz, ed., Birkhäuser, Boston, 2004.

[14] A. W. Conway, *Proc. Irish Acad.* **29** (1911), 1–9.

[15] A. W. Conway, *Phil. Mag.* **24** (1912), 208.

[16] L. Silberstein, *Phil. Mag.* **23** (1912), 790–809; **25** (1913), 135–144.

[17] L. Silberstein, *The Theory of Relativity*, Macmillan, London, 1914.

[18] C. Chu and T. Ohkawa, *Phys. Rev. Lett.* **48** (1982), 837.

[19] R. P. Feynman, *QED: The Strange Story of Light and Matter*, Princeton Science, 1985.

[20] A. H. Taub, *Phys. Rev.* **73** (1948), 786–798.

[21] D. Hestenes, *J. Math. Phys.* **15** (1974), 1768–1777; 1778–1786.

[22] W. E. Baylis and Y. Yao, *Phys. Rev. A* **60** (1999), 785–795.

[23] W. E. Baylis and J. Huschilt, *Phys. Lett.* A 301 (2002), 7–12.

[24] W. E. Baylis, *Phys. Rev. A* **45** (1992), 4293–4302.

[25] W. E. Baylis, *Adv. Appl. Clifford Algebras* **7(S)** (1997), 197.

[26] V. B. Berestetskii, E. M. Lifshitz, and L. P. Pitaevskii, *Quantum Electrodynamics* (Volume 4 of *Course of Theoretical Physics*), 2nd ed. (transl. from Russian by J. B. Sykes and J. S. Bell), Pergamon Press, Oxford, 1982.

[27] G. Trayling and W. E. Baylis, *J. Phys. A* **34** (2001), 3309–3324.

[28] J. S. R. Chisholm and A. K. Common, eds., *Clifford Algebras and their Applications in Mathematical Physics,* Riedel, Dordrecht, 1986.

[29] A. Micali, R. Boudet, and J. Helmstetter, eds., *Clifford Algebras and their Applications in Mathematical Physics,* Reidel, Dordrecht, 1992.

[30] Z. Oziewicz and B. Jancewicz and A. Borowiec, eds., *Spinors, Twistors, Clifford Algebras and Quantum Deformations,* Kluwer Academic, Dordrecht, 1993.

[31] J. Snygg, *Clifford Algebra, a Computational Tool for Physicists,* Oxford U. Press, Oxford, 1997.

[32] K. Gürlebeck and W. Sprössig, *Quaternions and Clifford Calculus for Physicists and Engineers,* J. Wiley and Sons, New York, 1997.

[33] R. Abłamowicz and B. Fauser, eds., *Clifford Algebras and their Applications in Mathematical Physics, Vol. 1: Algebra and Physics*, Birkhäuser, Boston, 2000.

[34] R. Abłamowicz, ed., *Clifford Algebras: Applications to Mathematics, Physics, and Engineering*, Birkhäuser, Boston, 2003.

[35] http://modelingnts.la.asu.edu/GC_R&D.html,
http://www.mrao.cam.ac.uk/~clifford,
http://www.uwindsor.ca/baylis-research

William E. Baylis
Department of Physics, University of Windsor
Windsor, ON, Canada N9B 3P4
E-mail: baylis@uwindsor.ca

Submitted: January 8, 2003; Revised: March 20, 2003.

5

Clifford Algebras in Engineering

J.M. Selig

ABSTRACT In this chapter, we look at some applications of Clifford algebra in engineering. These applications are geometrical in nature concerning robotics and vision mainly. Most engineering applications have to be implemented on a computer these days so we begin by arguing that Clifford algebra are well suited to modern microprocessor architectures.

Our first application is to satellite navigation and uses quaternions. The rotation of the satellite is to be found from observations of the fixed stars. The same problem occurs in many other guises throughout the natural sciences and engineering.

Next we use biquaternions to write down the kinematic equations of the Stewart platform. This parallel robot is used in aircraft simulators and in novel machine tools. The problem is to determine the position and orientation of the platform from the lengths of the hydraulic actuators.

In the next section, we introduce a less familiar Clifford algebra $C\ell(0,3,1)$. The homogeneous elements of this algebra can be used to represent points, lines and planes in three dimensions. Moreover, meets joins and orthogonal relationships between these linear subspaces can be modelled by simple formulas in the algebra. This algebra is used in the following sections to discuss a couple of problems in computer vision and the kinematics of serial robots.

5.1 Introduction

Traditionally in computing, methods involving many algebraic operations have not been favoured. In many applications such as graphics clipping, it has been possible to devise algorithms that use many comparison operations rather than take a more straightforward approach which would use a few polynomial evaluations. For simple processors these operations are much quicker and hence fast programs can be written. Today's processors are rather different and of course the ambitions of the programmers have developed too. Hence, even though the speed of processors has increased very rapidly over the last decade or so, the speed of operation is

---

This lecture was presented at "Lecture Series on Clifford Algebras and their Applications", May 18 and 19, 2002, as part of the 6th International Conference on Clifford Algebras and their Applications in Mathematical Physics, Cookeville, TN, May 20–25, 2002.

AMS Subject Classification: 15A66, 68T45, 68M01, 70B15.

Keywords: Clifford algebras, robotics, computer vision, satellite navigation, quaternions, biquaternions, Stewart platform, meet, join, kinematics.

still a crucially important factor. In modern processors the slowest operation turns out to be the 'fetch' operation, where the next instruction is retrieved from memory. To improve the efficiency therefore, possessors usually have a 'pipeline' of instructions which have been pre-fetched from memory. When a choice or comparison instruction is encountered the hardware must guess which branch will be taken after the instruction has been executed. Indeed much research has been directed to the problem of 'branch prediction' to improve the performance of these microprocessors. If the wrong guess is made the pipeline will have to be flushed and the instructions from the correct branch will be loaded. Hence, for modern processors comparison operations should be avoided if at all possible. Moreover multiplication, the traditional bottleneck, is now often supported in hardware, or at least the micro code for it will be cached on the processor chip. For these reasons algebraic methods are becoming more attractive.

Additionally, some modern processors have a lot of support for graphics. For instance, the Sony PlayStation 2 can perform a vector cross product faster than a scalar product. The PlayStation 2 (PS2) vector co-processors can perform a vector product in 2 instructions at one cycle each, giving the result after a total of 5 cycles. A scalar product in comparison takes 9 cycles (with 5 'free' wait cycles to schedule other instructions in), [5]. The problems of computer graphics can be expressed very neatly in terms of Clifford algebras. Indeed much of the early development of the subject was motivated by the desire to solve problems in 3-D geometry. Thus it is not too surprising that Clifford algebras are becoming more important in this area of computing, it is already common for graphics programs to use quaternions to represent rotations.

Since computer graphics is one of the main applications driving development in computer hardware, it does not seem unreasonable to expect even greater convergence between algebraic systems for representing geometry and the hardware to support it. A spin-off should be that the hardware will also support other Clifford algebras. Applications which can be represented in terms of Clifford algebras will find that this is the preferred route to computer implementation.

A major reason why Clifford algebras are so useful is that they contain all the exterior powers of a representation of some symmetry group. These representations usually have geometric significance, for example the points, lines and planes in 3 dimensions, see section 5.4 later. In the Clifford algebra we can combine elements from these representations and the group itself in the same algebra and they can be treated on an equal footing. Clifford algebras could be thought of as a way of hiding lots of determinant calculations. So above it has been assumed that fast methods can be found to compute these determinants as they have been already for the vector product.

Of course the benefits of using Clifford algebras in engineering applications are not solely computational. A major benefit of using Clifford algebras is the simplification of symbolic computation. Much of the power of any notation come from its conciseness. Clifford algebras are able to express relationships between elements with remarkably short equations. While it is uncommon for problems to be solvable by Clifford algebra methods alone, it is certainly true that as problems become more complicated it is easier to use the Clifford algebra as the equations are invariably more manageable. This argument naturally leads to consideration of symbolic algebra packages. Clifford algebras are ideally suited to implementation in such packages and indeed several implementations exist, notably CLICAL and CLIFFORD. These programs were intended as general purpose utilities for computations in any Clifford algebra.[1] I am not aware that they have been used in any particular engineering application, but this will surely come.

In this lecture I will attempt to give an account of some of the engineering applications where Clifford algebras have been used. The applications are linked by their use of geometry. That is the geometry of three-dimensional Euclidian space. The applications include computer vision and the kinematics of robots. There are other application of Clifford algebras not so closely tied to geometry, notably in signal and image processing but I have tried to present a coherent picture here rather than an encyclopedic review.

We begin by looking at a modern twist on the 'Clifford algebra' that started it all—Hamilton's quaternions.

5.2 Quaternions

The connection between quaternions and rotations has been explored in Lecture 1. As mentioned above, many computer graphics systems use quaternions to represent rotations, there are several reasons for this. Certainly storage requirements are less, quaternions are specified by four numbers while a 3×3 rotation matrix is determined by nine numbers. Another reason why quaternions are used is that there are well understood methods for interpolating unit quaternions. A more important reason for prefering quaternions is the simple way that errors can be handled. Rotations are represented by unit quaternions, hence after a computation the components of the quaternion may no longer satisfy this requirement. All that is necessary to recover a rotation though is a normalisation, that is division by the square root of the sum of the squared components of the quaternion. By contrast rotation matrices are 3×3 orthogonal matrices with unit determinants, to regain these properties after a computation, a time consuming Gramm-Schmidt orthogonalisation procedure is required.

In many branches of engineering, quaternions are used to solve practical problem involving rotations. Here we just look at one example, but see [12] for several others.

The problem considered here is that of satellite navigation. We wish to determine the orientation or attitude, of an earth orbiting artificial satellite relative to some fixed orientation. Many artificial satellites solve this problem using inertial navigation techniques, that is by carrying a gyroscope. Here we look at a differ-

[1]CLICAL and CLIFFORD are briefly reviewed in the Appendix in section 7.1.1, p. 190, and section 7.1.4, p. 196, respectively. *Edi.ors*

ent method which is also often used. For satellite navigation this problem can be traced back to Wahba [15] in the mid 1960s. However, the same problem occurs in the crystallography literature in 1957 [10], and in studies of paleomagnetism in the mid 1970s [11] and most recently in the late 1980s in the field of computer vision [8].

For this technique the satellite carries a vision system and can make observations of the fixed stars. The fixed stars can be assumed to be infinitely far away so that the measured direction to a star will be the same on the satellite as it would be on earth or at any position of the satellite. Any difference will be a result of the difference in the coordinate systems used and hence their relative orientations, see figure 5.1.

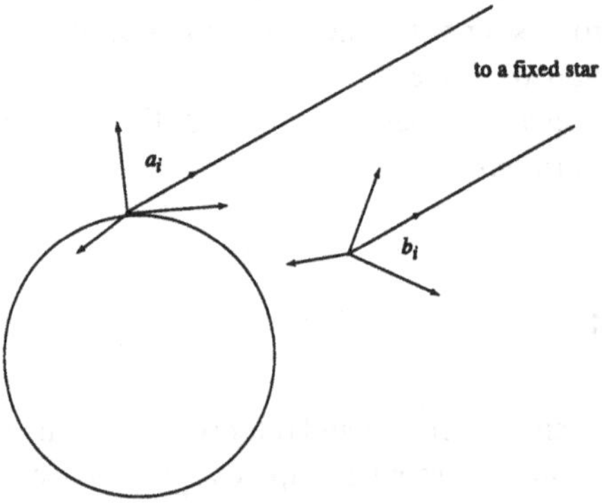

FIGURE 5.1. Fixed star observed from two different positions

Let a_i denote the unit direction vector to the ith fixed star in the fixed frame and b_i the unit vector to the same star in the satellite's coordinate frame. The relative rotation between the two frames will be represented by a unit quaternion h and hence the relationship between the two sets of vectors can be written as

$$b_i = ha_ih^-$$

where h^- denotes the quaternionic conjugate of h. For a single star this equation would have a 'circles-worth' of solutions, corresponding to the rotations which fix a_i. If several stars are observed then experimental errors will usually spoil the consistency of the set of these equations. Hence, the problem is to estimate the best h given several observations. The concept of 'best' here has usually been interpreted as the least squared distance. That is, h must be chosen so as to minimise the sum of the squares of the distances between the observed vector b_i and the estimated one ha_ih^-,

$$Q^2 = \sum_i |b_i - ha_ih^-|^2 .$$

Although this choice of objective function may seem arbitrary there are sound statistical reasons for using it, see [17].

Notice, this equation can be rewritten as,

$$Q^2 = \sum_i (b_i - ha_ih^-)(b_i^- - ha_i^-h^-) = \sum_i (2 - b_iha_i^-h^- - ha_ih^-b_i^-)$$

since a_i and b_i are assumed to be unit vectors.

The cost function Q^2 must be minimised subject to the condition that h remains a unit quaternion $hh^- = 1$. This can be done using a Lagrange multiplier, however, because the unit quaternions form a Lie group, differential geometry can be used to simplify the approach. The function Q^2 can be thought of a function defined on the 3-dimesional sphere defined by the unit quaternions, so a minimum of Q^2 will be a critical point. To find the critical points of Q^2 all that is needed is that its derivative along tangent vector fields vanish. These derivatives can be evaluated with the help of the exponential map. This maps Lie algebra elements to the group. A Lie algebra element here corresponds to a pure quaternion, $v = v_xi + v_yj + v_zk$. The exponential,

$$e^{\theta v} = 1 + (\theta v) + \tfrac{1}{2}(\theta v^2) + \cdots = \cos\theta + v\sin\theta$$

corresponds to a one-parameter subgroup, with parameter θ. Now to differentiate a quaternion along a left-invariant vector field we shift it along the one-parameter subgroup and take the difference with the original and divide by the parameter distance, finally we take the limit as the size of the shift decreases,

$$\partial_v h = \lim_{\theta \to 0} \frac{e^{\theta v}h - h}{\theta} = vh\,.$$

So the derivative of Q^2 along an arbitrary left-invariant vector field is,

$$\partial_v Q^2 = -\sum_i (b_ivha_i^-h^- + b_iha_i^-h^-v^- + vha_ih^-b_i^- + ha_ih^-v^-b_i^-)\,.$$

Now v, b_i and ha_ih^- are all pure quaternions, so that $v^- = -v$, $b_i^- = -b_i$ and so on. Further, for pure quaternions x, y we have the well know relation,

$$xy = -x \cdot y + x \times y$$

where $\cdot$ and $\times$ are the scalar and vector products of the corresponding vectors. Using these relations to simplify the derivative of Q^2 we obtain,

$$\partial_v Q^2 = -4v \cdot \left(\sum_i ha_ih^- \times b_i \right).$$

The critical points are given by the vanishing of this expression and since v is arbitrary we can write,

$$\sum_i ha_ih^- \times b_i = 0$$

as the condition for h to make Q^2 critical.

One way to make progress with this equation is to adopt a different representation for the rotation group. Suppose, $\mathbf{a}_i$ and $\mathbf{b}_i$ are the 3-vectors corresponding to the pure quaternions a_i and b_i. It is well known that 3-vectors can be represented by 3×3 antisymmetric matrices. A straightforward computation confirms that the antisymmetric matrix corresponding to the 3-vector $\mathbf{a}_i \times \mathbf{b}_i$ is given by the matrix product $\mathbf{b}_i\mathbf{a}_i^T - \mathbf{a}_i\mathbf{b}_i^T$. The effect of the rotation ha_ih^- can be represented by the product $R\mathbf{a}_i$ where R is the 3×3 rotation matrix corresponding to h and $-h$. Now the equation for a critical point reads,

$$\sum_i \left(\mathbf{b}_i\mathbf{a}_i^T R^T - R\mathbf{a}_i\mathbf{b}_i^T\right) = 0\,.$$

If we write $P = \sum_i \mathbf{a}_i\mathbf{b}_i^T$ then the equation simplifies to,

$$RP = P^T R^T\,.$$

This equation has 4 solutions in general, they are closely related to the polar decomposition of the matrix P. Let $P = R_p^T Q$ be the polar decomposition of P so that Q is a 3×3 symmetric matrix and R_p^T is a rotation matrix. Note that in the literature the polar decomposition of a matrix decomposes it into an orthogonal matrix and a positive definite symmetric matrix. Here we require the orthogonal matrix to be a rotation but it is not necessary that the symmetric matrix be positive. Hence if the classical polar decomposition yields a reflection we can multiply it by -1 to get a rotation and then we will have to multiply the symmetric matrix by -1 to get Q.

Now $R = R_p$ is certainly a solution, but we have 3 others, $R = R_1R_p, R_2R_p$ and R_3R_p. Where R_i is a rotation of π radians about the ith eigenvector of Q. That is, the R_is satisfy, $R_iQR_i = Q$. Finally the least squares solution we are seeking is given by the solution which minimises Q^2. It can be shown that if $\det(P) > 0$ then the minimising solution is $R = R_p$. If $\det(P) < 0$ then the minimising solution is $R = R_iR_p$ where R_i is the rotation by π about the eigenvector of Q which has the smallest magnitude.

Notice that, the quaternions were dispensed with after a certain point in the solution. This seems to be a characteristic of engineering approaches to problems: the solution is the important thing not the purity of the method used. No doubt there is a method of solving this problem which uses quaternions throughout or perhaps it is necessary to represent the vectors and rotations in a larger Clifford algebra to be able to produce a solution purely in terms of Clifford algebra. However the polar decomposition seems to have been little studied in the context of Clifford algebras, however see [14].

Finally here, the function Q^2 is essentially a function on the group of rotations, $SO(3)$. Above it was defined on the S^3 of unit quaternions however, it clearly has the same value for h and $-h$ and hence passes to the quotient $SO(3)$, which is topologically the projective space $\mathbb{P}\mathbb{R}^3$. In general, that is for most choices of stars, P will be non-singular, we need at least 3 stars though. Now, Q^2 will be a

Morse function on $\mathbb{PR}^3$ and hence from a knowledge of the topology of the space we have that the minimum number of critical points is 4. Moreover, it is possible to predict that these will be exactly one local maxima, one local minima and two types of saddle point.

5.3 Biquaternions

Clifford himself introduced the biquaternions [4] to represent rigid motions in three-dimensions. At the turn of the 20th Century Study used biquaternions to show that the group of rigid body motions $SE(3)$, consists of elements of a six dimensional projective quadric. These ideas seem to have been forgotten except by mechanical engineers working on the kinematics of mechanisms. Here, biquaternions are often used in conjunction with 'dual vectors' which represent lines in space. A general dual quaternion can be written as $\check{h} = h_0 + h_1\varepsilon$, here h_0 and h_1 are standard quaternions and ε represents the dual unit ε which squares to zero and commutes with the quaternions. Points in space can be represented by biquaternions of the form $\check{p} = 1 + p\varepsilon$, where $p = p_x i + p_y j + p_z k$ is the usual representation of a position vector by a pure quaternion. Rotations can be represented in exactly the same way as above by unit quaternions, $rr^- = 1$. Their effect on a point $\check{p}$ will be given by,

$$r(1 + p\varepsilon)r^- = (1 + rpr^-\varepsilon)\,.$$

Translations can be represented by dual quaternions of the form $(1+\frac{1}{2}t\varepsilon)$, where t is the translation vector written as a pure quaternion. The corresponding action of the translation on a point is given by,

$$(1 + \tfrac{1}{2}t\varepsilon)(1 + p\varepsilon)(1 - \tfrac{1}{2}t^-\varepsilon) = (1 + (p + t)\varepsilon)\,.$$

As usual, the group multiplication is modelled by Clifford multiplication, so that a general rigid body motion consisting of a rotation followed by a translation is given by a dual quaternion,

$$\check{h} = (1 + \tfrac{1}{2}t\varepsilon)r = (r + \tfrac{1}{2}tr\varepsilon)$$

Further, if we multiply two such group elements we get,

$$\begin{aligned}(r_1 + \tfrac{1}{2}t_1r_1\varepsilon)(r_2 + \tfrac{1}{2}t_2r_2\varepsilon) &= \\ &(r_1r_2 + \tfrac{1}{2}(r_1t_2r_2 + t_1r_1r_2)\varepsilon) = (r_1r_2 + \tfrac{1}{2}(r_1t_2r_1^- + t_1)r_1r_2\varepsilon)\end{aligned}$$

which shows the action of the rotations on the translations. Notice that, as with the rotations, this group is actually the double cover of the group of rigid transformations since $\check{h}$ and $-\check{h}$ give the same results on points.

If we extend the quaternion conjugate to the dual quaternions by defining,

$$\check{h}^- = (h_0^- + h_1^-\varepsilon)$$

then it is a simple matter to check that for a general rotation and translation we have,

$$\check{h}\check{h}^- = (r + \tfrac{1}{2}tr\varepsilon)(r + \tfrac{1}{2}tr\varepsilon)^- = 1.$$

Since, $rr^- = 1$ and $t^- = -t$, this equation splits into a pair of quaternion equations,

$$h_0h_0^- = 1, \qquad h_0h_1^- + h_1h_0^- = 0\,.$$

It is clear that any proper rigid motion, that is any combination of translations and rotations, can be represented as a dual quaternion satisfying these relations. It is also true, but a little harder to see, that any dual quaternion satisfying the above relations represents a proper rigid motion. Suppose we write a general dual quaternion as,

$$\check{h} = (x_0 + x_1 i + x_2 j + x_3 k) + (y_0 + y_1 i + y_2 j + y_3 k)\varepsilon\,.$$

The two relations for the dual quaternion to represent a proper rigid motion become,

$$x_0^2 + x_1^2 + x_2^2 + x_3^2 = 1$$

and

$$x_0y_0 + x_1y_1 + x_2y_2 + x_3y_3 = 0\,.$$

Since both $\check{h}$ and $-\check{h}$ represent the same motion we can identify these pairs by assuming $(x_0 : x_1 : x_2 : x_3 : y_0 : y_1 : y_2 : y_3)$ are homogeneous coordinates in the seven dimensional projective space $\mathbb{PR}^7$. This can be thought of as a projection which identifies pairs of elements in the double covering group. In $\mathbb{PR}^7$ the group elements must still satisfy the second relation which is homogeneous,

$$x_0y_0 + x_1y_1 + x_2y_2 + x_3y_3 = 0\,,$$

this defines a six dimensional quadric, known as the Study quadric. Each proper rigid motion corresponds to a point in this quadric and almost all the points of the quadric correspond to proper rigid motions. The exceptions comprise a special 3-plane,

$$x_0 = x_1 = x_2 = x_3 = 0\,.$$

Much of robotics is concerned with rigid motions and hence the geometry of the Study quadric is of fundamental importance. For example, the trajectory of the end-effector of a robot can be thought of as a curve in the Study quadric. The relative motions allowed by mechanical joints are subspaces of the Study quadric: revolute (or hinge joints) and prismatic (or sliding) joints allow motions corresponding to lines in the quadric while spherical and planar joints correspond to 3-planes. We will look at a more complex example in a moment but before that the dual quaternions will be related to the standard notations for Clifford algebras introduced in Lecture 2.

In Lecture 1 we saw that the quaternions could be thought of as the Clifford algebra $C\ell(0,2)$ which is also isomorphic to the even subalgebra of $C\ell(0,3)$. Here we show that the dual quaternions are isomorphic to the even subalgebra of the degenerate Clifford algebra $C\ell(0,3,1)$. This algebra has three generators which square to -1 and one which squares to 0, say $e_1^2 = e_2^2 = e_3^2 = -1$ and $e^2 = 0$. The isomorphism can be specified by giving the mapping on the generators,

$$i \mapsto e_2e_3, \quad j \mapsto e_3e_1, \quad k \mapsto e_1e_2, \quad \varepsilon \mapsto ee_1e_2e_3 .$$

Notice that the element $ee_1e_2e_3$ commutes with all the elements in the even subalgebra $C\ell^0(0,3,1)$. In this algebra, a typical biquaternion $\check{h}$ as above, would have the form,

$$\check{h} = x_0 + x_1e_2e_3 + x_2e_3e_1 + x_3e_1e_2 + y_0ee_1e_2e_3 + y_1e_1e + y_2e_2e + y_3e_3e .$$

As an example, we will look at the forward kinematics of a Stewart platform. This is a parallel manipulator, used originally to provide motion for aircraft simulators, but more recently this structure has been used in hexapod machine tools. The robot consists of a platform connected to the ground by six hydraulic cylinders. Each cylinder has a passive spherical joint at either end, see figure 5.2.

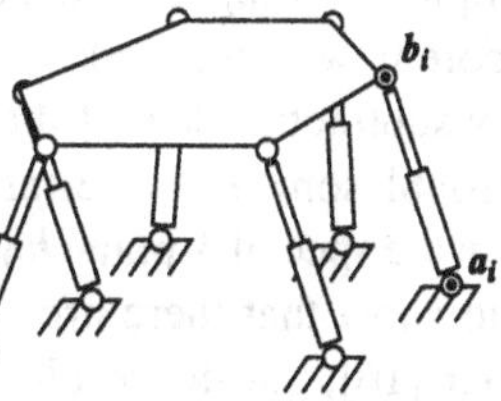

FIGURE 5.2. The Stewart platform

The forward kinematic problem is to determine the position and orientation of the platform given the 'leg-lengths' determined by the extension of the hydraulic cylinders. The mechanism is designed so that by actuating the six hydraulic cylinders the six degrees of freedom of the platform can be controlled.

Let the centres of the passive spherical joints on the base be located at points $b_1, b_2, \dots, b_6$. Further, suppose that in a standard position of the platform the centres of the spherical joints on the platform have position vectors $a_1, a_2, \dots, a_6$. Now suppose that the length of the ith leg, the one connecting b_i to a_i, is l_i. We seek a biquaternion $\check{h} = h_0 + h_1\varepsilon$ which takes the platform from its standard position to a position with the given leg-lengths. For each leg we get an equation,

$$l_i^2 = (h_0a_ih_0^- + t - b_i)(h_0a_ih_0^- + t - b_i)^-, \qquad i = 1, \dots, 6.$$

Here the translation $t = 2h_1h_0^- = -2h_0h_1^-$. Expanding these equations gives,

$$l_i^2 = |a_i|^2 + |b_i|^2 + h_0a_ih_0^-b_i + b_ih_0a_ih_0 - h_0a_ih_0^-t - th_0a_ih_0^- + b_it + tb_i - t^2,$$

$i = 1, \dots, 6$, where the fact that t and all the points are pure quaternions has been used to replace t^- by $-t$ for example. Next we may use the relations $t =$

$2h_1h_0^- = -2h_0h_1^-$ and the fact that $h_0h_0^- = 1$ to transform the equations to the form,

$$0 = (|a_i|^2 + |b_i|^2 - l_i^2)h_0h_0^- + h_0a_ih_0^-b_i + b_ih_0a_ih_0^- + \\ 2h_0a_ih_1^- - 2h_1a_ih_0^- + 2b_ih_1h_0^- - 2h_0h_1^-b_i + 4h_1h_1^-,$$

$i = 1, \dots, 6$. This is clearly a homogeneous quadratic equations in the coordinates, $(x_0 : x_1 : x_2 : x_3 : y_0 : y_1 : y_2 : y_3)$.

This gives us six quadratic equations which we must solve together with the quadratic equation for the Study quadric ($h_0h_1^- + h_1h_0^- = 0$), so seven quadratic equations in all. Thus, from Bézout's theorem, we might expect a maximum of $2^7 = 128$ solution but it is well known that there are at most 40. Looking at the equations above we see that each of the six 'leg-equations' contains the term $4h_1h_1^-$, now on the 3-plane $h_0 = 0$, or in coordinates $x_0 = x_1 = x_2 = x_3 = 0$, all of these equations reduce to $h_1h_1^- = 0$. That is all of our quadrics contain a 2-dimensional quadric which has nothing to do with the solution of our problem. However, it does mean that the seven quadratic equation do not form a complete intersection and hence simple counting arguments based on the degrees of the equations and Bézout's theorem, will not work.

Progress can be made by subtracting one of the equations above from the five others, this gets rid of the troublesome $h_1h_1^-$ terms. But now the five quadrics and the Study quadric all contain the special 3-plane $h_0 = 0$. Nevertheless, proceeding in this way it is possible to show that there are at most only 40 solutions. This approach is due to Wampler [16], but the result has also been demonstrated by more laborious means.

Finally in this section, note that there is a possible confusion in terminology, these biquaternions are also sometimes referred to as 'double quaternions'. However, these terms are also used to denote the algebra obtained by extending the quaternions with a basis element which squares to 1 rather than 0. The Clifford algebra obtained in this way is the even subalgebra of $C\ell(0,4)$, see lecture 2. This is the Clifford algebra for $SO(4)$, see [1] for example of the use of this algebra in robotics.

5.4 Points, lines, and planes

Above we have looked at the representation of the group of rigid body motions on the points in space $\mathbb{R}^3$. In order to extend these ideas to lines and planes in space it is necessary to use the Clifford algebra $C\ell(0,3,1)$. Now we saw above that the dual quaternions can be thought of as the even subalgebra of this algebra. So we might be tempted to use the same representation for the group and the points as before. This works well for the group, a group element

$$g = x_0 + x_1i + x_2j + x_3k + (y_0 + y_1i + y_2j + y_3k)\varepsilon$$

can simply be rewritten as,

$$g = x_0 + x_1 e_2 e_3 + x_2 e_3 e_1 + x_3 e_1 e_2 + y_0 e e_1 e_2 e_3 + y_1 e_1 e + y_2 e_2 e + y_3 e_3 e\,.$$

The representation of points cannot be treated in the same way, this is because we want the representation of the points to be compatible with the representation of the lines and the planes. It turns out that we can represent points as elements of grade 3 in the algebra. That is, a point with coordinates (p_x, p_y, p_z) will be represented by an algebra element of the form,

$$p = e_1 e_2 e_3 + p_x e_2 e_3 e + p_y e_3 e_1 e + p_z e_1 e_2 e\,.$$

Now it is not too difficult to see that the effect of a rigid transformation on such a point can be represented by the Clifford conjugation,

$$p' = g p g^{-}$$

where g is the Clifford element representing the transformation.

Lines in $\mathbb{R}^3$ can be represented by elements of grade 2. Notice that, these elements have even grade so if we were only interested in lines we could use the biquaternion algebra, that is lines can be represented as dual vectors. A line can be specified by its Plücker coordinates, these are given by a unit vector $\mathbf{v} = (v_x, v_y, v_z)$ in the direction of the line and a moment vector $\mathbf{u} = (u_x, u_y, u_z)$ where $\mathbf{u} = \mathbf{p} \times \mathbf{v}$ and $\mathbf{p}$ is the position vector of any point on the line. The Clifford algebra element corresponding to such a line is given by,

$$\ell = v_x e_2 e_3 + v_y e_3 e_1 + v_z e_1 e_2 + u_x e_1 e + u_y e_2 e + u_z e_3 e\,.$$

Not every grade 2 element of the Clifford algebra corresponds to a line. The form of the Plücker coordinates given above require that $\mathbf{v} \cdot \mathbf{v} = 1$ and $\mathbf{v} \cdot \mathbf{u} = 0$, both these conditions can be summarised as a single relation in the Clifford algebra as,

$$\ell \ell^{-} = 1\,.$$

Once again the action of a rigid transformation on a line is given by the conjugation,

$$\ell' = g \ell g^{-}\,.$$

Notice that these lines are directed lines, ℓ and $-\ell$ correspond to the same line but with opposite directions.

As might be expected by now, planes can be represented by elements of grade 1. Every plane can be specified by a unit normal vector $\mathbf{n} = (n_x, n_y, n_z)$, and a perpendicular distance from the origin d. A typical plane is represented by the algebra element,

$$\pi = n_x e_1 + n_y e_2 + n_z e_3 + d e\,.$$

The requirement that the normal must be a unit vector can be neatly written in the algebra as,

$$\pi \pi^{-} = 1$$

and the action of the group of rigid motions is again given by the conjugation,

$$\pi' = g\pi g^- .$$

The advantage of having all the linear elements represented in the same algebra is that it is now possible to look at the meets and joins of these spaces. We begin by looking at some incidence relations. The condition for a point $\mathbf{p}$ to lie on a line with direction $\mathbf{v}$ and moment $\mathbf{u}$ is simply that the moment vector is given by the cross product of the point and the direction vector, $\mathbf{u} - \mathbf{p} \times \mathbf{v} = 0$. In the Clifford algebra the point is represented by an element,

$$p = e_1e_2e_3 + p_xe_2e_3e + p_ye_3e_1e + p_ze_1e_2e$$

and the line is represented by,

$$\ell = v_xe_2e_3 + v_ye_3e_1 + v_ze_1e_2 + u_xe_1e + u_ye_2e + u_ze_3e .$$

A direct computation shows that,

$$\begin{aligned} p\ell^- + \ell p^- = {} & 2(u_x - p_yv_z + p_zv_y)e_2e_3e + \\ & 2(u_y - p_zv_x + p_xv_z)e_3e_1e + 2(u_z - p_xv_y + p_yv_x)e_2e_1e . \end{aligned}$$

Hence, the Clifford algebra condition for a point p to lie on a line ℓ is simply,

$$p\ell^- + \ell p^- = 0 .$$

For a point p to lie in a plane, we must have that $\mathbf{n} \cdot \mathbf{p} - d = 0$, where $\mathbf{n}$ is the unit normal to the plane and d the perpendicular distance of the plane to the origin. In the Clifford algebra this condition becomes, a point p lies in a plane π if and only if,

$$p\pi^- + \pi p^- = 0 .$$

In a similar manner it is possible to show that the condition for a line ℓ to lie in a plane π is,

$$\ell\pi^- + \pi\ell^- = 0 .$$

These incidence relations can be used to help verify the relations for the meet of a pair of linear subspaces. For example, suppose we have line ℓ and a plane π. Consider the Clifford algebra expression,

$$\rho = \ell\pi^- + \pi\ell^- .$$

Now since π is homogeneous of grade 1 and ℓ is homogeneous of grade 2, ρ can only contain elements of grades 1 or 3. But it is easy to see that $\rho^- = \rho$ and hence we may conclude that ρ is homogeneous of grade 3. The coefficient of $e_1e_2e_3$ in ρ is simply, $-2\mathbf{n} \cdot \mathbf{v}$ where $\mathbf{n}$ is the unit normal vector to the plane and $\mathbf{v}$ is the unit vector in the direction of the line. Hence we see that this coefficient is zero only if the line is parallel to the plane. If we divide ρ by this coefficient we produce a point, this point is the intersection point of the line and the plane. This can be seen by considering the incidence relations, showing that the point lies on both the plane and the line. It is not necessary to consider the constant factor here, the element ρ can be used. So the point lies on the line since,

$$\begin{aligned}\rho\ell^- + \ell\rho^- &= (\ell\pi^- + \pi\ell^-)\ell^- + \ell(\ell\pi^- + \pi\ell^-)^- \\ &= \ell\pi^-\ell^- + \pi\ell^2 + \ell^2\pi^- + \ell\pi\ell^- = 0\,.\end{aligned}$$

Notice that the relations $\ell^- = -\ell$, $\pi^- = -\pi$ and $\ell^2 = -1$, have been used to simplify this.

Also the point lies on the plane because,

$$\begin{aligned}\rho\pi^- + \pi\rho^- &= (\ell\pi^- + \pi\ell^-)\pi^- + \pi(\ell\pi^- + \pi\ell^-)^- \\ &= \ell\pi^2 + \pi\ell^-\pi^- + \pi\ell\pi^- + \pi^2\ell^- = 0\,.\end{aligned}$$

Here the relation $\pi^2 = -1$ has also been used.

In a similar fashion it is possible to show that the line determined by the intersection of a pair of planes π_1 and π_2 is given by finding $\lambda = \pi_1\pi_2^- - \pi_2\pi_1^-$ and then dividing by the "magnitude" of λ, that is, $\lambda/\sqrt{\lambda\lambda^-}$.

These relations can be neatly summarised using the exterior product. In a Clifford algebra this exterior product can be defined by the relation,

$$x \wedge u = \tfrac{1}{2}(xu + (-1)^k ux)$$

where x is a grade 1 element of the algebra and u is an element of grade k. Juxtaposition of symbols on the right of this definition denotes the Clifford product. The exterior product can then be extended to the rest of the algebra by requiring it to be linear and associative, for more details see, [9]. Using the exterior product, the meet of a line ℓ, and a plane π, becomes,

$$p = \ell \wedge \pi / \pm\sqrt{(\ell\wedge\pi)(\ell\wedge\pi)^-}\,.$$

The sign of the square root is chosen so that the coefficient of $e_1e_2e_3$ is $+1$. Similarly the meet of a pair of planes π_1 and π_2 is given by the line,

$$\ell = \pi_1 \wedge \pi_2 / \sqrt{(\pi_1\wedge\pi_2)(\pi_1\wedge\pi_2)^-}\,.$$

Further meet operations are given by this normalised exterior product.

To find the join of a pair of linear elements we need a new idea. One approach is to use the Hodge star, this is a dualising operation which depends on the metric. The operation is linear, so we need only give the mapping of the basis elements in the algebra

$$\begin{array}{rclcrcl}
\star 1 &=& e_1e_2e_3e & & \star e_1e_2e_3e &=& 1 \\
\star e_1 &=& e_2e_3e & & \star e_2e_3e &=& -e_1 \\
\star e_2 &=& e_3e_1e & & \star e_3e_1e &=& -e_2 \\
\star e_3 &=& e_1e_2e & & \star e_1e_2e &=& -e_3 \\
\star e &=& -e_1e_2e_3 & & \star e_1e_2e_3 &=& e \\
\star e_1e_2 &=& e_3e & & \star e_3e &=& e_1e_2 \\
\star e_2e_3 &=& e_1e & & \star e_1e &=& e_2e_3 \\
\star e_3e_1 &=& e_2e & & \star e_2e &=& e_3e_1
\end{array}$$

So, for example, if $p = e_1e_2e_3 + p_xe_2e_3e + p_ye_3e_1e + p_ze_1e_2e$ then $\star p = e - p_xe_1 - p_ye_2 - p_ze_3$. Notice that this operation is not coordinate invariant, a change of basis will change the operation.

In a non-degenerate Clifford algebra this operation can be represented by multiplication by the basis element of highest grade, (the unit pseudoscalar). In our degenerate $C\ell(0,3,1)$, the basis element of highest grade will be $e_1e_2e_3e$ and multiplying this by any other basis element containing an e will annihilate it. So the above mapping approach is forced on us.

Now the join of a pair of linear elements can be found by dualising the elements, finding the meet and then dualising the result. For example the join of a pair of points p_1 and p_2 should be the line joining these two points. Dualising, we get a pair of grade 1 elements, $(\star p_1)$ and $(\star p_2)$. The meet of these elements is the grade 2 element,

$$\lambda = (\star p_1)(\star p_2)^- - (\star p_2)(\star p_1)^- .$$

The line joining the points is the dual of this divided by the magnitude,

$$\ell = (\star\lambda)/\sqrt{(\star\lambda)(\star\lambda)^-} .$$

The alternative approach is to introduce a new product, the shuffle product. This is taken directly from the theory of Grassmann-Cayley algebras, see for example [18]. We will write the shuffle product as $\vee$ and define it on exterior products of grade 1 elements. So for two such elements, $a = a_1 \wedge a_2 \wedge \cdots \wedge a_j$ and $b = b_1 \wedge b_2 \wedge \cdots \wedge b_k$ in a general Clifford algebra, with $j + k \geq n$ the dimension of the algebra, we define,

$$a \vee b = \sum_{\sigma} \operatorname{sign}(\sigma) \det(a_{\sigma(1)}, \dots, a_{\sigma(n-k)}, b_1, \dots, b_k) a_{\sigma(n-k+1)} \wedge \cdots \wedge a_{\sigma(j)} .$$

The sum is taken over all permutations σ of $1, 2, \dots, j$ such that$\sigma(1) < \sigma(2) < \cdots < \sigma(n-k)$ and $\sigma(n-k+1) < \sigma(n-k+2) < \cdots < \sigma(j)$.

Each a_i can be written as a sum of basis elements,

$$a_i = a_{i1}e_1 + a_{i2}e_2 + \cdots + a_{in}e_n .$$

So the determinant in the above definition is the determinant of the matrix whose columns are the coefficients $a_{\sigma(1)i}, a_{\sigma(2)i}, \dots, b_{\sigma(k)i}$. The shuffle product is then extended to the entire Clifford algebra by demanding that it distributes over addition. As an example, consider the shuffle product $e_1e_2e_3 \vee e_2e_3e$. For an orthogonal basis we can confuse the Clifford product of basis elements with the exterior product, so the above formula gives,

$$\begin{aligned} e_1e_2e_3 \vee e_2e_3e = {} & \det(e_1, e_2, e_3, e)e_2e_3 \\ & - \det(e_2, e_2, e_3, e)e_1e_3 + \det(e_3, e_2, e_3, e)e_1e_2 \end{aligned}$$

where we have taken e to be the fourth basis element "e_4". Only the first term on the right hand side of this equation is non-zero since the determinants vanish in the other two terms, and hence we can conclude that,

$$e_1e_2e_3 \vee e_2e_3e = e_2e_3 \,.$$

Using this approach the join of two points is given by the formula,

$$\ell = p_1 \vee p_2 / \sqrt{(p_1 \vee p_2)(p_1 \vee p_2)^-} \,.$$

The two approaches are related by the relation,

$$\star(a \vee b) = (\star a \wedge \star b) \,.$$

An advantage of the Clifford algebra is that we can also discuss Euclidian distances and perpendicularity. For example, the condition for a line ℓ to be perpendicular to a plane π is simply $\ell\pi^- - \pi\ell^- = 0$. This can be demonstrated by a direct computation, assuming,

$$\ell = v_x e_2e_3 + v_y e_3e_1 + v_z e_1e_2 + u_x e_1e + u_y e_2e + u_z e_3e$$

and

$$\pi = n_x e_1 + n_y e_2 + n_z e_3 + de$$

then

$$\begin{aligned} \ell\pi^- - \pi\ell^- = {} & -2(v_y n_z - v_z n_y)e_1 - 2(v_z n_x - v_x n_z)e_2 \\ & - 2(v_x n_y - v_y n_x)e_3 - 2(n_x u_x + n_y u_y + n_z u_z)e \end{aligned}$$

and if $\mathbf{v} = \pm\mathbf{n}$ this clearly vanishes and conversely if this vanishes it can be seen that the unit vectors $\mathbf{v}$ and $\mathbf{n}$ must be either parallel or anti-parallel.

We can also use perpendicularity to construct new linear elements. For instance, suppose we are given a line ℓ and a point p then the plane perpendicular to the line passing through the point is given by,

$$\pi^\perp = \tfrac{1}{2}(p\ell^- - \ell p^-) \,.$$

This can be checked using the condition for perpendicularity with ℓ and the incidence relation with p.

In the next couple of sections these ideas are used to look at some practical problems in robotics and computer vision. Note that there are other approaches to the problem of representing the geometry of points, lines and planes using Clifford algebras, see for example [6] and [2].

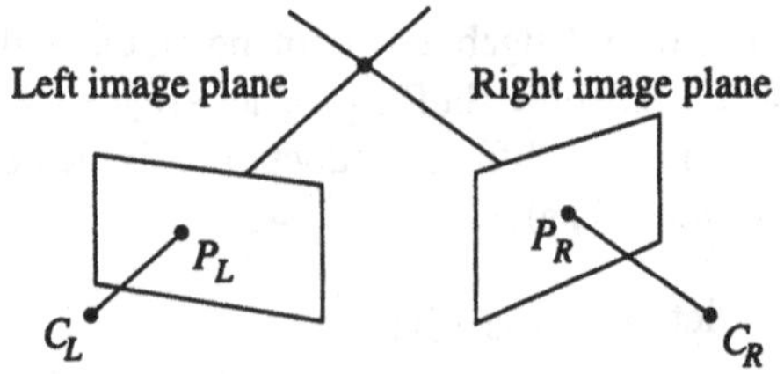

FIGURE 5.3. Two cameras observing a single point

5.5 Computer vision example

A central problem in Computer Vision is the correspondence problem. Suppose we have two or more cameras looking at the same scene. If we can find points in the images which correspond to the same point in the scene then we have solved the correspondence problem. The difficulty of this problem can be reduced by realising that there are geometric constraints on image points which are the projection of the same object point. To keep things simple we will keep to just two cameras, that is stereo vision. Now given an image point in the right hand camera say, we know that the object point could be anywhere along the line joining the image point to the centre of projection in that camera. The image of this line in the left hand camera is called the epipolar line and clearly the corresponding image point in this camera must lie on this line, see figure 5.3. In a general situation we will have four points, c_L, c_R, the centres of projection for the left and right cameras respectively and p_L, p_R which are the image points, that is the points on the right and left image planes respectively. We require a simple algebraic condition to determine whether or not the points are consistent with a single object point. The condition required is simply,

$$p_L \vee c_L \vee c_R \vee p_R = 0\,.$$

To see why this is so consider the following relation between a point p and a plane π,

$$p \vee \pi = d - p_x n_x - p_y n_y - p_z n_z\,.$$

This is the negative of the perpendicular distance between the point and the plane. Hence we can use $p \vee \pi = 0$ as the incidence relation which ensures that the point lies on the plane. In fact we have that,

$$\tfrac{1}{2}(p\pi^- + \pi p^-) = -(p \vee \pi)e_1e_2e_3e\,.$$

Now the relation for four points to lie on a plane can be found from the above relation by assuming that the plane is the mutual join of three of the points. Notice that this relation also caters for the case where the object point is at infinity, the four points will still be coplanar in this situation.

The above relation is very neat but it requires that we know the coordinates of all four points in some global coordinate frame. It is more usual to have the coordinates of the image points in local coordinate frames. Moreover, the image coordinate are usually taken to be homogeneous, that is the image plane is assumed to be a projective plane. So if we know the global coordinates for three non-collinear points in a plane, say p_1, p_2 and p_3 then any other point in the plane can be written as,

$$p = \alpha_1 p_1 + \alpha_2 p_2 + \alpha_3 p_3 .$$

The α_is here are the homogeneous coordinates. The grade 3 element p here will not be a point, in the sense that the coefficient of $e_1e_2e_3$ will not necessarily be 1. However, dividing by this coefficient will not affect the homogeneous coordinates and further it will not affect the condition for the four points to be coplanar since the condition is also homogeneous. Hence if we write the left image point as,

$$p_L = \alpha_1 p_{1L} + \alpha_2 p_{2L} + \alpha_3 p_{3L}$$

and the right image point as,

$$p_R = \beta_1 p_{1R} + \beta_2 p_{2R} + \beta_3 p_{3R}$$

then the consistency condition can be written,

$$(\alpha_1 p_{1L}+\alpha_2 p_{2L}+\alpha_3 p_{3L}) \vee c_L \vee c_R (\beta_1 p_{1R}+\beta_2 p_{2R}+\beta_3 p_{3R}) = \sum_{i,j=1}^{3} \alpha_i F_{ij} \beta_j = 0 .$$

The matrix F_{ij} is known as the fundamental matrix of the system and clearly its elements are given by,

$$F_{ij} = p_{iL} \vee c_L \vee c_R \vee p_{jR}, \qquad i,j = 1,\ 2,\ 3 .$$

These ideas can be extended to the cases where there are three or more cameras, see [13]. There are also many other geometric problems in computer vision which benefit from this sort of algebraic approach. For example, if we assume that we can solve the correspondence problem, then using a single camera we could capture successive images. Using the correspondences we could try to find the rigid motion undergone by the camera between the two images. This is a similar problem to the one studied in section 5.2 above, however this problem doesn't usually have a unique solution.

5.6 Robot kinematics

Kinematics is a key problem in robotics. In section 5.3 above we looked at the forward kinematics for a parallel robot. The inverse kinematics of a serial manipulator will be studied here. The problem here is to determine the joint angles of the robot that will place the end-effector in a prescribed position. For a six joint industrial robot this is a formidable problem and solutions have been found for only a few particular structures.

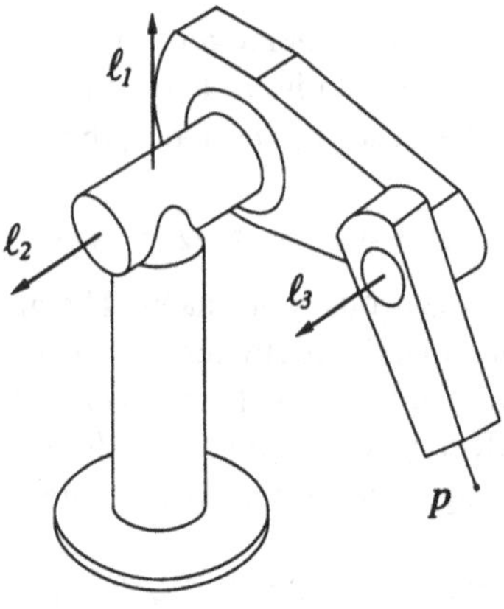

FIGURE 5.4. A three-joint robot

Here we will look at a three joint machine placing a point at a predetermined position, see figure 5.4. In a full six joint robot the orientation of the end-effector as well as its position, would be important.

The joints of the robot will be assumed to be revolute, that is simple hinge joint allowing rotation about an axis. If ℓ is the line determining the axis of the joint then the possible motions allowed by the joint are given by the exponential,

$$e^{\frac{\theta}{2}\ell} = \cos\tfrac{\theta}{2} + \ell \sin\tfrac{\theta}{2} = a(\theta)\,.$$

Recall that $\ell^2 = -1$. We will label these one-parameter subgroups $a(\theta)$, as an allusion to the usual A-matrices of standard robotics.

Now suppose that we have the coordinates of a point on the end-effector. In the home position of the robot, where all the joint angles are zero, this point is represented by the Clifford algebra element p_0. After the robot has moved the position of this point on the end-effector becomes p. This final position can be found from the initial position by rotating about each of the joints in turn. The axes of the joints in the home position will be labelled, ℓ_1, ℓ_2 and ℓ_3 and the corresponding joint angles will be θ_1, θ_2 and θ_3, see figure 5.4. So the final position of the point is given by,

$$p = a_1(\theta_1)a_2(\theta_2)a_3(\theta_3)p_0a_3^-(\theta_3)a_2^-(\theta_2)a_1^-(\theta_1)\,.$$

Notice that we rotate about the distal joints first, that is those furthest from the fixed base of the robot. Now we must solve this equation for the joint angles θ_1, θ_2 and θ_3.

Let us introduce two new points, $p_\alpha = a_1^- p a_1$ and $p_\beta = a_3 p_0 a_3^-$, with these definitions the kinematic equation becomes,

$$p_\alpha = a_2 p_\beta a_2^-\,.$$

The next step is to eliminate θ_2 from this equation to leave equations involving θ_1 and θ_3 only.

To do this, we look at some generalities first. Consider the action of a general rotation, $a(\theta) = \cos(\frac{1}{2}\theta) + \ell\sin(\frac{1}{2}\theta)$ on a general point p. If we write $p = \frac{1}{2}(p - \ell p\ell) + \frac{1}{2}(p + \ell p\ell)$ then using standard trigonometric relations and the fact that $\ell^2 = -1$, we obtain the relation,

$$a(\theta)pa^-(\theta) = \tfrac{1}{2}(p - \ell p\ell) + \tfrac{1}{2}(p + \ell p\ell)\cos\theta + \tfrac{1}{2}(\ell p - p\ell)\sin\theta\,.$$

The terms here have useful interpretations, the point $\frac{1}{2}(p - \ell p\ell)$ is the orthogonal projection of the point p onto the line ℓ. The terms $\frac{1}{2}(p + \ell p\ell)$ and $\frac{1}{2}(\ell p - p\ell)$ are element of grade 3 but are not points, they lack the necessary $e_1e_2e_3$ term. However, they can be thought of as free vectors, under a rigid transformation they transform according to the rotations only, they are unaffected by translations. Moreover, these free vectors can be thought of as the differences between pairs of points. So now the free vector $\frac{1}{2}(p + \ell p\ell)$ is the vector joining p to its orthogonal projection on ℓ. The free vector $\frac{1}{2}(\ell p - p\ell)$, has the same length as the first one but is orthogonal to it and to the line ℓ.

The point $\frac{1}{2}(p - \ell p\ell)$ is clearly independent of the angle of rotation θ. Hence, we can write the following equation between p_α and p_β not involving θ_2,

$$p_\alpha - \ell_2 p_\alpha \ell_2 = p_\beta - \ell_2 p_\beta \ell_2\,.$$

Alternatively, we could use the idea that the planes perpendicular to ℓ_2 and passing through the points p_α and p_β should coincide. In fact this turns out to be a simple rearrangement of the above equation,

$$p_\alpha \ell_2 + \ell_2 p_\alpha = p_\beta \ell_2 + \ell_2 p_\beta\,.$$

Either of these equations will give us just one linear condition for the coordinates of the points p_α and p_β. To see this, note that the we know that the normal to the perpendicular plane, it must be the direction of the line ℓ_2. So the only information that the equation gives us concerns the distance of the plane from the origin.

We can get another equation from the length of the free vector $\frac{1}{2}(p + \ell p\ell)$ or $\frac{1}{2}(\ell p - p\ell)$. This is, just the perpendicular distance from the point to the rotation axis, so clearly independent of the rotation angle. Using the Hodge star, if we let $q_\alpha = \star(\ell_2 p_\alpha - p_\alpha \ell_2)$ and $q_\beta = \star(\ell_2 p_\beta - p_\beta \ell_2)$, then the second equation can be written,

$$q_\alpha q_\alpha^- = q_\beta q_\beta^-\,.$$

This equation is quadratic in the components of p_α and p_β, however, the variables we are really interested in are $\cos\theta_1$, $\sin\theta_1$, $\cos\theta_2$ and $\sin\theta_2$. It is possible to simplify the equation using the first equation $p_\alpha \ell_2 + \ell_2 p_\alpha = p_\beta \ell_2 + \ell_2 p_\beta$, and the trigonometric identity $\cos^2\theta + \sin^2\theta = 1$ to show that the equation above is actually linear in the variables $\cos\theta_1$, $\sin\theta_1$, $\cos\theta_2$ and $\sin\theta_2$.

Hence, we have two linear equations in the four variables, $\cos\theta_1$, $\sin\theta_1$, $\cos\theta_2$ and $\sin\theta_2$. From the trigonometric identities $\cos^2\theta_1 + \sin^2\theta_1 = 1$ and $\cos^2\theta_3 + \sin^2\theta_3 = 1$ we obtain two more equations. The system now reduces to a pencil of conic curves.

Clearly if we were given a particular mechanism, that is, home positions for the lines ℓ_1, ℓ_2, ℓ_3 and coordinates of the point p_0, then it would be a reasonably straightforward matter to give a solution for the joint angles in terms of the coordinates of the point p. However, with this symbolic approach there is some hope that we can address more general problems concerning this robot. For example, how do the singularities of the robot depend on its design parameters? The singularities are configurations where there are less than 4 solutions to the inverse kinematics problems. All these robots will have singularities, in particular at the edge of the robot's workspace, with only revolute joints the robot can't reach points that are a long way from the first joint. The boundary of the set of reachable points will be a set of singularities. But it is the singularities inside the workspace that are more interesting.

Another interesting problem here concerns what are known as cuspidal robots. The different solutions to the inverse kinematics represent different configurations of the robot which place the point at the same position. It has been claimed that for some 3-joint robots it is possible for the robot to move from one of these configurations to another without passing through a singularity. This is clearly a topological property of the map from the 3-torus of joint space, that is the space coordinatised by the joint variables, to the robot's workspace. In particular, the set of regular points in the robot's work space would have to have non-trivial fundamental group.

5.7 Concluding remarks

It would be useful if the methods presented above could be extended to the dynamics of robots and other systems of rigid bodies. To do this the algebra would need to include the coadjoint representation of group of rigid body motions. This is because we would need to use wrenches, these are combinations of forces and torques. Evaluating a wrench on a Lie algebra element or screw should result in a scalar, the work done. This evaluation map needs to be modelled in the algebra too. A possible solution has be proposed by Hestenes [6, 7]. This uses the Clifford algebra $C\ell(4,1)$, that we met in Lecture 2 in connection with the conformal group.

In general it would be helpful if we could write all of robotics using a single algebra. We could then do the modelling, analysis and control in a single formalism, and when we were ready to implement our algorithms it would simple if could write efficient code using the same formalism again. Moreover, if we needed to use information from cameras, so called visual servoing, this could come in the same formalism too. At a lower level here, there are applications of Clifford analysis to image processing, see [3]. Ideally these low level vision tasks could also be integrated into the same algebraic scheme.

So far, the best candidate for such a universal language would appear to be a Clifford algebra. However, there are several competing flavours each with advantages and disadvantages.

Exercises

1. The Frobenius inner product of two 3 rotation matrices, R_1 and R_2 is defined as,

$$< R_1, R_2 >_F = \mathrm{Tr}(R_1^T R_2)$$

where Tr() is the trace of the matrix. Show that if h_1 and h_2 are unit quaternions corresponding to the rotation matrices then,

$$\mathrm{Tr}(R_1^T R_2) = 4\,\mathrm{Re}(h_1^- h_2)^2 - 1$$

where Re() denotes the real part of the quaternion.

2. Biquaternions can be used to represent lines in space as follows. A line with direction $\mathbf{v} = (v_x, v_y, v_z)^T$ and moment $\mathbf{u} = (u_x, u_y, u_z)^T$ is represented by a biquaternion of the form,

$$\check{\ell} = (v_x i + v_y j + v_z k) + \varepsilon(u_x i + u_y j + u_z k)\,.$$

Given two lines $\check{\ell}_1$ and $\check{\ell}_2$, investigate the geometrical significance of the biquaternion expressions, $\check{\ell}_1\check{\ell}_2^- + \check{\ell}_2\check{\ell}_1^-$ and $\check{\ell}_1\check{\ell}_2^- - \check{\ell}_2\check{\ell}_1^-$.

3. Using the Clifford algebra $C\ell(0,3,1)$ as above, find expressions for the following,

   (i) the condition for two lines to be co-planar.

   (ii) the square of the perpendicular distance between a point and a line.

   (iii) the line along the common perpendicular to a pair of lines.

   (iv) the condition for two planes to be mutually perpendicular.

4. Extend the analysis in section 5.5 above to the case where there are three cameras, hence obtain an expression for the trifocal tensor.

5.8 REFERENCES

[1] S.G. Ahlers and J.M. McCarthy. The Clifford algebra and the optimization of robot design. In E. Bayro Corrochano and G. Sobczyk (eds.) *Geometric Algebra with Applications in Science and Engineering*, chap. 12, Birkhäuser, Boston, 2001.

[2] E. Bayro Corrochano and G. Sobczyk. Applications of Lie algebra and the algebra of incidence. In E. Bayro Corrochano and G. Sobczyk (eds.) *Geometric Algebra with Applications in Science and Engineering*, chap. 13, Birkhäuser, Boston, 2001.

[3] T. Bülow and G. Sommer Local hypercomplex signal representations and applications. In G. Sommer (ed.) *Geometric Computing with Clifford Algebras*, chap. 11, Springer, Berlin, 2001.

[4] W.K. Clifford, Preliminary sketch of the biquaternions. *Proceedings of the London Mathematical Society* iv(64/65):381–395, 1873.

[5] C. Ericson. Sony Computer Entertainment, Santa Monica. Personal communication. 2001.

[6] D. Hestenes. Old wine in new bottles: a new algebraic framework for computational geometry. In E. Bayro Corrochano and G. Sobczyk (eds.) *Geometric Algebra with Applications in Science and Engineering*, chap. 1, Birkhäuser, Boston, 2001.

[7] D. Hestenes and E.D. Fasse. Homogeneous Rigid Body Mechanics with Elastic Coupling, in *Applications of Geometric Algebra in Computer Science and Engineering*, eds. L. Dorst, C. Doran and J. Lasenby, Birkhäuser, Boston 2002, pp. 197–212.

[8] B.K.P Horn. Closed-form solution of absolute orientation using unit quaternions. *Journal of the Optical Society of America*, A-**4**:629–642, 1987.

[9] P. Lounesto. Marcel Riesz's work on Clifford algebra. in *Clifford Numbers and Spinors* by M. Riesz, edited by E. F. Bolinder and P. Lounesto, Kluwer, Dordrecht, 1993.

[10] J.K. MacKenzie. The estimation of an orientation relationship. *Acta. Cryst.*, **10**:62–62, 1957.

[11] P.A.P. Moran. Quaternions, Harr measure and estimation of paleomagnetic rotation. In J. Gani (ed.), *Perspectives in Prob. and Stats.*, 295–301, Applied Probability Trust, Sheffield,,1975.

[12] L. Meister. Quaternion optimization problems in engineering. In E. Bayro Corrochano and G. Sobczyk (eds.) *Geometric Algebra with Applications in Science and Engineering*, chap. 19, Birkhäuser, Boston, 2001.

[13] C.B.U. Perwass and J. Lasenby. A unified description of multiple view geometry. In G. Sommer (ed.) *Geometric Computing with Clifford Algebras*, chap. 14, Springer, Berlin, 2001.

[14] G. Sobczyk. Universal geometric algebra. In E. Bayro Corrochano and G. Sobczyk (eds.) *Geometric Algebra with Applications in Science and Engineering*, chap. 2, Birkhäuser, Boston, 2001.

[15] G. Wahba. Section on problems and solutions: A least squares estimate of satellite attitude. *SIAM rev*, **8**:384–385, 1966.

[16] C. Wampler. Forward displacement analysis of general six-in-parallel SPS (Stewart) platform manipulators using soma coordinates. *General Motors Technical Report R&D*–8179, May 1994.

[17] G.S. Watson. Statistics of Rotations, In *Lecture Notes in Mathematics*, Springer Verlag, **1379**:398–413, 1989.

[18] N. White. Grassmann-Cayley algebra and robotics. *J. Intell. Robot Syst.*, **11**:91–107, 1994.

Jon Selig,
Faculty of Business, Computing and Information Management
London South Bank University
Borough Rd.
London SE1 0AA, U.K.
E-mail: seligjm@sbu.ac.uk

Received: January 31, 2002; Revised: March 15, 2003.

6

Clifford Bundles and Clifford Algebras

Thomas Branson

ABSTRACT In view of General Relativity, it is necessary to study physical fields, including solutions of the Dirac equation, in curved spacetimes. It is generally believed that the study of Riemannian (positive definite) metrics (infinitesimal distance functions) will ultimately be relevant to the more directly physical problem of Lorentz signature metrics, via principles of analytic continuation in signature. This adds impetus to the natural mathematical pursuit of studying spin structure, the Dirac operator, and other related operators on Riemannian manifolds. This lecture is a biased attempt at an introduction to this subject, with an emphasis on fundamental ideas likely to be important in future work, for example, Stein–Weiss gradients, Bochner–Weitzenböck formulas, the Hijazi inequality, and the Penrose local twistor idea. This provides at least a framework for the study of advanced topics such as spectral invariants and conformal anomalies, which are not treated here.

6.1 Spin geometry

The starting point for the study of the spinor bundle and the Dirac operator on manifolds is the realization that, like differential forms and other species of tensors that people are more familiar with, spinors are an *associated bundle*. This means the following. The *structure group* of Riemannian geometry is $O(n)$. Roughly speaking, this means that there is a copy of $O(n)$ acting on each tangent space. In fact, these groups and actions fit together into a *principal bundle* [32].

6.1.1 Associated bundles

With *oriented* Riemannian geometry (in which we can give ordered local orthonormal frames, consistent up to even permutation), the structure group reduces

---

This lecture was presented at "Lecture Series on Clifford Algebras and their Applications", May 18 and 19, 2002, as part of the 6th International Conference on Clifford Algebras and their Applications in Mathematical Physics, Cookeville, TN, May 20–25, 2002.

AMS Subject Classification: 53C27, 53A30.

Keywords: Spinor, Dirac operator, Bochner–Weitzenböck formula, Hijazi inequality, local twistors, spannors, plyors, tractors, Stein–Weiss gradients.

to $SO(n)$. It's natural to ask what we get when we now go to the simply connected double cover $\mathbf{Spin}(n)$ of $SO(n)$, and the answer, in brief is spin geometry.

In fact, we should really be speaking of covers of $SO(p,q)$, in order to cover pseudo-Riemannian spin geometry of all signatures. For the most part, we shall ignore this here, however. It turns out that in spin geometry, one has to take a little more care in changing signature than in most of (tensorial) pseudo-Riemannian geometry.

An *associated bundle* is made from two ingredients: a principal bundle $\mathcal{F}$ with structure group H on a manifold M, and a representation (λ, V) of H. The associated bundle is a vector bundle $\mathbf{V} = \mathcal{F} \times_\lambda V$, whose section space is fairly easy to describe: the sections may be naturally identified with smooth functions $\varphi : M \to V$ with the special property that

$$\varphi(x \cdot h) = \lambda(h)^{-1} f(x), \quad x \in M, \ h \in H.$$

With this, we can get at least some understanding of vector bundles by a kind of mental shorthand: the tangent bundle "is" the defining representation of $SO(n)$; the two-form bundle $\Lambda^2 M$ "is" just the alternating two-tensor representation, and so on. Even without having the details of the associated bundle construction near at hand, one can believe in them and do computations based on their fundamental properties. That is, we use the associated bundle construction to promote linear algebra to vector bundle theory.

In fact, much of this thinking and calculating can be done on the level of the Lie algebra $\mathfrak{h}$, as opposed to the group H. Part of the genesis of spinors was the realization that, when one examines the irreducible, finite-dimensional representations of the classical Lie algebra $\mathfrak{so}(n,\mathbb{C})$ using the usual classification by dominant integral weights, one gets more than just the irreducible constituents of the tensor algebra over the defining representation. The new wrinkle is the *spin representation* Σ, which is irreducible for odd n, and breaks into two irreducible summands $\Sigma_\pm$ when n is even. Now everything can be realized in the tensor algebra over Σ; for example, when n is even, $\Sigma \otimes \Sigma \cong_{\mathfrak{so}(n)} \Lambda$ (the right side being alternating tensors of all orders), so in particular we recover all tensors, since these are generated by Λ^1. The representation Σ can be integrated to $\mathbf{Spin}(n)$, but not to $SO(n)$.

Something is being swept under the rug in the above; however, there is a topological obstruction to having a principal $\mathbf{Spin}(n)$ bundle over a given manifold M, analogous to the orientability obstruction that governs reduction of $O(n)$ structure to $SO(n)$ structure. Roughly speaking, granting orientability, the spin obstruction has to do with having local orthonormal frames for the tangent bundle which are split into pairs: (X_1, X_2), (X_3, X_4), ... We need to be able to arrange the overlaps so that pairs go to pairs (though not necessarily pairs with corresponding subscripts). From now on, we shall assume that this condition is satisfied, so that our manifolds are *spinnable*. An additional warning to heed is that there are generally many different spin structures — i.e., choices of a principal $\mathbf{Spin}(n)$ bundle covering the $SO(n)$ bundle of orthonormal frames. For the most part, statements we make will be true for any choice of the spin structure.

In any case, the old $SO(n)$ bundles are still with us as we do spin geometry; they are just the ones associated to representations $\lambda : \mathbf{Spin}(n) \to V$ which factor through $SO(n)$ via the covering homomorphism $\mathbf{Spin}(n) \to SO(n)$.

Where does Clifford algebra come in? Computing some elementary tensor products reveals a rich structure. First, we find that

$$\Lambda^1 \otimes \Sigma \cong_{\mathbf{Spin}(n)} \Sigma \oplus \mathbf{Tw},$$

where Σ and $\mathbf{Tw}$ have *no joint content*—i.e., no isomorphic $\mathbf{Spin}(n)$ summands. This means there is a nonzero $\mathbf{Spin}(n)$-invariant map – the projection

$$\Lambda^1 \otimes \Sigma \overset{\cong}{\to} \Sigma \oplus \mathbf{Tw} \overset{\mathrm{Proj}}{\to} \Sigma.$$

We may view this, by the usual trick of identifying $\mathrm{End}(V)$ with $V \otimes V^*$, as an invariant map

$$\begin{array}{ll} \Lambda^1 & \to \Sigma \otimes \Sigma^* \cong_{\mathbf{Spin}(n)} \mathrm{End}(\Sigma), \\ \eta & \qquad \mapsto \gamma(\eta). \end{array}$$

This brings *Clifford multiplication* into view. In fact, γ is a section of the vector bundle $T \otimes \mathrm{End}(\Sigma)$, and is called the *Clifford section*.

6.1.2 The Clifford relation

The famous *Clifford relation* is forced upon us as follows. Symmetrizing two Clifford multiplications,

$$\tfrac{1}{2}(\eta \otimes \alpha + \alpha \otimes \eta) \mapsto \tfrac{1}{2}(\gamma(\eta)\gamma(\alpha) + \gamma(\alpha)\gamma(\eta))$$

gives a well-defined invariant map from the *symmetric 2-tensors* $\odot^2 T^*$ to $\mathrm{End}(\Sigma)$. (The reader should check this.)

By the reverse of the trick above, this gives $\mathbf{Spin}(n)$-invariant maps

$$\odot^2 T^* \to \Sigma \otimes \Sigma^*, \qquad \odot^2 T^* \otimes \Sigma \to \Sigma.$$

But there is only one such thing (up to a constant factor), since

$$\odot^2 T^* \otimes \Sigma \cong_{\mathbf{Spin}(n)} \Sigma \oplus \mathcal{M},$$

where $\mathcal{M}$ has no joint content with Σ. This is a fairly special property of Σ; there are few modules with this property.

Furthermore, $\odot^2 T^*$ splits as $\mathbb{R}g \oplus \mathrm{TFS}^2$, where g is the metric (or inner product), and TFS^2 are the *trace-free* symmetric 2-tensors. Our map above must emanate from just one of these; otherwise we could restrict and contradict uniqueness. But

$$\mathbb{C}g \cong_{\mathbf{Spin}(n)} \Lambda^0$$

is the trivial representation, and of course $\Lambda^0 \otimes \Sigma = \Sigma$. So the map we are looking at must be

$$\tfrac{1}{2}(\gamma(\eta)\gamma(\alpha) + \gamma(\alpha)\gamma(\eta))\psi = c g^\sharp(\eta, \alpha)\psi$$

for some constant c, where $g^\sharp$ is the metric on T^*M determined by the metric g on TM. By renormalizing γ, we recover our favorite normalization for the Clifford relation; for example, $c = -2$.

In *abstract index notation* [38], we have $\gamma(\eta) = \gamma^a \eta_a$. The above argument may be rephrased as

$$\gamma^a\gamma^b + \gamma^b\gamma^a = \underbrace{\left(\gamma^a\gamma^b + \gamma^b\gamma^a - \frac{2}{n}\gamma^c\gamma_c g^{ab}\right)}_{\mathrm{TFS}^2 \to \mathrm{End}(\Sigma),\ \therefore\ 0} + \frac{2}{n}\underbrace{\gamma^c\gamma_c}_{-n} g^{ab}.$$

The value $-n$ in the last underbrace is a consequence of our choice of normalization just above.

The Clifford section γ now plays a role analogous to that of the metric tensor g; it is a section which encodes information about the geometry, allowing us to compute without always referring to the Lie theoretic constructs that gave birth to the theory. This is underlined by the fact that we may now describe the even-dimensional splitting $\Sigma = \Sigma_+ \oplus \Sigma_-$ using the Clifford section. We take the volume form $E_{i_1 \dots i_n}$ and contract with the tensor argument on the gammas to form a *chirality operator*

$$\tilde{\chi} := \frac{1}{n!} E_{i_1 \dots i_n} \gamma^{i_1} \cdots \gamma^{i_n}.$$

One computes that $\tilde{\chi}^2 = (-1)^{\left[\frac{n+1}{2}\right]}$. If n is odd, $\tilde{\chi} = \mathrm{const} \cdot \mathrm{Id}_\Sigma$ by irreducibility (Schur's Lemma). If n is even, there are two eigenmodules for $\tilde{\chi}$, and

$$\chi := \mathrm{i}^{n/2}\tilde{\chi} = \begin{pmatrix} 1 & 0 \\ 0 & -1 \end{pmatrix}$$

in the corresponding block decomposition. This gives exactly the $\Sigma_+ \oplus \Sigma_-$ splitting. We could also have chosen $(-\mathrm{i})^{n/2}\tilde{\chi}$ as our definition of χ, but the choice above turns out to make better contact with weight theoretic considerations for the Lie algebra $\mathfrak{so}(n)$.

6.1.3 Bundles with Spin(n) *structure*

The spinor bundle(s) only generate the additional **Spin**(n) bundles that are "out there". One can capture these additional bundles in a few different ways. First, one could take bundles associated to other irreducible **Spin**(n) modules; there are infinitely many of these which are not already $SO(n)$ modules. Second, we could take iterated tensor products of Σ, and pick out the irreducible) summands. (We get infinitely many duplicate copies of each module in this way.) Or, finally, we could take the transition functions relating trivializations of the bundle Σ; these will be valued in **Spin**(n). Now compose these transition functions with various representations of **Spin**(n) to get new sets of transition functions:

$$\mathcal{O}_\alpha \cap \mathcal{O}_\beta \xrightarrow{\tau_{\beta\alpha}} \mathbf{Spin}(n) \xrightarrow{\lambda} \mathrm{Aut}(V).$$

Here we are relying on the fact that (σ, Σ) is *faithful* to be assured that we are getting everything.

These three processes are really equivalent. For an example of approach **(2)**, take

$$\Lambda^1 \otimes \Sigma \cong_{\mathbf{Spin}(n)} \Sigma \oplus \underbrace{\mathbf{Tw}}_{\text{new}} . \tag{1.1}$$

This allows us to discover the *twistor bundle* $\mathbf{Tw}$. It also suggests a tensor-spinor *realization* of this bundle. First take spinor-1-forms φ_a, where the index indicates the Λ^1 content. From a general such thing, we may make a spinor $\gamma^a \varphi_a$; this corresponds to the Σ summand in (1.1).

It also hints that we can get our hands on the other summand by writing a counter term which will assure that this derived object vanishes:

$$\varphi_a = \underbrace{\varphi_a + \frac{1}{n}\gamma_a \gamma^b \psi_b}_{\mathbf{Tw}\text{ part}} \underbrace{- \frac{1}{n}\gamma_a \gamma^b \varphi_b}_{\Sigma \text{ part}} . \tag{1.2}$$

Thus *twistors* are spinor-1-forms φ_a with $\gamma^a \varphi_a = 0$. Like the spinors, these are irreducible for odd n, and split into two pieces (another chiral decomposition) when n is even.

There are some obvious generalizations of this construction, some of which are under intense scrutiny in current research. For example, one could start with the spinor-k-forms $\Sigma \otimes \Lambda^k$ as in [7]. The generalization of the map used just above is the *interior Clifford multiplication*

$$\begin{aligned} \operatorname{int}(\gamma) : \Sigma \otimes \Lambda^k &\to \Sigma \otimes \Lambda^{k-1}, \\ \varphi_{a_1 \ldots a_k} &\mapsto \gamma^{a_1} \varphi_{a_1 \ldots a_k} . \end{aligned} \tag{1.3}$$

Since it is possible for powers of $\operatorname{int}(\gamma)$ to annihilate a spinor-form without $\operatorname{int}(\gamma)$ itself annihilating, $\Sigma \otimes \Lambda^k$ will generally break up into several irreducible pieces [22]. We shall denote the "top piece," i.e., the null bundle for $\operatorname{int}(\gamma)$, by $\mathbf{Y}_k$. Alternatively, we could replace Λ^k above with the symmetric tensors $\odot^k$, and define $\operatorname{int}(\gamma)$ in formally the same way. The null bundle $\mathbf{Z}_k$ for $\operatorname{int}(\gamma)$ will automatically consist of spinor-*trace-free*-symmetric k-tensors, as

$$\gamma^{a_1} \varphi_{a_1 \ldots a_k} = 0 \Rightarrow 0 = \gamma^{a_2} \gamma^{a_1} \varphi_{a_1 \ldots a_k} = -g^{a_1 a_2} \varphi_{a_1 \ldots a_k} , \tag{1.4}$$

by symmetry and the Clifford relation. By symmetry again, if the $a_1 a_2$ trace vanishes, all traces must vanish.

6.1.4 Connections

It is sometimes said that a differential geometer is someone who knows what a connection is. In spin geometry, as in other branches of geometry, an understanding of connections is indeed a kind of gateway to the theory.

The Levi-Civita connection ∇ on the orthonormal frame bundle $\mathcal{F}$ (see, e.g., [26]) lifts to a connection on the bundle S of spin frames. Transferring to associated bundles, or proliferating the connection to duals and tensor products, we get a connection ∇ on all $\mathbf{Spin}(n)$ bundles, satisfying

$$\nabla g = 0, \;\; \nabla E = 0, \;\; \nabla\gamma = 0.$$

On the spinor bundle, there is a family of so-called *Friedrich connections* [25]

$$\nabla_a + \text{const} \cdot \gamma_a \tag{1.5}$$

which are arguably just as distinguished as the spin connection, and which are quite useful in problems involving Dirac eigenspinors.

Exercise 1. *Show that among the Friedrich connections, only the spin connection annihilates the Clifford section* γ.

6.1.5 Stein–Weiss gradients

There is a universal construction of first-order differential operators in Riemannian spin geometry due to Stein and Weiss [39]. First note that if $\mathbf{V}$ is a natural $\mathbf{Spin}(n)$ bundle, then

$$\nabla : \mathbf{V} \to T^*M \otimes \mathbf{V}.$$

One way to think of this is as follows: if a section φ of $\mathbf{V}$ has index structure φ_I, then $\nabla\varphi$ has index structure $\nabla_a\varphi_I$. That is, an extra cotangent index has been added, signaling tensor product with the cotangent bundle.

Now as it happens, the defining representation T of $\mathfrak{so}(n)$ has a multiplicity free *selection rule* governing its tensor products with other $\mathfrak{so}(n)$-modules. This means that if V is irreducible, then $T \otimes V$ breaks up as $W_1 \oplus \cdots \oplus W_N$, for some N, with the various W_u coming from distinct isomorphism classes. On the bundle level,

$$\begin{aligned} &T^*M \otimes \mathbf{V} \cong_{\mathbf{Spin}(n)} \mathbf{W}_1 \oplus \cdots \oplus \mathbf{W}_{N(\mathbf{V})}\,, \\ &\mathbf{W}_u \cong_{\mathbf{Spin}(n)} \mathbf{W}_v \;\Rightarrow u = v. \end{aligned} \tag{1.6}$$

Here we write $N = N(\mathbf{V})$ to highlight the fact that N depends on $\mathbf{V}$ — in fact it is a very useful numerical invariant.

This allows us to speak of projections

$$\text{Proj}_u := \text{Proj}_{\mathbf{W}_u}$$

on the bundle $T^* \otimes \mathbf{V}$, and thus of *gradients*

$$G_u = \text{Proj}_{\mathbf{W}_u} \circ \nabla.$$

These are universal, $\mathbf{Spin}(n)$-equivariant operators, since ∇ and the $\text{Proj}_{\mathbf{W}_u}$ are. These operators are unique up to isomorphic realization of bundles and up to constant factors; because of this, they actually give a *classification* of equivariant first-order operators.

For example, take the spinor bundle Σ, in odd dimensions for now. The decomposition (1.2) exactly realizes the decomposition in (1.6), and the corresponding gradients are

$$\begin{aligned} \mathbf{T} = G_{\mathbf{Tw}} : \psi &\mapsto \nabla_a \psi + \tfrac{1}{n}\gamma_a \gamma^b \nabla_b \psi, \\ G_\Sigma : \psi &\mapsto -\tfrac{1}{n}\gamma_a \gamma^b \nabla_b \psi. \end{aligned} \tag{1.7}$$

In the formula for G_Σ, we are just finding a different realization of the operator

$$\nabla\!\!\!/ : \psi \mapsto \gamma^b \nabla_b \psi;$$

i.e., of the *Dirac operator*. (In fact, to get to the standard realization, just left-multiply the formula for G_Σ by γ^a.) $\mathbf{T}$ is called the *twistor operator*. In this sense $\nabla\!\!\!/$ and $\mathbf{T}$ are the two gradients emanating from the spinor bundle, and in particular $N(\Sigma) = 2$.

In even dimensions, we have the chiral decomposition, so we get two Dirac operators and two twistor operators:

$$\nabla\!\!\!/ : \Sigma_\pm \to \Sigma_\mp\,, \quad \mathbf{T} : \Sigma_\pm \to \mathbf{Tw}_\pm\,.$$

That $\nabla\!\!\!/$ should reverse chirality is clear from the observation that ∇_b (resp. γ^b) should preserve (resp. reverse) chirality.

Other examples of gradients, or (when reducible bundles are involved) operators with a block decomposition whose entries are gradients, are the following. The *exterior derivative* on differential forms is

$$d : \Lambda^k \to \Lambda^{k+1}, \quad (d\varphi)_{a_0 a_1 \ldots a_k} = (k+1)\nabla_{[a_0}\varphi_{a_1 \ldots a_k]}\,. \tag{1.8}$$

Thus the usual normalization of d is $k+1$ times the gradient from Λ^k to Λ^{k+1}. Note that in abstract index notation, an expression like $\nabla_a \varphi_b$ means $(\nabla\varphi)_{ab}$. The *conformal Killing operator* S carrying Λ^1 to the trace-free symmetric 2-tensors TFS^2 is

$$(S\eta)_{ab} = \nabla_{(a}\eta_{b)} - \frac{1}{n} g_{ab} \nabla_c \eta^c\,.$$

As usual, the parentheses around the indices indicate symmetrization. The name of this operator derives from the fact that its null space (viewed as a space of vector fields) consists exactly of the *conformal vector fields* X; i.e., those vector fields for which

$$\mathcal{L}_X g = 2\omega g \tag{1.9}$$

for some smooth function ω, where $\mathcal{L}$ is the Lie derivative. In index notation, (1.9) translates to $\nabla_{(a}X_{b)} = \omega g_{ab}$, and contracting, we find that $\omega = \frac{1}{n}\nabla_c X^c$.

By the basic uniqueness result, the formal adjoint G^* of a gradient G must be a gradient. If G is realized as above, i.e., as going from $\mathbf{V}$ to a subbundle of $T^*M \otimes \mathbf{V}$, then G^* is always a divergence in the additional T^*M tensor argument:

$$G^* = (\mathrm{Proj}_u \nabla)^* = \nabla^* \mathrm{Proj}_u^* = \nabla^* \mathrm{Inj}_u\,,$$

where Inj_u injects the copy of $\mathbf{W}_u$ into $T^*M \otimes \mathbf{V}$. For example, we have the interior derivative $\delta : \Lambda^{k+1} \to \Lambda^k$, defined by

$$(\delta\varphi)_{a_1 \dots a_k} = -\nabla^{a_0}\varphi_{a_0 \dots a_k} .$$

Exercise 2. *The alert reader will notice an apparent normalization problem when comparing with (1.8). This is because δ is conventionally computed in the form of inner product $(k!)^{-1}\varphi^{a_1 \dots a_k}\psi_{a_1 \dots a_k}$, whereas the above general discussion uses an inner product on $T^*M \otimes \mathbf{V}$ with the property that $|\xi \otimes \varphi|^2 = |\xi|^2|\varphi|^2$. This "default" inner product, when applied to building differential form inner products, leads to an inner product without the $(k!)^{-1}$ formula above. Show that d and δ are formal adjoints in the form inner product, and that the gradient $(k+1)^{-1}d$ and δ are formal adjoints in the default inner product.*

The above implies (as the reader may check explicitly) that

$$(S^*\tau)_b = -\nabla^a\tau_{ab}, \quad (\mathbf{T}\varphi)^* = -\nabla^a\varphi_a \tag{1.10}$$

are the formal adjoints of the conformal Killing and twistor operators, respectively.

Exercise 3. *Verify equations (1.10).*

Note that when speaking of formal adjoints, a statement like $\nabla\!\!\!/^* = \nabla\!\!\!/$, while true, hides some subtleties. First, recall from above that $\nabla\!\!\!/$ exchanges the two chiralities:

$$\Sigma_+ \underset{\nabla\!\!\!/_-}{\overset{\nabla\!\!\!/_+}{\rightleftarrows}} \Sigma_- .$$

In dimension $4k$, each of Σ_+ and Σ_- is self-dual, and $\nabla\!\!\!/_+$ and $\nabla\!\!\!/_-$ are formal adjoints of each other. In dimension $4k+2$, Σ_+ and Σ_- are duals, so each of $\nabla\!\!\!/_+$ and $\nabla\!\!\!/_-$ is formally self-adjoint.

6.2 Conformal structure

The Dirac operator is deeply related to *conformal structures* on manifolds. A conformal structure is an equivalence class of Riemannian metrics, with g equivalent (or *conformal*) to $\hat{g}$ iff $\hat{g} = \Omega^2 g$ for some positive smooth function Ω. In fact, not just the Dirac operator, but *every* gradient depends only on conformal structure, in a suitable sense.

6.2.1 Conformal covariance and weights

Let g and $\hat{g} = \Omega^2 g$ be two conformal metrics. Since the volume element and Clifford section have the metric-related properties ($|E|^2 = 1$ and the Clifford relation), they need to be given a compatible pointwise scaling:

$$\hat{E} = \Omega^n E, \quad \hat{\gamma} = \Omega^{-1}\gamma.$$

Theorem 1. [24] *Each gradient* $G : \mathbf{V} \mapsto \mathbf{W}$ *is* conformally covariant*: there are numbers* a, b *with*

$$\hat{G}\varphi = \Omega^{-b} G(\Omega^{a} \varphi)$$

for all sections φ *of* $\mathbf{V}$.

The *conformal biweight* (a, b) moves around depending on what realizations of the bundles we use. For example, viewing the conformal Killing operator S as acting on tangent, as opposed to cotangent, vector fields results in different (a, b). To understand and control this, we need to take a broader view: there's really a structure group larger than $\mathbf{Spin}(n)$ at work in statements like these, namely $\mathbb{R}_{+} \times \mathbf{Spin}(n)$. The $\mathbb{R}_{+}$ part has something to do with the pointwise scaling. To make precise versions of these statements, we need to know something about *weights* of modules for (some) Lie algebras.

Irreducible $\mathbf{Spin}(n)$ modules (thus bundles on spinnable manifolds) are parameterized by the *dominant* $\mathfrak{so}(n)$ *weights*

$$\lambda = (\lambda_1, \ldots, \lambda_\ell) \in \mathbf{Z}^{\ell} \cup \left(\tfrac{1}{2} + \mathbf{Z}\right)^{\ell},$$

where $\ell = [n/2]$. The dominance condition is

$$\begin{array}{ll} \lambda_1 \geq \cdots \geq |\lambda_\ell|, & n \text{ even}, \\ \lambda_1 \geq \cdots \geq \lambda_\ell \geq 0, & n \text{ odd}. \end{array}$$

λ is the *highest weight* of the module, which we may call $V(\lambda)$. We shall call the corresponding bundle $\mathbf{V}(\lambda)$. In general, *weights* parameterize the possible irreducible representations of a (fixed choice of) maximal abelian subalgebra $\mathfrak{t}$. Such irreducible representations are necessarily one-dimensional, and are given by

$$m^1 X_1 + \cdots + m^{\ell} X_{\ell} \mapsto (\mu_1 m^1 + \cdots + \mu_\ell m^{\ell})\mathrm{Id}.$$

The ℓ-tuple μ is the *weight* of the representation.

In the case of $\mathfrak{so}(n)$, this plays out as follows [39, 40]. In the usual matrix presentation, $\mathfrak{so}(n)$ is the Lie algebra of skew-symmetric $n \times n$ matrices. A basis for this consists of the L_{ab} for $a < b$, where L_{ab} has a 1 in the (a, b) entry, and a -1 in the (b, a) entry. A good choice for $\mathfrak{t}$ is the span of $L_{12}, L_{34}, \ldots, L_{2\ell-1,2\ell}$.

Each irreducible $\mathfrak{so}(n)$ module has a lexicographically maximal weight, which is necessarily dominant, and necessarily occurs with multiplicity 1; this is the highest weight. Some Lie algebra theory forces *integrality* conditions on the weights, which, in the present case, work out to the $\mathbf{Z}^{\ell} \cup \left(\tfrac{1}{2} + \mathbf{Z}\right)^{\ell}$ condition above.

Some familiar modules/bundles and their highest weights are:

$\Lambda^k,\ k<\ell$	$(\overbrace{1,\dots,1}^{k},0,\dots,0)$
$\Lambda^\ell_\pm$ (n even)	$(1,\dots,1,\pm 1)$
TFS^p	$(p,0,\dots,0)$
Σ (n odd)	$(\frac{1}{2},\dots,\frac{1}{2})$
$\Sigma_\pm$ (n even)	$(\frac{1}{2},\dots,\frac{1}{2},\pm\frac{1}{2})$
$\mathbf{Tw}$ (n odd)	$(\frac{3}{2},\frac{1}{2},\dots,\frac{1}{2})$
$\mathbf{Tw}_\pm$ ($4\le n$ even)	$(\frac{3}{2},\frac{1}{2},\dots,\frac{1}{2},\pm\frac{1}{2})$
$\mathbf{Tw}_\pm$ ($n=2$)	$(\pm\frac{3}{2})$
algebraic Weyl tensors ($n\ge 5$)	$(2,2,0,\dots,0)$
algebraic Weyl$_\pm$ tensors ($n=4$)	$(2,\pm 2)$

Recall there is a gradient $\mathbf{V}(\lambda)\to\mathbf{V}(\mu)$ when $T^*M\otimes\mathbf{V}(\lambda)$ has a $\mathbf{V}(\mu)$ summand, i.e., if

$$(1)\otimes\lambda\cong_{\mathbf{Spin}(n)}\mu\oplus\cdots$$

on the level of representations. This happens iff

$$\begin{gathered}\mu=\lambda\pm e_a,\quad\text{some } a\in\{1,\dots,\ell\},\\ \underline{\textbf{or}}\\ n\text{ is odd},\ \lambda_\ell\neq 0,\ \mu=\lambda.\end{gathered}\tag{2.11}$$

Some important gradients in odd dimensions are the *self-gradients* — operators associated to the exceptional case in (2.11). Among these are the Dirac operator, the *Rarita–Schwinger operator* S on $\mathbf{Tw}$ (see below), and the operator $\star d$ on Λ^ℓ. In this last example, $\star$ is the *Hodge star operator* taking k-forms to $(n-k)$-forms:

$$(\star\varphi)_{a_{k+1}\dots a_n}:=E_{a_1\dots a_n}\varphi^{a_1\dots a_k}.$$

In even dimensions, there will be gradients

$$\mathbf{V}(\sigma,\pm\tfrac{1}{2})\to\mathbf{V}(\sigma,\mp\tfrac{1}{2})$$

whenever σ is an $(\ell-1)$-tuple with $(\sigma,\frac{1}{2})$ dominant and integral. By the above table of weights, the Dirac operators will be among these. It is then natural to speak of the Dirac *operator* on $\mathbf{V}(\frac{1}{2},\dots,\frac{1}{2})\oplus\mathbf{V}(\frac{1}{2},\dots,\frac{1}{2},-\frac{1}{2})$.

We can see that there will be an operator on twistors, $\mathcal{R}:\mathbf{Tw}\to\mathbf{Tw}$ in odd dimensions, and for $4\le n$ even,

$$\mathbf{Tw}_+\ \underset{\mathcal{R}_-}{\overset{\mathcal{R}_+}{\rightleftarrows}}\ \mathbf{Tw}_-.$$

This is the *Rarita–Schwinger* operator. This also makes it clear why there should be no Rarita–Schwinger operator in dimension 2, since this would have to try to take $\mathbf{V}(\frac{3}{2})$ to itself or $\mathbf{V}(-\frac{3}{2})$.

In fact, we may make an elementary construction of the Rarita-Schwinger operator S by forming the Dirac symbol on φ_a, that is $\gamma^b\nabla_b\varphi_a$, and then adding a *counterterm* designed to attain the condition $\gamma^a(S\varphi)_a = 0$:

$$(S\varphi)_a = \gamma^b\nabla_b\varphi_a - \frac{2}{n}\gamma_a\nabla^b\varphi_b.$$

A somewhat better known set of gradients is that emanating from $\mathbf{V}(1) \cong \Lambda^1$. (Here we adopt the convention of omitting terminal strings of zeros in weights.) By the selection rule (2.11), these should go to $\mathbf{V}(2)$, $\mathbf{V}(1,1)$, and $\mathbf{V}(0)$ for $n \geq 5$. These bundles are exactly TFS^2, Λ^2, and Λ^0, and the corresponding operators are (up to constants and isomorphic realizations) S, d, and δ. More generally, there is an S-like operator on Λ^k [9], in addition to the familiar d and δ.

Returning to our discussion of conformal structure, note that we have less to work with than we did in the Riemannian regime — g is determined only up to a positive function multiple. This may be reflected by taking a *larger* number of trivializing neighborhoods $(\mathcal{O}_\alpha, g_\alpha)$; here g_α is a metric on $\mathcal{O}_\alpha$. On an overlap $\mathcal{O}_\alpha \cap \mathcal{O}_\beta$, we have $g_\beta = \Omega^2_{\beta\alpha} g_\alpha$, and we may build a bundle $\mathbf{V}(w|\lambda)$ with transition functions

$$\Omega^w_{ba}\tau_{ba}$$

where τ_{ba} are (Riemannian spin) transitions $(\mathcal{O}_\alpha, g_\alpha) \to (\mathcal{O}_\beta, g_\alpha)$. What we get out of this are tensor-spinor *density* bundles — corresponding to each old (Riemannian) bundle $\mathbf{V}(\lambda)$, we now have a whole family $\mathbf{V}[w|\lambda]$ of bundles, parameterized by a complex number w. Since $\mathbf{V}[w|\lambda]$ is naturally isomorphic to $\mathbf{V}[w|0] \otimes \mathbf{V}[0|\lambda]$, we could have gotten away with defining the *density bundles* $\mathcal{E}[w] := \mathbf{V}[w|0]$, and the tensor-spinors of *conformal weight* $w = 0$.

Exercise 4. *Show that the tangent bundle is naturally isomorphic to* $\mathbf{V}[1|1]$, *while the cotangent bundle is* $\mathbf{V}[-1|1]$.

In the case of tensors, this means that we can do some of our weight accounting by using index position. For example, if $\mathcal{E}^a$ (resp. $\mathcal{E}_a$) is the tangent (resp. cotangent) bundle, and we put, for example, $\mathcal{E}_a{}^b[w] := \mathcal{E}[w] \otimes \mathcal{E}_a \otimes \mathcal{E}^b$, then

$$\mathcal{E}_a{}^b[w] \cong \mathcal{E}_{ab}[w+2] \cong \mathcal{E}^{ab}[w-2].$$

In fact, this explains why conformal weights and biweights move around as we move among different isomorphic realizations of bundles.

To describe the conformal biweights of the gradients, and many other operators as well, consider the *rho-shift* of a weight $[w|\lambda]$:

$$(\tilde{w}|\tilde{\lambda}) = (w + n/2|\lambda + \rho_n), \text{ where } 2\rho_n = (n-2, n-2, \ldots, n-2\ell).$$

The *affine Weyl group* acts on such shifted weights, according to permutations and

an even number of sign changes,	n even,
any number of sign changes,	n odd.

The equivalence relation of lying in the same affine Weyl orbit will be denoted by $\sim$.

The first necessary condition for having a conformally invariant differential operator $\mathbf{V}[w|\lambda] \to \mathbf{V}[u|\mu]$ is that

$$(\tilde{w}|\tilde{\lambda}) \sim (\tilde{u}|\tilde{\mu}).$$

That is, the bundles must have the same *central character* under the algebra $\mathfrak{so}(n+1,1)$. The relevance of this algebra will be explored in somewhat more detail later, in Section 6.3. In addition, the drop in conformal weight, $w - u$, must be the order of the differential operator. Note that our "operators" are really functors that assign operators to conformal structures. We adopt the convention that our operators must be nonzero on flat space, so that (for example) $K\Delta$ is not an operator. The condition about the drop in conformal weight is then just a homogeneity condition under uniform scaling of the metric. If our prospective conformally invariant operator is a gradient, we must thus have $u = w - 1$.

Suppose there is a gradient $G : \mathbf{V}(\lambda) \to \mathbf{V}(\mu)$. If we choose any conformal weight w_0 and view G as carrying $\mathbf{V}[w_0|\lambda]$ to $\mathbf{V}[w_0 - 1|\mu]$, some invariant theory shows that the conformal variation of G along the curve of metrics $e^{2\varepsilon\omega}g_0$ must take the form

$$(w_0 + c)[G, m_\omega],$$

where m_ω is multiplication by ω, and c is a constant. Thus we have an invariant operator when $w_0 = -c$. Knowing now that gradients are conformally invariant for *some* choice of the initial conformal weight, we may determine *which* choice does the trick using the above central character considerations. For example, if we have a self-gradient $\mathbf{V}(\bullet|\lambda) \to \mathbf{V}(\bullet|\lambda)$ in odd dimensions (where the bullets $\bullet$ represent numbers we do not necessarily know), then we must have

$$\mathbf{V}(\tfrac{1}{2}|\lambda) \to \mathbf{V}(-\tfrac{1}{2}|\lambda).$$

Exercise 5. *Show that if our gradient takes* $\mathbf{V}(\bullet|\cdots\lambda_i\cdots)$ *to* $\mathbf{V}(\bullet|\cdots\lambda_i + \varepsilon\cdots)$, *where* $\varepsilon = \pm 1$ *and the corresponding lists* $\cdots$ *agree, we must have*

$$\mathbf{V}(\varepsilon(\lambda_i + \varepsilon)|\cdots\lambda_i\cdots) \to \mathbf{V}(\varepsilon\lambda_i|\cdots\lambda_{i+1}\cdots).$$

For example, to work out the conformal biweight of the conformal Killing operator, we reason as follows:

$$\begin{aligned}
\mathbf{V}[\bullet|1] &\to \mathbf{V}[\bullet|2], \\
\mathbf{V}\left(\bullet|\tfrac{n}{2}, \tfrac{n-4}{2}, \ldots, \tfrac{n-2\ell}{2}\right) &\to \mathbf{V}\left(\bullet|\tfrac{n+2}{2}, \tfrac{n-4}{2}, \ldots, \tfrac{n-2\ell}{2}\right), \\
\mathbf{V}\left(\tfrac{n+2}{2}|\tfrac{n}{2}, \tfrac{n-4}{2}, \ldots, \tfrac{n-2\ell}{2}\right) &\to \mathbf{V}\left(\tfrac{n}{2}|\tfrac{n+2}{2}, \tfrac{n-4}{2}, \ldots, \tfrac{n-2\ell}{2}\right), \\
\mathbf{V}[1|1] &\to \mathbf{V}[0|2], \\
\mathcal{E}^a &\to \mathcal{E}_{(ab)_0}[2],
\end{aligned}$$

where $(\ )_0$ is trace-free symmetrization.

When spinors are in play, conventions vary on the treatment of indices. One choice is to suppress spin indices altogether, and another is to write them, assigning them weights of $\pm 1/2$ (depending on position) in analogy with the weight ± 1 of a tensor index. We shall have occasion to consider this briefly later, in Subsection 6.3.1.

Remark 1. *Weyl's dimension formula* [30] gives a formula for the dimension of the module $V(\lambda)$, and thus for the fiber dimension of $\mathbf{V}(\lambda)$, in terms of the entries of λ. For example in dimension 4,

$$\dim V(\lambda) = \tilde{\lambda}_1^2 - \tilde{\lambda}_2^2. \tag{2.12}$$

Exercise 6. *Use (2.12) to find the dimensions of the half-spinor bundles $\Sigma_\pm$; the form bundles Λ^1, $\Lambda^2_\pm$, and Λ^3; the bundles $\mathcal{W}_\pm$ of algebraic Weyl tensors; and the bundles TFS^p of trace-free symmetric p-tensors in dimension 4.*

Though it is not obvious, the above information is almost enough to compute the spectra of all natural operators on bundles over the round sphere S^n; in particular the G^*G when G is a gradient [10]. Implicit in this are the spectra of self-gradients [11]. This recovers old results on, for example, the spectrum of the Dirac operator. But it is general, so that one may plug in weight parameters and get, for example, the spectrum of the self-gradient on $\mathbf{V}(p+\frac{1}{2}, \frac{1}{2}, \dots, \frac{1}{2}, \pm\frac{1}{2})$; in fact, this is the bundle described in (1.4) above. This operator has been an object of considerable recent study.

This in turn says something about classifying Bochner–Weitzenböck (BW) formulas on *general* Riemannian spin manifolds [10]. There are deep connections between this and estimates that can be made on solutions of first-order elliptic differential systems [12, 19], with consequences for decay rates of these solutions on open manifolds. Indeed, this last question actually inspired the original invention of gradients on the part of Stein and Weiss.

6.2.2 *Bochner–Weitzenböck formulas and a discrete leading symbol*

Here is a partial illustration of all this in a "small" situation. The ideas are closely related to those of the general calculation, but no heavy machinery is required. One may compute (from the conformal variations of the gradients and scalar curvature) that for certain explicit constants k_u,

$$\frac{K}{2(n-1)} - \sum_u k_u G_u^* G_u \tag{2.13}$$

is conformally invariant

$$\mathbf{V}\left(\frac{2-n}{2}\middle|\lambda\right) \to \mathbf{V}\left(\frac{-2-n}{2}\middle|\lambda\right),$$

for each λ [9]. In fact, if G_u carries $\mathbf{V}(\lambda)$ to $\mathbf{V}(\sigma_u)$,

$$k_u = 2(1 + \langle \lambda + \sigma_u + 2\rho_n , \lambda - \sigma_u \rangle)^{-1}.$$

(In case k_u is undefined, our invariant operator is $G_u^* G_u$.) In the most elementary example, $\lambda = 0$, we get (up to a factor) the *Yamabe operator* or *conformal Laplacian*

$$Y := \Delta + \frac{n-2}{4(n-1)} K. \tag{2.14}$$

The operator (2.13) need not be second-order in general – it could be zero, or at least have order zero, for some bundles $\mathbf{V}(\lambda)$. In fact, it turns out to have order 2 exactly when the numerical invariant $N(\mathbf{V}(\lambda)) =: N(\lambda)$ (recall (1.6) and the ensuing discussion above) is *odd.* By some invariant theory, being of order less than 2 and being conformally invariant for $N(\lambda)$ even, expression (2.13) must be an action of the Weyl tensor C. In particular, this discussion applies if our bundle is Σ (n odd) or $\Sigma_\pm$ (n even), since then $N(\lambda) = 2$. Furthermore, the Weyl tensor cannot act, by weight considerations. Thus the expression (2.13) vanishes.

As a result, we have two equations on the two $G_u^* G_u$:

$$\begin{gathered} \frac{K}{2(n-1)} + \frac{2}{n-1} G_1^* G_1 - 2 G_2^* G_2 = 0, \\ G_1^* G_1 + G_2^* G_2 = \nabla^* \nabla. \end{gathered} \tag{2.15}$$

(The second of these follows directly from the definition of gradients.) Here and below, we shall list gradients according to the lexicographical ordering of their target weights σ_u, the largest weights coming first. (In particular, σ_1 is always $\lambda + e_1$.) Solving the system (2.15) for the $G_u^* G_u$, we get

$$\begin{aligned} n G_2^* G_2 &= \nabla^* \nabla + \frac{K}{4}, \\ \frac{n}{n-1} G_1^* G_1 &= \nabla^* \nabla - \frac{K}{4(n-1)}. \end{aligned} \tag{2.16}$$

These are, respectively, the *Lichnerowicz formula* and *Friedrich formula.* After taking account of the normalization change resulting from realizing the spinor bundle in different ways (as Σ and in $T^* \otimes \Sigma$), the Lichnerowicz formula becomes

$$\nabla\!\!\!\!/^{\,2} = \nabla^* \nabla + \frac{K}{4}. \tag{2.17}$$

The left side of the Friedrich formula is $T^* T$, where T is the twistor operator.

To do this in a more general setting, we use a device called the *discrete leading symbol* [1]. To use this, we need a complete understanding of the model case of round S^n. Here the section space of $\mathbf{V}(\lambda)$ is (a completion of) a direct sum of modules,

$$\mathcal{V}(\alpha; \lambda) \cong_{\mathbf{Spin}(n+1)} \overline{\mathbf{V}}(\alpha)$$

for $\mathbf{Spin}(n+1)$, where we use an overline to signal the transition from n to $n+1$. The $\mathbf{Spin}(n+1)$ module labeled by α occurs (with multiplicity 1) iff $\alpha|_{\mathbf{Spin}(n)}$

has a λ summand — this is an instance of *Frobenius reciprocity*. If α occurs, we say that $\alpha \downarrow \lambda$ or $\lambda \uparrow \alpha$. In fact, $\downarrow$ is an example of a *branching rule*, in this case the one describing restriction of $\mathbf{Spin}(n+1)$ representations to a $\mathbf{Spin}(n)$ subgroup (which is imbedded in the standard way). In terms of weight arithmetic, this branching rule is given by the *interlacing* condition

$$\begin{aligned} &\alpha_1 \geq \lambda_1 \geq \alpha_2 \geq \lambda_2 \geq \cdots \alpha_\ell \geq |\lambda_\ell|, && n \text{ even},\\ &\alpha_1 \geq \lambda_1 \geq \alpha_2 \geq \lambda_2 \geq \cdots \geq \lambda_\ell \geq |\alpha_{\ell+1}|, && n \text{ odd}. \end{aligned}$$

We have

$$\Gamma(\mathbf{V}(\lambda)) = \bigoplus_{\alpha \downarrow \lambda} \mathcal{V}(\alpha;\lambda) \cong_{\mathbf{Spin}(n+1)} \bigoplus_{\alpha \downarrow \lambda} \overline{\mathbf{V}}(\alpha).$$

A curious but useful aspect of the interlacing rule reveals itself upon branching twice, from $\mathbf{Spin}(n+1)$ to $\mathbf{Spin}(n)$ to $\mathbf{Spin}(n-1)$. Let $L := [(n+1)/2]$. If

$$\mathring{\alpha} := (\alpha_2, \ldots, \alpha_L),$$

then

$$\alpha \downarrow \lambda \iff \lambda \downarrow \mathring{\alpha} \text{ and } \alpha_1 - \lambda_1 \in \mathbf{N}.$$

Note that $\mathring{\alpha}$ runs through a finite set, say $\mathcal{A}_\lambda$, for $\alpha \downarrow \lambda$, while α_1 runs through an infinite set.

For a *natural differential operator* on $\mathbf{V}(\lambda)$, there are two descriptions of the discrete leading symbol, one microlocal (or at least pointwise), and one global:

(1) Pick a point (x, ξ) in the cotangent bundle of an *arbitrary* M (with the Riemannian spin structure), with $\xi \neq 0$. This choice of a distinguished direction ξ leaves us with a symmetry group $\mathbf{Spin}(n-1)_\xi$ isomorphic to $\mathbf{Spin}(n-1)$. Take the leading symbol $\sigma_{\text{lead}}(D)(x,\xi) \in \text{End}(\mathbf{V}(\lambda)_x)$, and decompose relative to the $\mathbf{Spin}(n-1)_\xi$-splitting:

$$\mathbf{V}(\lambda)_x = \underline{\mathbf{V}}(\beta_1)_\xi \oplus \cdots \oplus \underline{\mathbf{V}}(\beta_{b(\lambda)})_\xi\,,$$

where

$$\underline{\mathbf{V}}(\beta_i)_\xi \cong_{\mathbf{Spin}(n-1)} \underline{\mathbf{V}}(\lambda).$$

(The underline signals the transition from n to $n-1$.) We get

$$\sigma_{\text{lead}}(D)(x,\xi) = \sum_i \mu_i(D)\, \text{Id}_{\underline{\mathbf{V}}(\beta_i)_\xi}$$

for some eigenvalues $\mu_i(D)$. A main point is that *the list of* $(\beta_i, \mu_i(D))$ *is independent of* x, ξ, *and* M. In fact the β_i just list the $\beta \uparrow \lambda$. So do the $\mathring{\alpha}$ for $\alpha \downarrow \lambda$, so we can speak of $\mu(D, \mathring{\alpha})$.

(2) Take the special case of the sphere, and take the leading asymptotics of spec(D), in the sense

$$\text{eig}(D, S^n, \alpha) = \mathbf{k}(D, \mathring{\alpha})\alpha_1^d + o(\alpha_1^d) \text{ as } \alpha_1 \to \infty.$$

(As part of the game, one has to show there *are* such asymptotics.)

Theorem 2. [1] *These are the same:* $\mu(D, \mathring{\alpha}) = \mathbf{k}(D, \mathring{\alpha})$.

Remark 2. The calculation of the spectrum makes essential use of conformal covariance. The noncompact group $G = \mathbf{Spin}(n+1, 1)$ gets involved; see [10, 18].

Remark 3. To oversimplify a bit, the G^*G for gradients G are the basic "building blocks" of natural differential operators. However, these are not usually conformally covariant. The first-order operators G and G^*, on the other hand, are conformally covariant, but don't usually have spectra, since they generally go from one bundle to another. Nevertheless, the calculation works, as G and G^* can be made to live in a diagram of operators which are "simultaneously diagonalizable" (in a suitable loose sense). This is slightly more involved than it looks at first blush, as some of these operators act on subquotients (under the conformal group) of section spaces, rather than on full section spaces.

Remark 4. By the interlacing rule, the α and $\mathring{\alpha}$ may be identified with lattice points in rectangular boxes. The box $\mathbf{A}$ of α is infinite on one end, and its base is the box $\mathring{\mathbf{A}}$ of all $\mathring{\alpha}$.

For example, the spinor bundle Σ for n odd gives rise to a two-point set $\mathring{\mathbf{A}}$, and the discrete leading symbol $\mathbf{k}(\nabla\!\!\!\!/\,)$ of the Dirac operator takes the values ± 1 on these points. The square of the Dirac operator has $\mathbf{k}(\nabla\!\!\!\!/\,^2) \equiv 1$, and in fact the discrete leading symbol takes compositions to products:

$$\mathbf{k}(D_1 \circ D_2)(\beta) = \mathbf{k}(D_1)(\beta) \cdot \mathbf{k}(D_2)(\beta).$$

In particular, this shows how to construct an operator D_1 with $D_1 D_2$ of the form $(\nabla^*\nabla)^k + \text{(lower order)}$, if such exists.

Generalizing the example of $\nabla\!\!\!\!/\,^2$, the discrete leading symbol of any operator of the form

$$(\nabla^*\nabla)^k + \text{(lower order)}$$

on any bundle $\mathbf{V}(\lambda)$ is the constant function 1.

Ellipticity is detected by the discrete leading symbol: D is elliptic iff $\mathbf{k}(D)$ is nowhere zero. D is positively elliptic (has positive definite leading symbol) iff $\mathbf{k}(D)$ is a positive function. In fact, the best ellipticity constant c for an operator of order $2p$,

$$\sigma_{2p}(D)(\xi) \geq c|\xi|^{2p}\mathrm{Id}$$

in the sense of endomorphisms, is the minimum value of the discrete leading symbol.

The characteristic function of a point in $\mathring{\mathbf{A}}$ corresponds to a fundamental projection in the algebra of leading symbols. Here the prototypical example is the differential form bundle Λ^k for $k < (n-2)/2$, where the operators δd and $d\delta$ give rise to the fundamental projections. This example is somewhat misleading in its simplicity however, as generally the operators representing fundamental projections are of higher order, and need not commute.

If $G_1, \ldots, G_N$ are the gradients emanating from $\mathbf{V}(\lambda)$, then

$$\min_{\beta \in \mathcal{B}(\lambda)} \mathbf{k}(G_{a_1}^* G_{a_1} + \cdots + G_{a_p}^* G_{a_p})(\beta)$$

is also the best Kato constant k for solutions of the first-order system $G_{a_1}\varphi = \cdots = G_{a_p}\varphi = 0$, in the sense that

$$\Big|d|\varphi|\Big|^2 \leq c|\nabla\varphi|^2 \text{ off } \{\varphi_\xi = 0\}. \tag{2.18}$$

(See [12, 19].)

The presence of a Bochner–Weitzenböck (BW) formula

$$a_1 G_1^* G_1 + \cdots + a_N G_N^* G_N = (0^{\underline{\text{th}}} \text{ order})$$

is signaled by the discrete leading symbol relation

$$a_1 \mathbf{k}(G_1^* G_1) + \cdots + a_N \mathbf{k}(G_N^* G_N) = 0.$$

Given a knowledge (from [10]) of the $\mathbf{k}(G_u^* G_u)$, we can algebraically determine the left sides of BW formulas. Specifically, there are numbers s_u, $\tilde{c}_u$ (built from conformal weights) such that

$$\begin{aligned} & a_1 \mathbf{k}(G_1^* G_1) + \cdots + a_N \mathbf{k}(G_N^* G_N) = 0 \iff \\ & a_1 \tilde{c}_1 s_1^{2j} + \cdots + a_N \tilde{c}_N s_N^{2j} = 0 \\ & \text{for } j = 0, 1, \ldots, \left[\frac{N(\lambda) - 1}{2}\right]. \end{aligned}$$

BW formulas are key ingredients in the business of relating local geometry and spectral data, since

$$\underbrace{\sum_{a_u > 0} a_u G_u^* G_u}_{\text{nonneg. op.}} = \underbrace{\sum_{-a_u > 0} (-a_u) G_u^* G_u}_{\text{nonneg. op.}} + \text{Curv.}$$

The above shows there are $[N(\lambda)/2]$ independent BW formulas on $\mathbf{V}(\lambda)$. If the curvature term is positive (resp. negative) definite, so is the operator on the left (resp. the first term on the right). This is only the most obvious of a sophisticated bag of tricks for proving vanishing theorems and eigenvalue estimates.

The classification of BW left sides depends on the classification theory of second-order conformally covariant operators, which in turn depends on a complete understanding of second-order contributors to $\nabla^*\nabla$:

$$T^* \otimes \underbrace{(T^* \otimes \mathbf{V}(\lambda))}_{G_u \text{ targets}} \cong \underbrace{(T^* \otimes T^*)}_{\Lambda^0 \oplus \Lambda^2 \oplus \mathrm{TFS}^2} \otimes \mathbf{V}(\lambda).$$

An example of a bundle with $N(\lambda) = 4$ (so that there are 2 independent BW formulas) is the bundle introduced above in (1.1). The gradient target bundles are Σ (or the copy of Σ contained in $T^* \otimes \mathbf{Tw}$), $\mathbf{Tw}$ itself, and the bundles $\mathbf{Y}_2$ and $\mathbf{Z}_2$. (Recall the discussion around (1.3) and (1.4) above.) Denoting the corresponding gradients by G_Σ, $G_{\mathbf{Tw}}$, $\Gamma_{\mathbf{Y}}$, and $G_{\mathbf{Z}}$, the discrete leading symbols of the $G_u^* G_u$ on the two-point space $\mathcal{B}(\lambda)$ are:

$G_\Sigma^* G_\Sigma$	$\frac{1}{n}$ •	0 •
$G_{\mathbf{Tw}}^* G_{\mathbf{Tw}}$	$\frac{n-2}{n(n+2)}$ •	$\frac{n}{(n+2)(n-2)}$ •
$G_{\mathbf{Y}}^* G_{\mathbf{Y}}$	0 •	$\frac{n-3}{2(n-2)}$ •
$G_{\mathbf{Z}}^* G_{\mathbf{Z}}$	$\frac{n}{n+2}$ •	$\frac{n+1}{2(n+2)}$ •

The corresponding BW formulas are derived in [17]; a basis is:

$$\begin{aligned}
&\frac{(n-3)(n-2)}{2n} G_\Sigma^* G_\Sigma - \frac{(n-3)(n+2)}{2n} G_{\mathbf{Tw}}^* G_{\mathbf{Tw}} + \mathrm{G}_{\mathbf{Y}}^* \mathrm{G}_{\mathbf{Y}} \\
&\quad = -\frac{(n-2)(n-3)}{8n(n-1)} K + \frac{n-3}{2(n-2)} b \cdot + \frac{1}{8} C : , \\
&-\frac{(n-1)(n+2)}{2n} G_\Sigma^* G_\Sigma - \frac{(n-2)(n+1)}{2n} G_{\mathbf{Tw}}^* G_{\mathbf{Tw}} + \mathrm{G}_{\mathbf{Z}}^* \mathrm{G}_{\mathbf{Z}} \\
&\quad = -\frac{(n+2)(n+1)}{8n(n-1)} K - \frac{n+1}{2(n-2)} b \cdot + \frac{3}{8} C : .
\end{aligned}$$

Here b is the trace-free Ricci tensor, and the curvature actions $b\cdot$ and $C :$ are:

$$\begin{aligned}
(b \cdot \varphi)_\mu &= \frac{1}{n} \left((n-2) b_{\mu\lambda} - b_{\alpha\lambda} \gamma^\alpha \gamma_\mu \right) \varphi^\lambda , \\
(C : \varphi)_\mu &= C_{\alpha\beta}{}^\lambda{}_\mu \gamma^\alpha \gamma^\beta \varphi_\lambda .
\end{aligned}$$

These are the *unique* invariant actions of TFS^2 (where b lives) and of the algebraic Weyl tensor bundle $\mathcal{W}$ (where C lives) on twistors.

Here is a sample theorem demonstrating what one can do with the resulting estimates:

Theorem 3. [17] *Suppose* $n = 8$, *and* g *is an Einstein manifold (i.e.,* $b = 0$*). Then any twistor* φ *in* $\ker(\mathrm{G}_\Sigma) \cap \ker(\mathrm{G}_{\mathbf{Y}})$ or *in* $\ker(\mathrm{G}_{\mathbf{T}}) \cap \ker(\mathrm{G}_{\mathbf{Z}})$ *is parallel (i.e.,* $\nabla\varphi = 0$*).*

6.2.3 The Hijazi inequality

One of the major touchstones of spin geometry has been the *Hijazi inequality* [27], which related the bottom eigenvalue of the conformal Laplacian Y (recall (2.14)) with that of the square of the Dirac operator. To explain how this comes about, recall the first equation of (2.15), which integrates to

$$0 = \int \left(\frac{K}{2(n-1)} |\psi|^2 - 2|G_2\psi|^2 + \frac{2}{n-1}|G_1\psi|^2 \right) dv_g$$
$$= \int \left(\frac{K}{2(n-1)} |\psi|^2 - \frac{2}{n}|\nabla\!\!\!/\,\psi|^2 \right) dv_g + (\text{nonnegative}). \quad (2.19)$$

(Recall the normalization change implicit in (2.16,2.17).) The Yamabe operator governs the conformal change of the scalar curvature K according to

$$\frac{K_\omega}{2(n-1)} = \frac{2}{n-2} e^{-(n+2)\omega/2} Y e^{(n-2)\omega/2}, \quad (2.20)$$

where K_ω is the scalar curvature of the metric $g_\omega = e^{2\omega} g$, and the operator Y is computed in the metric g. A well-known symmetrization argument shows that the bottom eigenfunction of Y does not change sign, so we may assume that it is a positive function, which we may then write as $e^{(n-2)\omega_1/2}$ for some smooth function ω_1. With $\omega = \omega_1$ in (2.20), and μ_1 the bottom eigenvalue of Y, we have

$$\frac{K_{\omega_1}}{2(n-1)} = \frac{2}{n-2} e^{-(n+2)\omega_1/2} \cdot \mu_1 Y e^{(n-2)\omega_1/2} = \frac{2}{n-2} \mu_1 e^{-2\omega_1}.$$

Substituting into the basic inequality (2.19) at the metric $g_1 := g_{\omega_1}$, we have

$$0 = \int \left(\frac{2}{n-2} \mu_1 e^{-2\omega_1} |\psi|^2 - \frac{2}{n} |\nabla\!\!\!/_{\omega_1} \psi|^2 \right) dv_{g_1} + (\text{nonnegative}). \quad (2.21)$$

The obvious thing to do now is to take advantage of the conformal covariance relation for the Dirac operator, in the form

$$\nabla\!\!\!/_{\omega_1} \psi = e^{-(n+1)\omega_1/2} \nabla\!\!\!/ \left(e^{(n-1)\omega_1/2} \psi\right).$$

If $e^{(n-1)\omega_1/2}\psi$ is an eigenspinor for $\nabla\!\!\!/$, with eigenvalue λ, then

$$0 = \int e^{-2\omega_1} |\psi|^2 \left(\frac{2}{n-2} \mu_1 - \frac{2}{n} \lambda^2 \right) dv_{g_1} + (\text{nonnegative}),$$

and we have:

Theorem 4. [27] $\lambda^2 \geq n\mu_1/(n-2)$.

It is clear how one would use the derivation above to study the case of equality. There are various improvements in the case of Kähler manifolds, quaternionic Kähler manifolds, and the like; see, for example, [28], [31], and [33].

There are two main obstacles to the attempt to extend this line of reasoning to other bundles. First, we depended on the fact that, in some BW formula, we could have just *one* $G_u^* G_u$ appear with a coefficient of a given sign. In general, the coefficient signs tend to try to split about evenly, as one might expect. Another is that it is difficult to keep the other curvatures (b and C) out of the picture. The spinor and scalar bundles are special in that weight considerations prevent b and C from acting; this simplified state of affairs never occurs again as we go up into the space of bundles. There *is* a BW formula omitting b, whenever $N(\lambda)$ is even, the combination (2.13). Effective estimation using this, however, will depend on having some control over C. See [15, 16] for some attempts in this direction.

Like central results in other fields, the Hijazi inequality is closely related to other fundamental phenomena. For example, an argument of Bär and Moroianu explained in [20] ties it closely to the sharp Kato estimate for solutions of the Dirac equation, namely

$$\nabla\!\!\!/\,\psi = \lambda\psi \Rightarrow \left|d|\psi|\right|^2 \leq \frac{n-1}{n}|\tilde{\nabla}\psi|^2 \text{ off } \{\psi_\xi = 0\},$$

where $\tilde{\nabla}_a = \nabla_a + \lambda\gamma_a/n$. (Recall (1.5) and (2.18) above.) We compute

$$\left(Y - \frac{n-2}{n}\lambda^2\right)|\psi|^{(n-2)/(n-1)}$$
$$= \frac{n(n-2)}{(n-1)^2}\left(\left|d|\psi|\right|^2 - \frac{n-1}{n}|\tilde{\nabla}\psi|^2\right)|\psi|^{-n/(n-1)}. \quad (2.22)$$

Modulo some analytic niceties having to do with the possible zero set of ψ, the nonnegativity of the right side of (2.22) establishes the Hijazi inequality.

6.3 Tractor constructions

The constructions of spin-tractors and form-tractors below is joint work of the author and Rod Gover [14].

6.3.1 Enlarging the structure group

To some extent, the use of the word *twistor* to describe the bundle $\mathbf{Tw}$ is a mistake that has been solidified by tradition. The Penrose *local twistor bundle* is an object whose generalizations are sure to be much-studied in the near future. Recall that we are working in the setting of *Riemannian* spin geometry. The constructions we are about to introduce generalize more or less readily to the pseudo-Riemannian case, including the Lorentzian regime where Penrose's construction finds its most natural home.

A first step is to realize that there is a larger group lurking at the edges of the picture, namely

$$G = \mathbf{Spin}(n+1, 1),$$

and that the structure groups we have dealt with thus far, as well as some others, are natural subgroups of G. When the *Iwasawa decomposition* and *Langlands decomposition* of a semisimple Lie group,

$$G = KAN = \bar{N}MAN,$$

are applied to our G, the groups $\bar{N}$ and N are copies of $\mathbb{R}^n$, and the group A is a copy of $\mathbb{R}_+$. (The *conformal weight* lives in the dual of this.) The group M is $\mathbf{Spin}(n)$, and K is $\mathbf{Spin}(n+1)$. The *stereographic injection* $\mathbb{R}^n \to S^n$ is

$$\bar{N} \hookrightarrow \bar{N}MAN \to \bar{N}MAN/MAN.$$

The introduction of conformal weights carried us from M-bundles to MA-bundles. But there's a further step, suggested by Cartan's work: $P = MAN$-bundles — not just on homogeneous spaces for G, but on conformal (spin) manifolds. That is, there is a P-structure to these manifolds that is not usually exploited. It is not generally known that there is a rich supply of MAN-bundles, in which the N factor acts **nontrivially** — in contrast to the situation for the classical tensor-spinor-density bundles. These bundles exist on MA-manifolds, i.e., manifolds endowed with a conformal structure. That is, our manifolds need only have MA structure, and we get bundles with the richer MAN structure free of charge.

The basic concept is one that seems to be rediscovered in different forms at irregular intervals. Besides the work of Penrose [38], pioneering work was done by T.Y. Thomas [42] in the 1920s, but lacking the language of bundles, it was largely forgotten until the explanation of [2] in modern terminology. I.E. Segal's *spannors* and *plyors* [36, 37], introduced to explain fundamental physical phenomena, are examples of such objects, although Segal treated them only on homogeneous model spaces.

One important difference feature of MAN-bundles, as opposed to M- or MA-bundles, is the lack of *complete reducibility*. That is, MAN-bundles have composition lattices, or (dropping some information) Jordan–Hölder series of the form

$$\mathbf{V}_s \oplus \mathbf{V}_{s-1} \oplus \cdots \oplus \mathbf{V}_r. \tag{3.23}$$

The $\oplus$ notation means that each $\mathbf{V}_q \oplus \cdots \oplus \mathbf{V}_r$ for $s \geq q \geq r$ is an invariant subbundle, but there is no invariant direct complement; that is, any attempt to get just the information from $\mathbf{V}_s, \ldots, \mathbf{V}_{q+1}$ involves taking a quotient. If we reduce the structure group to MA, we may replace the $\oplus$ in (3.23) with $\oplus$, but the action of N carries us from $\mathbf{V}_q$ to $\mathbf{V}_{q-1}$ for each q.

On the face of it, it would appear to be a formidable task to find lists of MA-representations that can be "glued together" using actions of N. But in fact, there is an abundant source of such things, and they are geometrically very natural. First, we can always restrict representations of G to MAN. When we do this for the (half) spin representation in dimension 4, we obtain Penrose local twistors and Segal spannors. When we use the full differential form representation, we get Segal plyors. But there is really not too much that is special about dimension 4 from this point of view — the general setup only really recognizes a difference between even and odd dimensions.

With these restrictions of G-representations in hand, we can always take MAN-*subquotients*. This means we iterate and alternate the processes of taking subrepresentations and quotients. Following the standard (after [2]) modern parlance, we shall call these *tractor* (or tractor-density) bundles. The *full* tractor bundles play a central role in the theory — these are those obtained from restricting a (usually irreducible) G-representation. One benefit of having all these bundles around is a cure for the *composition problem* for conformally invariant operators. Such operators on conventional tensor-spinor-density bundles almost never compose, since forming the composition AB would mean that the weights a and d in

$$\hat{A} = \Omega^{-b} A \Omega^{a}, \quad \hat{B} = \Omega^{-d} B \Omega^{c}$$

must match up — in practice, this "almost never" happens. With all these MAN-bundles though, there are many compositions, and in fact this is a good way of constructing new operators. In particular, one may start in a "conventional" tensor-spinor-density bundle (where N acts trivially), travel through some tractor bundles, and emerge at a conventional bundle. In fact, doing so is necessary for the curved version of *Jantzen–Zuckerman translation*, or *curved translation* for short [23].

To concentrate on local twistors and spannors, or what we might call *spin-tractors* from the current viewpoint: If we realize the positive spin representation of $\mathbf{Spin}(n+2, \mathbb{C})$ in the real form $\mathbf{Spin}(n+1, 1)$, then restrict to $M = \mathbf{Spin}(n)$, we get an object "containing" both positive and negative spinors:

$$\Sigma_{+}^{M}[\tfrac{1}{2}] \oplus \Sigma_{-}^{M}[-\tfrac{1}{2}].$$

Here the first summand is an MAN-quotient, and the second is an MAN-submodule; recall that this is the meaning of the semidirect sum sign $\oplus$. The numbers in brackets encode the conformal weights. A more Penrosian notation would be

$$\mathcal{E}^{P} \oplus \mathcal{E}^{P'}[-1].$$

Here the M-modules involved are indicated by the indices adorning the generic bundle symbol $\mathcal{E}$, the P indicating positive spinors, and the P' negative. In this notation, the index positions carry conformal weights. This is also true in the tensor case, where, for example, the cotangent bundle $\mathcal{E}_a$ is isomorphic to $\mathcal{E}^a[2]$ (the vector field 2-densities) as an MA-bundle. The difference is that reversal of a spinor index position incurs only half the "penalty" (or "reward") of a tensor index move. One point that must be observed is that in dimension $4k$, the positive and negative spinors are self-contragradient (their own dual modules), while in dimension $4k+2$, they are dual to each other. The upshot is that, for example,

$$\mathcal{E}_P = \mathcal{E}^P[-1], \quad n = 4k,$$
$$\mathcal{E}_P = \mathcal{E}^{P'}[-1], \quad n = 4k+2.$$

Let us denote spin-tractor indices by $\mathbf{A}, \mathbf{B}, \cdots$. Then, as a consequence of the discussion immediately above,

$$\mathcal{E}^{\mathbf{B}} \cong_{MAN} \mathcal{E}_{\mathbf{B}} \cong \begin{cases} \mathcal{E}^{P} \dotplus \mathcal{E}_{P'}, & n = 4k, \\ \mathcal{E}^{P} \dotplus \mathcal{E}_{P}, & n = 4k+2. \end{cases} \tag{3.24}$$

What do form-tractors (plyors) look like? To see, one could use the relation to spinors in even dimensions: as M-bundles,

$$\Lambda \cong \Sigma \otimes \Sigma.$$

On the tractor level, we can take the tensor square of the spin-tractors $\mathcal{E}^{\mathbf{B}}$. In dimension $4k$, this gives

$$\mathcal{E}^{PQ} \dotplus (\mathcal{E}^{P}{}_{Q'} \oplus \mathcal{E}^{P'}{}_{Q}) \dotplus \mathcal{E}_{P'Q'}.$$

This reflects the usual notational convention on tensor products; for example, $\mathcal{E}^{PQ} = \mathcal{E}^{P} \otimes \mathcal{E}^{Q}$. Thus we are getting forms of (reduced) weights 1, 0, and -1. This is different from what we get in the de Rham complex with its natural conformal weights:

$$\mathcal{E}_{[a_1 \cdots a_k]} = \Lambda^k[-k].$$

In dimension 4, the subquotients of the above are:

$$\mathcal{E}^{P}{}_{Q'} \cong \mathcal{E}^{P'}{}_{Q} \cong \mathcal{E}_a[1],$$

$$\mathcal{E}^{PQ} \cong \mathcal{E}^{\mathrm{SD}}_{ab}[3] \oplus \mathcal{E}[1],$$

$$\mathcal{E}_{P'Q'} \cong \mathcal{E}^{\mathrm{ASD}}_{ab}[1] \oplus \mathcal{E}[-1],$$

where SD (resp. ASD) means self-dual (resp. anti-self-dual). The first line of this is one of the things that makes 2-component spinor calculus in dimension 4 so appealing: a pair of spinor indices, one primed and one unprimed, is "equivalent" to a tensor index.

This does not quite tell the whole story of the MAN-structure of the form-tractors, even in dimension 4. In fact, what one is getting is a direct sum of 1-form tractors and self-dual 3-form tractors:

$$\mathcal{E}^{\mathbf{BC}} \cong \mathcal{E}_{A} \oplus \mathcal{E}^{\mathrm{SD}}_{[ABC]}.$$

The duality referred to here is under the structure group $G = \mathbf{Spin}(5,1)$. This basically comes down to the fact that the full tractor bundle based on a tensor product of G-modules is the tensor product of the relevant full tractor bundles — tractorization and tensor product commute.

A form-tractor bundle $\mathcal{E}_{[A_1 \cdots A_k]}$ (for k small enough to safely avoid considerations of duality in the middle order of form) looks like

$$\begin{aligned} &\Lambda^{k-1}[1] \dotplus (\Lambda^k[0] \oplus \Lambda^{k-2}[0]) \dotplus \Lambda^{k-1}[-1] \\ &\quad \cong \mathcal{E}_{[a_1 \cdots a_{k-1}]}[k] \dotplus (\mathcal{E}_{[a_1 \cdots a_k]}[k] \oplus \mathcal{E}_{[a_1 \cdots a_{k-2}]}[k-2]) \dotplus \mathcal{E}_{[a_1 \cdots a_{k-1}]}[k-2]. \end{aligned}$$

In particular, *standard tractors* are 1-form-tractors:

$$\mathcal{E}_A \cong \mathcal{E}[1] \mathbin{+\!\!\!\!\supset} \mathcal{E}_a[1] \mathbin{+\!\!\!\!\supset} \mathcal{E}[-1], \tag{3.25}$$

and *adjoint tractors* are 2-form tractors:

$$\mathcal{E}_{[AB]} \cong \mathcal{E}^a \mathbin{+\!\!\!\!\supset} (\mathcal{E}_{[ab]}[2] \oplus \mathcal{E}) \mathbin{+\!\!\!\!\supset} \mathcal{E}_a\,.$$

The name "adjoint" comes from the fact that the adjoint representation of $\mathfrak{so}(n+2,\mathbb{C})$ is isomorphic to the 2-form representation of that algebra. One may also compute, for example, that trace-free symmetric p-tensor tractors, i.e., the full tractor bundle based on the trace-free symmetric representation of $\mathfrak{so}(n+2,\mathbb{C})$, look like

$$\mathcal{E}_{(A_1\cdots A_p)_0} \cong$$
$$\mathbin{+\!\!\!\!\supset}_{q=-p}^{p} \oplus \{\mathcal{E}_{(a_1\cdots a_r)_0}[q+r] \mid (-1)^r = (-1)^p(-1)^q,\ 0 \le r \le p-|q|\}.$$

There are undoubtedly several conditions in standard Riemannian geometry that have natural equivalents in tractor geometry. For example:

Theorem 5 (Gover). *A (Riemannian signature) conformal manifold admits a parallel tractor if and only if the conformal structure contains an Einstein metric.*

6.3.2 Spin-tractors

To get back to spin-tractors: Suppose we are given just a conformal class of metrics, that is, the set of metrics of the form $\Omega^2 g$, where g is a given metric and Ω is a positive smooth function. Note that it is not really necessary to specify a "background metric" g; we can formulate things in terms of equivalence classes of metrics.

To describe spin-tractors, it is useful to alter our convention on the normalization of the Clifford symbol γ. For present purposes, we switch to a Clifford symbol $\beta = \gamma/\sqrt{2}$, so that the Clifford relation becomes

$$\beta^a\beta^b + \beta^b\beta^a = -g^{ab}\mathrm{Id}. \tag{3.26}$$

We shall call the associated Dirac operator $\not{D} := \beta^a\nabla_a$.

We start with an ordinary pair of spinors,

$$\begin{pmatrix}\psi\\ \phi\end{pmatrix}, \tag{3.27}$$

of opposite chirality. In fact, for the sake of definiteness, let us say that ψ is a positive spinor ψ^P, and ϕ is a negative spinor $\phi^{P'}$. This pair is meant to represent the spin tractor at the choice of metric g; at the conformally related choice $\hat{g} = \Omega^2 g$, we get (suppressing indices and their weights)

$$\begin{pmatrix} \hat{\psi} \\ \hat{\phi} \end{pmatrix} = \begin{pmatrix} \Omega^{w+1/2}\psi \\ \Omega^{w-1/2}(\phi + \omega_c\beta^c\psi) \end{pmatrix}. \tag{3.28}$$

Here $\omega = \log\Omega$, and ω_c is an abbreviation for $\omega_{|c} = \nabla_c\omega$. (We shall propagate this abbreviation to higher derivatives, so that $\omega_{ab} = \nabla_b\nabla_a\omega$.)

If we had a set of trivializations for the spinor bundle parameterized by some set B, what we now have is a really big set of trivializations, parameterized by $B \times \{\Omega\}$. In fact, what we have defined is $\mathcal{E}^{\mathbf{B}}[w]$, the bundle of spin-tractor-w-densities, where w is a complex number. One could also start with spinor-$(w \pm \frac{1}{2})$-densities ψ and ϕ; the effect would be to eliminate the powers of Ω from (3.28).

In order to avoid dealing with the difference between dimension $4k$ and $4k+2$ continually, it is useful to work without a chirality assumption on ψ or φ in (3.27). This means that we are really taking a direct sum of two spin-tractor bundles:

$$(\mathcal{E}^P \oplus\!\!\!\!+ \mathcal{E}_{P'}) \oplus (\mathcal{E}^{P'} \oplus\!\!\!\!+ \mathcal{E}_P), \quad n = 4k, \tag{3.29a}$$

$$(\mathcal{E}^P \oplus\!\!\!\!+ \mathcal{E}_P) \oplus (\mathcal{E}^{P'} \oplus\!\!\!\!+ \mathcal{E}_{P'}), \quad n = 4k+2. \tag{3.29b}$$

In fact, with this convention, we can also handle the odd-dimensional case in unified calculations. In odd dimensions, the spin-tractor bundle is a direct sum of two copies of $\Sigma[w + \frac{1}{2}] \oplus\!\!\!\!+ \Sigma[w - \frac{1}{2}]$. A distinguished splitting comes from the action of $\mathbf{Pin}(n)$ (which doubly covers $O(n)$). This is analogous to the splitting, in the differential form bundle Λ, of $\Lambda^{(n\pm1)/2}$, which are isomorphic under $SO(n)$, but not under $O(n)$. Using primed and unprimed indices to distinguish the two $\mathbf{Pin}(n)$-modules in question, the splitting looks typographically just like the case $n = 4k$ in (3.29).

Though it is not completely obvious, an easy calculation shows that (3.28) is an action of the group of positive functions Ω, and thus is consistent. That is, if we make the change of trivialization forced by the scale change $g \to \Omega^2 g$, and then the change by $\Omega^2 g \to \Xi^2\Omega^2 g$, this is the same as if we make the scale change $g \to (\Omega\Xi)^2 g$ all at once.

Remarkably, there is a conformally invariant connection on $\mathcal{E}^{\mathbf{B}}[w]$ — this is quite unlike our experience with "conventional" tensor-spinor-density bundles. At a scale g, this is built from the spin connection by

$$\nabla_b \begin{pmatrix} \psi \\ \varphi \end{pmatrix} = \begin{pmatrix} \nabla_b\psi + \beta_b\varphi \\ \nabla_b\varphi + \mathsf{P}_{ab}\beta^a\psi \end{pmatrix}. \tag{3.30}$$

Here P is the normalization of the Ricci tensor that is used in conformal geometry,

$$\mathsf{J} := \frac{\text{Scal}}{2(n-1)}, \qquad \mathsf{P} = \frac{\text{Ric} - \mathsf{J}g}{n-2}.$$

In fact, the appearance of the Ricci tensor signals the presence of the *Cartan connection* — curvature coefficients for the M-structure appear as connection coefficients for the MAN-structure.

The problem with this conformally invariant connection, or with analogous connections for other tractor bundles, is that one cannot iterate it. That is, one cannot form $\nabla_b \nabla_a \Psi^{\mathrm{B}}$ in a conformally invariant way, because of the tensor content of $\nabla_a \Psi^{\mathrm{B}}$, a section of $\mathcal{E}_a \otimes \mathcal{E}^{\mathrm{B}}[w]$. To do so would require a conformally invariant connection on the contangent bundle $\mathcal{E}_a$, which of course we don't have. This state of affairs is cured by the introduction of tractor-indexed replacements for ∇_a, notably the *tractor-D* operator.

If f_* is a tractor of weight w (where $*$ denotes any number of full tractor indices of any type, in up and/or down positions), let

$$D_A f_* = \begin{pmatrix} f_* \\ w^{-1}\nabla_a f_* \\ -\frac{1}{2}[w(w+\frac{1}{2}(n-2))]^{-1}(\nabla^a\nabla_a + w\mathrm{J})f_* \end{pmatrix}. \tag{3.31}$$

The vertical stack on the right side reflects the standard tractor content (recall (3.25); the rightward direction is now downward). On the left side, standard tractor content is indicated by the index A. Though this formula may not make elementary sense for some values of w, it does make perfect sense over the field of *rational functions of* w; that is, over $\mathbb{C}(w)$. Evaluation at "forbidden" values of w sometimes makes sense (and in fact is very productive), when viewed in this way. For example, a zero and a pole may neutralize each other, or it may actually be a residue we are after in a given calculation.

In fact, looking back at (3.31), the residue at $w = 0$ (if $n \neq 2$) is

$$\begin{pmatrix} 0 \\ \nabla_a f_* \\ \star \end{pmatrix}.$$

This gives the invariance of the exterior derivative d on scalar functions (or full tractors of weight 0). If $n = 2$,

$$D_A f_* = w^{-2}\begin{pmatrix} 0 \\ 0 \\ -\frac{1}{2}\nabla^a\nabla_a f_* \end{pmatrix} + w^{-1}\begin{pmatrix} 0 \\ \nabla_a f_* \\ 0 \end{pmatrix} + \begin{pmatrix} f_* \\ 0 \\ 0 \end{pmatrix}$$

gives the conformal invariance of the scalar Laplacian. With this in place, we may then conclude the invariance of d at the w^{-1} level, and of the identity map at the w^0 level.

Back in the case $n \neq 2$, the residue at $w = \frac{1}{2}(2-n)$ is

$$\frac{1}{2}\begin{pmatrix} 0 \\ 0 \\ -\nabla^a\nabla_a + \frac{1}{2}(n-2)\mathrm{J})f_* \end{pmatrix},$$

showing the invariance of the Yamabe operator

$$Y : \mathcal{E}\left[\frac{2-n}{2}\right] \to \mathcal{E}\left[\frac{-2-n}{2}\right].$$

Back to spin-tractors again: Where does the Dirac operator appear? Consider the problem of *tractor extension*. The D_A construction, in fact, solves a certain tractor extension problem for f_*, namely to extend it to a tractor with f_* in the top slot. As indicated by formulas like (3.28), the top slot is an invariant function of the tractor stack — one can *project* onto this slot. This is not true of the other slots (again, see (3.28)), but one may invariantly *inject* into the bottom slot (writing 0 above). This is, of course, due to the semidirect sum nature of tractors; for example (3.24).

A computationally productive way to think of this is to define *positional sections*, so that, for example, a standard (one-form) tractor is

$$\begin{pmatrix} \sigma \\ \mu_a \\ \rho \end{pmatrix} = Y_A \sigma + Z_A{}^a \mu_a + X_A \rho.$$

Because of (3.25),

$$Y_A \in \mathcal{E}_A[-1], \qquad Z_A{}^a \in \mathcal{E}_A{}^a[1], \qquad X_A \in \mathcal{E}_A[1].$$

The above remarks about projecting slots translate to the statement that X_A is an invariant section, but $Z_A{}^a$ and Y_A depend on the choice of metric (within the conformal class). Positional sections make it clear how we can compute (even with a computer!) with sections having several tractor indices — rather than trying to work with multidimensional arrays, we just have long formulas with several indexed positional operators. Such representations may not be very evocative, but they reduce computation (when this is necessary) to a purely mechanical matter.

Let's write a spin-tractor as

$$\begin{pmatrix} \psi \\ \varphi \end{pmatrix} = \mathbf{y}\psi + \mathbf{x}\varphi.$$

Here we have suppressed the spin-tractor index, as well as the spin indices, so that our symbols are now noncommutative. By (3.28), $\mathbf{y}$ has weight $-\frac{1}{2}$, and $\mathbf{x}$ has weight $\frac{1}{2}$. By (3.30),

$$\nabla_b \mathbf{y} = \mathbf{x} \mathsf{P}_{ab} \beta^a, \qquad \nabla_b \mathbf{x} = \mathbf{y} \beta_b .$$

Using a prime for the conformal variation, that is, letting the metric run through a conformal curve $g_\varepsilon = e^{2\varepsilon\omega} g_0$ for ω a smooth function, and letting $' = (d/d\varepsilon)|_{\varepsilon=0}$, we have (from (3.28))

$$\mathbf{x}' = 0, \;\; \mathbf{y}' = -\omega_c \mathbf{x} \beta^c.$$

Having $\mathbf{y}\psi+\mathbf{x}\varphi$ as a tractor extension means that $(\mathbf{y}\psi+\mathbf{x}\varphi)'=0$; this is satisfied by

$$\varphi=\frac{2}{n+2w}\not{D}\psi. \tag{3.32}$$

Thus the Dirac operator has appeared, but what about its conformal invariance? This is just the observation that

$$\begin{pmatrix}\psi\\ (w+\frac{1}{2}n)^{-1}\not{D}\psi\end{pmatrix}$$

has a pole at $w=-n/2$, with residue

$$\begin{pmatrix}0\\ \not{D}\psi\end{pmatrix}.$$

It is not hard to see that the first (topmost) *nonzero* slot in such an invariant expression projects out to an invariant operator; the result is the conformal invariance of

$$\not{D}:\Sigma[(1-n)/2]\to\Sigma[(-1-n)/2],$$

or with (weighted) indices attached,

$$\not{D}:\mathcal{E}^{P}[-n/2]\to\mathcal{E}^{P'}[-n/2-1],$$
$$\not{D}:\mathcal{E}^{P'}[-n/2]\to\mathcal{E}^{P}[-n/2-1].$$

In this tractor extension problem, we implicitly reasoned that the expression $\not{D}=\beta^a\nabla_a$ was, up to a constant, the only quantity with the right valence, invariance and homogeneity properties to land in the bottom slot; we put it there with an undetermined coefficient, and pinned down the coefficient by conformal variation. There are also ways to do a direct computation of the tractor extension, using fundamental tractor operators applied to tractor quantities with undetermined slots, like

$$\begin{pmatrix}\psi\\ \star\end{pmatrix},$$

which we shall not explain in detail here. That is, it is not necessary to input the Dirac operator with an undetermined coefficient in order to get the Dirac operator as output. In the unlikely event that someone (from another planet, say) were to discover spin-tractors before discovering the Dirac operator, he, she, or it would inexorably be led to the Dirac operator anyway.

Applying the tractor-D to our spin-tractor, we obtain

$$\begin{aligned}D_A\Psi&=Y_A(\mathbf{y}\psi+\mathbf{x}\varphi)+w^{-1}Z_A{}^a(\mathbf{y}(\nabla_a\psi+\beta_a\varphi)+\mathbf{x}(\nabla_a\varphi+\mathsf{P}_{ab}\beta^b\psi))\\&\quad-\tfrac{1}{2}[w(w+(n-2)/2)]^{-1}X_A(\mathbf{y}(\nabla^a\nabla_a\psi+2\not{D}\varphi+(w-\tfrac{1}{2})\mathsf{J}\psi)\\&\qquad+\mathbf{x}(\nabla^a\nabla_a+2\mathsf{P}_{ab}\beta^b\nabla^a\psi+(w-\tfrac{1}{2})\mathsf{J}\varphi+\mathsf{J}_b\beta^b\psi)).\end{aligned} \tag{3.33}$$

The residue at $w = 0$ is

$$Y_A \cdot 0 + Z_A{}^a(\mathbf{y}(\nabla_a \psi + \beta_a \varphi) + \mathbf{x} \cdot \star) + X_A \cdot \star.$$

In the obvious array notation, this is

$$\begin{pmatrix} 0 & 0 \\ \nabla_a \psi + \beta_a \varphi & \star \\ \star & \star \end{pmatrix};$$

we are allowed to project out an entry if there are no nonzero entries to its left or above it. If we take (ψ, φ) to be the tractor extension of ψ as in (3.32), we have

$$Y_A \cdot 0 + Z_A{}^a(\mathbf{y}\underbrace{(\nabla_a + \frac{2}{n}\beta_a \not{D}\psi)}_{(\mathbf{T}\psi)_a} + \mathbf{x} \cdot \star) + X_A \cdot \star,$$

showing the invariance of the *twistor operator* $\mathbf{T} : \mathcal{E}^P \to \mathcal{E}_{\underline{a}}{}^P$ (or similarly with P replaced by P'). Here the underline means that we take only the sections τ_a annihilated by interior Clifford multiplication:

$$\beta^a \tau_a = 0.$$

Exercise 7. *Getting φ from ψ by the tractor extension and take the residue at $w = (2-n)/2$, construct a third-order conformally invariant operator D_3 from $\mathcal{E}^P[(2-n)/2] \to \mathcal{E}^{P'}[(-4-n)/2]$ (or the same with P and P' interchanged). Use the Lichnerowicz formula to write this in the form $D_3 = \not{\nabla}^3 +$ (lower order).* Hint: *In array form as above, we have (up to a constant factor)*

$$\begin{pmatrix} 0 & 0 \\ 0 & 0 \\ 0 & D_3\psi \end{pmatrix}.$$

A tractor analogue of the Clifford section is

$$\alpha^A = X^A \mathbf{y}\mathbf{y}^* + Z^A{}_a(\mathbf{y}\beta^a\mathbf{x}^* - \mathbf{x}\beta^a\mathbf{y}^*) - Y^A\mathbf{x}\mathbf{x}^*.$$

This satisfies

$$\alpha^A\alpha^B + \alpha^B\alpha^A = -h^{AB},$$

where h^{AB} is the tractor metric

$$h_{AB} = Z_A{}^a Z_{Ba} + X_A Y_B + Y_A X_B\,.$$

Note that h is indefinite. The tractor metric and tractor Clifford section are parallel in the connections given.

To compute $\alpha^A D_A$, we need the identities

$$Y_A Y^A = 0, \quad X_A X^A = 0, \quad Y_A X^A = 1, \quad Z_{Aa} Z^A{}_b = g_{ab},$$
$$\mathbf{y}\mathbf{x}^* + \mathbf{x}\mathbf{y}^* = \mathrm{Id}, \quad \mathbf{x}^*\mathbf{y} = \mathrm{Id}, \quad \mathbf{y}^*\mathbf{x} = \mathrm{Id},$$
$$\mathbf{x}^*\mathbf{x} = 0, \quad \mathbf{y}^*\mathbf{y} = 0.$$

With this, if we abbreviate the formula (3.33) as

$$D_A \Psi = Y_A(\mathbf{y}\psi + \mathbf{x}\varphi) + Z_A{}^a(\mathbf{y}\eta_a + \mathbf{x}\theta_a) + X_A(\mathbf{y}\pi + \mathbf{x}\xi),$$

we have

$$\alpha^A D_A \Psi = y(\varphi + \beta^a \eta_a) + \mathbf{x}(\pi - \beta^a \theta_a).$$

If Ψ takes the form $\mathbf{y} \cdot 0 + \mathbf{x}\varphi$, this is

$$(1 - \tfrac{1}{2} n w^{-1})\mathbf{y}\varphi - w^{-1}\mathbf{x}(1 + (w + \tfrac{1}{2}(n-2))^{-1}) \not{D} \varphi,$$

or in vertical notation,

$$\begin{pmatrix} (1 - \tfrac{1}{2} n w^{-1})\varphi \\ -w^{-1}(1 + (w + \tfrac{1}{2}(n-2))^{-1}) \not{D} \varphi \end{pmatrix}.$$

The Dirac operator emerges as the residue of this at $w = \frac{1}{2}(2-n)$. Note that since the tractor-D lowers weights by 1, this is an operator from

$$\Sigma[(1-n)/2] \to \Sigma[(-1-n)/2].$$

6.4 REFERENCES

[1] I. Avramidi and T. Branson, A discrete leading symbol and spectral asymptotics for natural differential operators, *J. Funct. Anal.* **190** (2002), 292–337.

[2] T.N. Bailey, M.G. Eastwood and A.R. Gover, The Thomas structure bundle for conformal, projective and related structures, *Rocky Mountain Journal of Math.* **24** (1994), 1–27.

[3] H. Baum, *Spin-Strukturen und Dirac-Operatoren, ber pseudo-Riemannsche Mannigfaltigkeiten,* Teubner-Verlag, Leipzig, 1981.

[4] H. Baum, T. Friedrich, R. Grunewald, and I. Kath, *Twistor and Killing Spinors on Riemannian Manifolds,* Teubner–Verlag, Stuttgart/Leipzig, 1991.

[5] J.-P. Bourguignon, The magic of Weitzenböck formulas, in *Variational Methods* (Paris 1988), H. Berestycki, J.-M. Coron, I. Ekeland, eds., PNLDE, vol. 4, Birkhäuser, 1990, pp. 251–271.

[6] J.-P. Bourguignon, O. Hijazi, J.-L. Milhorat, A. Moroianu, *A Spinorial Approach to Riemannian and Conformal Geometry*, Monograph (in preparation).

[7] T. Branson, Intertwining differential operators for spinor-form representations of the conformal group, *Advances in Math.* **54** (1984), 1–21.

[8] T. Branson, Group representations arising from Lorentz conformal geometry, *J. Funct. Anal.* **74** (1987), 199–291.

[9] T. Branson, Nonlinear phenomena in the spectral theory of geometric linear differential operators, *Proc. Symp. Pure Math.* **59** (1996), 27–65.

[10] T. Branson, Stein–Weiss operators and ellipticity, *J. Funct. Anal.* **151** (1997), 334–383.

[11] T. Branson, Spectra of self-gradients on spheres, *J. Lie Theory* **9** (1999), 491–506.

[12] T. Branson, Kato constants in Riemannian geometry, *Math. Research Letters* **7** (2000), 245–261.

[13] T. Branson and A.R. Gover, Form-tractors and a conformally invariant computation of cohomology, in preparation.

[14] T. Branson and A.R. Gover, Spin-tractors, in preparation.

[15] T. Branson and O. Hijazi, Vanishing theorems and eigenvalue estimates in Riemannian spin geometry, *International J. Math.* **8** (1997), 921–934.

[16] T. Branson and O. Hijazi, Improved forms of some vanishing theorems in Riemannian spin geometry, *International J. Math.* **11** (2000), 291–304.

[17] T. Branson and O. Hijazi, Bochner–Weitzenböck formulas associated with the Rarita-Schwinger operator, *International J. Math.* **13** (2002), 137–182.

[18] T. Branson, G. Ólafsson, and B. Ørsted, Spectrum generating operators, and intertwining operators for representations induced from a maximal parabolic subgroup, *J. Funct. Anal.* **135** (1996), 163–205.

[19] D. Calderbank, P. Gauduchon, and M. Herzlich, Refined Kato inequalities and conformal weights in Riemannian geometry, *J. Funct. Anal.* **173** (2000), 214–255.

[20] D. Calderbank, P. Gauduchon, and M. Herzlich, On the Kato inequality in Riemannian geometry, in *Global Analysis and Harmonic Analysis* (Marseille-Luminy, 1999), Séminaires & Congrès, Collection Société Mathématique de France, num. 4 (2000), eds. J.-P. Bourguignon, T. Branson, and O. Hijazi, pp. 95–113.

[21] A. Čap, A.R. Gover, Tractor bundles for irreducible parabolic geometries, in *Global Analysis and Harmonic Analysis* (Marseille-Luminy, 1999), Séminaires & Congrès, Collection Société Mathématique de France, num. 4 (2000), eds. J.-P. Bourguignon, T. Branson, and O. Hijazi, pp. 129–154.

[22] R. Delanghe, F. Sommen, and V. Soucek, *Clifford Algebra and Spinor-Valued Functions,* Kluwer Academic Publishers, Dordrecht, 1992.

[23] M.G. Eastwood and J.W. Rice, Conformally invariant differential operators on Minkowski space and their curved analogues, *Commun. Math. Phys.* **109** (1987), 207–228. Erratum, *Commun. Math. Phys.* **144** (1992), p. 213.

[24] H. Fegan, Conformally invariant first order differential operators, *Quart. J. Math.* (Oxford) **27** (1976), 371–378.

[25] T. Friedrich, Der erste Eigenwert des Dirac-Operators einer kompakten Riemannschen Mannigfaltigkeit nichtnegativer Skalar-krümmung, *Math. Nachr.* **97** (1980), 117–146.

[26] S. Helgason, *Differential Geometry, Lie Groups, and Symmetric Spaces,* Academic Press, New York, 1978.

[27] O. Hijazi, A conformal lower bound for the smallest eigenvalue of the Dirac operator and Killing spinors, *Commun. Math. Phys.* **104** (1986), 151–162.

[28] O. Hijazi and J-L. Milhorat, Twistor operators and eigenvalues of the Dirac operator on compact quaternionic spin manifolds, *Ann. Global Anal. Geom.* **2** (1997), 117–131.

[29] N. Hitchin, Harmonic spinors, *Adv. in Math.* **14** (1974), 1–55.

[30] J. Humphreys, *Introduction to Lie Algebras and Representation Theory,* Springer-Verlag, Berlin, 1972.

[31] K.-D. Kirchberg, An estimation for the first eigenvalue of the Dirac operator on closed Kähler manifolds of positive scalar curvature, *Ann. Global Anal. Geom.* **3** (1986), 291–325.

[32] S. Kobayashi and K. Nomizu, *Foundations of Differential Geometry,* vols. I and II, Interscience, New York, 1963 and 1969.

[33] W. Kramer, U. Semmelmann, and G. Weingart, The first eigenvalue of the Dirac operator on quaternionic Kähler manifolds, *Commun. Math. Phys.* **199** (1998), 327–349.

[34] B. Lawson and M.-L. Michelson, *Spin Geometry,* Princeton University Press, 1989.

[35] A. Lichnerowicz, Spineurs harmoniques, *C.R. Acad. Sci. Paris* **257** (1963), 7–9.

[36] B. Ørsted and I.E. Segal, A pilot model in two dimensions for conformally invariant particle theory, *J. Funct. Anal.* **83** (1989), 150–184.

[37] S. Paneitz, I.E. Segal, and D. Vogan, Analysis in space-time bundles IV: Natural bundles deforming into and composed of the same invariant factors as the spin and form bundles, *J. Funct. Anal.* **75** (1987), 1–57.

[38] R. Penrose and W. Rindler, *Spinors and Space-time*, vols. 1,2, Cambridge University Press, Cambridge, 1984 and 1986.

[39] E. Stein and G. Weiss, Generalization of the Cauchy–Riemann equations and representations of the rotation group, *Amer. J. Math.* **90** (1968), 163–196.

[40] R. Strichartz, Linear algebra of curvature tensors and their covariant derivatives, *Canad. J. Math.* **40** (1988), 1105–1143.

[41] S. Sulanke, *Die Berechnung des Spektrums des Quadrates des Dirac-Operators auf der Sphäre*, Doktorarbeit, Humboldt-Universität zu Berlin, 1979.

[42] T.Y. Thomas, On conformal geometry, *Proc. Natl. Acad. Sci. USA* **12** (1926), 352–359.

Thomas Branson
Department of Mathematics
The University of Iowa
Iowa City IA 52242 USA
E-mail: thomas-branson@uiowa.edu

Received: January 31, 2002; Revised: March 15, 2003.

Appendix

In this appendix we briefly review existing software for computations with Clifford geometric algebras. We believe that being able to perform quick and reliable computations with these algebras leads to a greater understanding of the theory and applications. The software provides tools for easily generating examples, checking conjectures, finding counterexamples, and advancing understanding via experimentation. for these purposes [34, 36]. In fact, many of the examples and exercises that appeared in [35], and most recently in Lecture 1 of this volume, were derived or checked by Lounesto with CLICAL.

7.1 Software for Clifford (geometric) algebras

Here we follow in the footsteps of the late Pertti Lounesto who, to the best of our knowledge, was the first to design and use the computer software called CLICAL.

It was the advent of personal computers that made CLICAL, originally written in PASCAL for a mainframe computer, more popular. In 1987 its first successful PC version appeared [37] with the most recent version 4 being available for downloading from [38]. Both editors of this volume are indebted to Lounesto for introducing us to CLICAL. While CLICAL has semi-symbolic abilities, its code was optimized for computations with Clifford algebras $C\ell(Q)$ of a quadratic nondegenerate form Q, and was not designed to deal with Clifford algebras $C\ell(B)$ of an arbitrary bilinear form B. While in 1987 a paper [22] on deformations of Clifford algebras appeared, the properties of Clifford algebras $C\ell(B)$ where B has an antisymmetric part really came into focus in the early 1990s (see [5] and [33]). The need to compute with $C\ell(B)$ for an arbitrary B gave a stimulus to the development of CLIFFORD, a Maple package which was first presented in 1996 (see [1]). Fundamental to the development of this package was Chevalley's 1954 recursive definition of Clifford multiplication for an arbitrary bilinear form such that $B(\mathbf{x}, \mathbf{x}) = Q(\mathbf{x})$ (see [33] and references therein). Most recently, a new non-recursive definition of Clifford multiplication based on the Rota–Stein cliffordization process [3] has been introduced to CLIFFORD.

CLIFFORD is not the only program capable of symbolic computations. There exist quite a number of other packages, some based upon computer algebra systems such as Maple or Mathematica, and some standalone programs based on C++ libraries or Matlab. We will review a few of the available packages here, many of which are freely downloadable from the web. Our review is intended not only as a brief software guide for the reader, but in some cases as an introduction to applications of Clifford (geometric) algebras not covered in the previous chapters. Since these algebras have applications in physics, robotics, computer vision, image processing, signal processing and space dynamics, it should not be surprising that a number of special and general-purpose packages exist and that their number continues to grow. Some of these packages, and results derived with them, have already been presented in [6] seven years ago. Since then, new attempts, especially in the area of visualization, have yielded very interesting new programs. The reader is strongly encouraged to download and explore one or more of these programs.

7.1.1 Standalone software

We will begin by describing standalone programs that do not require any additional software such as Maple or Mathematica.

CLICAL – Complex Number, Vector Space and Clifford Algebra Calculator

It was around 1982 when a group of scientists from the Helsinki University of Technology, directed by Pertti Lounesto, wrote the first version of CLICAL for a mainframe computer. In 1987 a PC version that ran in a DOS window became available. The most recent version 4, from 1992, can be downloaded from [38]. CLICAL is powerful calculator for quickly performing computations in non-degenerate[1] algebras $C\ell_{p,q}$ for any signature (p, q) where $n = p + q < 10$. For example, to work in $C\ell_{1,3}$ one sets the dimension and the signature at the program prompt as follows:

```
>dimension 1,3
```

The following information about $C\ell_{p,q}$ can be now displayed:

```
>info

Algebra Cl(1,3)
- isomorphic with M2(H)
- a simple algebra
- square of the oriented volume j = e1234 is -1
- primitive idempotent ip = 1/2*(1+e14)
- automorphism groups of scalar products of spinors
  for reversion Sp(2,2) and for conjugation Sp(2,2)
```

The above display says that $C\ell_{1,3} \cong \mathrm{Mat}(2, \mathbb{H})$ which is in agreement with Table 1.1 page 35 (Lecture 2). Furthermore, the Clifford element $i_p = \frac{1}{2}(1 + \mathbf{e}_{14})$ where $\mathbf{e}_{14} = \mathbf{e}_1\mathbf{e}_4 = \mathbf{e}_1 \wedge \mathbf{e}_4$ is a primitive idempotent that can be used to generate a minimal left ideal (a spinor module) $S = C\ell_{1,3}i_p$. In $S \times S$ we have two

[1] By a non-degenerate Clifford algebra we mean a Clifford algebra $C\ell(Q)$ of a non-degenerate quadratic form Q.

bilinear forms[2] $\beta^{(+)}(\psi,\varphi)$ and $\beta^{(-)}(\psi,\varphi)$ such that

$$(\psi,\varphi) \to p\beta^{(+)}(\psi)(\varphi) = \lambda i_p \quad \text{and} \quad (\psi,\varphi) \to p'\beta^{(-)}(\psi)(\varphi) = \lambda' i_p$$

for some "pure spinors" $p, p' \in C\ell_{1,3}$ and $\lambda, \lambda' \in \mathbb{K} \simeq \mathbb{H}$. It is precisely these two bilinear forms, $\beta^{(+)}$ with reversion and $\beta^{(-)}$ with conjugation that have $Sp(2,2)$ as their automorphism group [32].

CLICAL can handle computations with complex numbers, quaternions, octonions, vectors, in addition to elements of $C\ell_{p,q}$. For the latter, Clifford (geometric), wedge, and dot products are available. Several transcendental functions including exponential and logarithmic are built in. Elements of $C\ell_{p,q}$ must be entered with numeric coefficients (real or complex) but the multivectors are entered symbolically as shown in the display above. Depending on whether $\mathbf{e}_i^2 = 1$ or $\mathbf{e}_j^2 = -1$, the indices i and j appearing in basis Grassmann monomials (blades) are displayed in different colors. Examples and tutorials are accessible from within the program. The manual [37] explains computations in various low dimensional geometries, 3D vector algebra, electromagnetism and Lorentz transformations, spin groups, the Cayley transform and the outer exponential, Möbius transformations and Dirac spinors. There are also external files with tutorials on octonions, spinors, Dirac equation, etc., that can be read into the program. There is also a file on triality, one of the favorite topics of Pertti Lounesto.

GAIGEN – A C++ code optimized generator

GAIGEN is a special purpose program for implementation of geometric algebras. It generates C++ and C source codes which produces any Clifford algebra $C\ell_{p,q,r}$, including degenerate (see Lecture 2), in dimensions up to 8 as might be requested by the user. The following operations are available: geometric and outer products, left and right contractions, scalar product, (modified) Hestenes inner product, the outermorphism operator, and the delta product. Several useful functions such as factorization, meet and join are built in. GAIGEN has been used in computer graphics to write a simple scanline renderer using different geometric algebras. Several test movies and images that have been rendered with the program are displayed at [21].

7.1.2 C++ Template classes for Geometric Algebras

At [56] one can find a source code in C++ containing template classes for implementing the geometric algebras $C\ell_{p,q}$, and documentation. The main template class is `GeometricAlgebra`. It implements the addition of a multivector and a scalar, the subtraction of a scalar from a multivector, and (left or right) multiplication of a multivector by a scalar. Next, it implements the negation, addition, subtraction, multiplication, accessing coefficients, grade-involution, reversion, and

[2] See Section 6.4 in Lecture 1 where Lounesto discusses two bilinear forms in the spinor space $S = C\ell_3 f \simeq \mathbb{C}(2)f$.

conjugation of an arbitrary element. It requires three template parameters: the data type for scalars, the value of p, and the value of q. The data type for scalars must support addition, subtraction, multiplication, assignment from another member of the same type, and assignment from the integer 0. The multiplication need not be commutative. According to the documentation, this `GeometricAlgebra` template class inherits from the `GeomMultTable` template class which maintains lookup tables that are used in the multiplication, grade involution, reversion, and conjugation. Both classes use a `GeomGradeTable` template class. There are also some external data files available. The `geoma.h` file contains the declarations and implementations of each of the classes. The `geomaData.h` file contains all of the static variable declarations needed by the template classes in `geoma.h`. A C++ code for a fractal generator that uses these files and the `GeometricAlgebra` template class is posted on the web page. For more information, we suggest visiting the page and downloading the C++ source files.

7.1.3 *Visualization software*

There have been two very interesting projects similar to GAIGEN that attempt to visualize elements of geometric algebras and the operations that can be performed on them, such as geometric multiplication. As described in Lecture 1, multivectors can represent lines, planes, volume elements, and other geometric quantities.

The CLU Project - A C++ library with OPEN-GL graphics

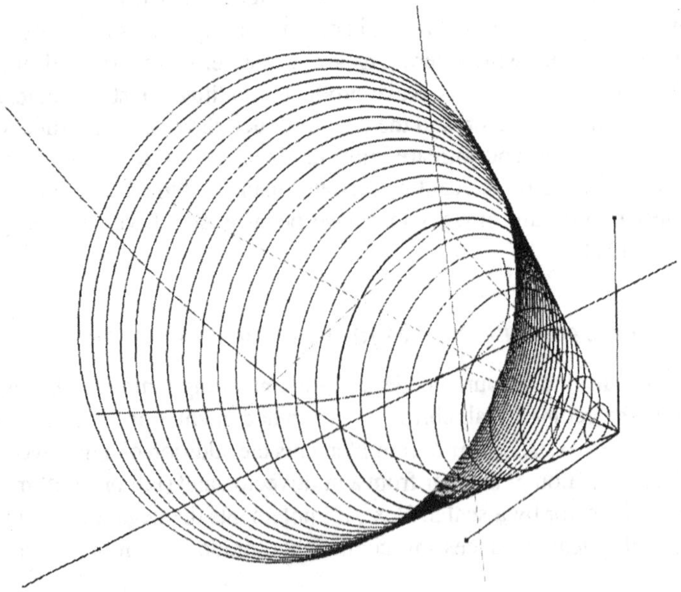

FIGURE 7.1. Visualization of the embedding of a 1D-Euclidean space in the corresponding conformal space using CLUCALC

Project CLU [43] consists of several parts. The CLU library encodes Clifford algebra operations in C++ language, CLUDRAW offers visualization of Grassmann multivectors through OpenGL, and CLUCALC visualizes multivector calculations (see figures (7.1) and (7.2)). Thus, while CLICAL was dubbed by Lounesto "Clifford algebra calculator", CLUCALC is a "visual Clifford algebra calculator". Grassmann basis multivectors are realized as *blades*, with an ability to mix blades coming from different spaces. The program can perform computations in projective space (PGA), the spacetime algebra (STA)[3], the geometric algebra of the standard Euclidean 3D space[4], and the Clifford algebra $C\ell_{1,2}$ useful in modeling a Clifford neural network. The meet and join of blades can be computed, general multivector equations can be solved, and the null space and the inverse of a multivector can be found. CLUDRAW can visualize points, lines, planes, circles, spheres, rotors, motors and translators, represented by multivectors. It contains various tutorial programs useful for the user. The entire CLU project is described in [43] from where CLUCALC and a manual can be downloaded.

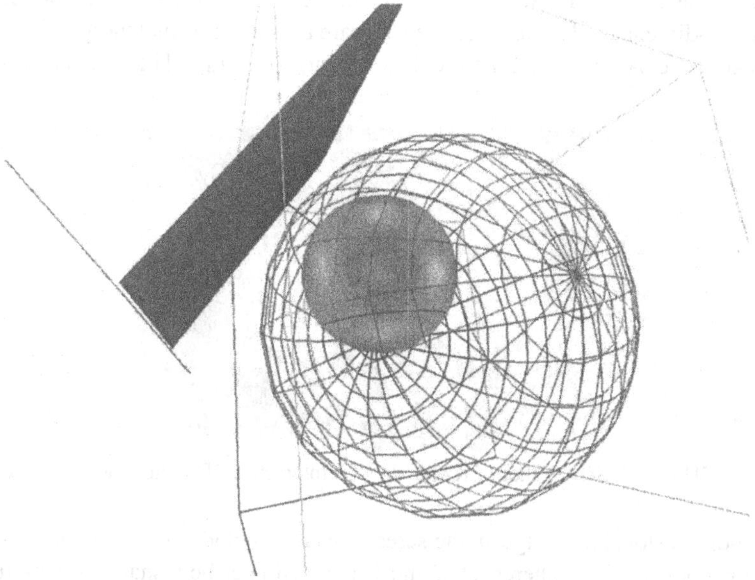

FIGURE 7.2. Inversion of a plane and a line on a sphere using CLUCALC

[3]See Lecture 4 where the STA is described.

[4]See Lecture 1 where the Clifford algebra $C\ell_3$ of the Euclidean 3D space is introduced.

GABLE – A Matlab package with visualization

GABLE (Geometric AlgeBra Learning Environment) is a Matlab package for learning Geometric Algebra $C\ell_3$ of the Euclidean 3D-space. It can be downloaded from [18], along with a tutorial [39] that describes its features. This tutorial, according to its authors, *"is aimed at the sophomore college level, although it may provide a gentle introduction to anyone interested in the topic."* The aim of GABLE is to visualize the geometric product, the outer product, the inner product, and the geometric operations that may be formed with them. The user can perform various computations using the geometric algebra, including projections and rejections, orthogonalization, interpolation of rotations, and the intersection of linear offset spaces such as lines and planes. As in CLUCALC, blades are used to represent subspaces, while meet and join are used to manipulate them. An example of the Euclidean geometry of 2-dimensional space represented in the 3-dimensional homogeneous model is presented in the tutorial. The author of GABLE has used this model in teaching his college geometry courses.

KAMIWAAI – Interactive 3D sketching with Java

In [27] one can find KAMIWAAI, a very interesting interactive Java application for three-dimensional sketching. The software uses a conformal model of the Euclidean space based on the Clifford (de Sitter) algebra $C\ell_{4,1}$. The user can perform

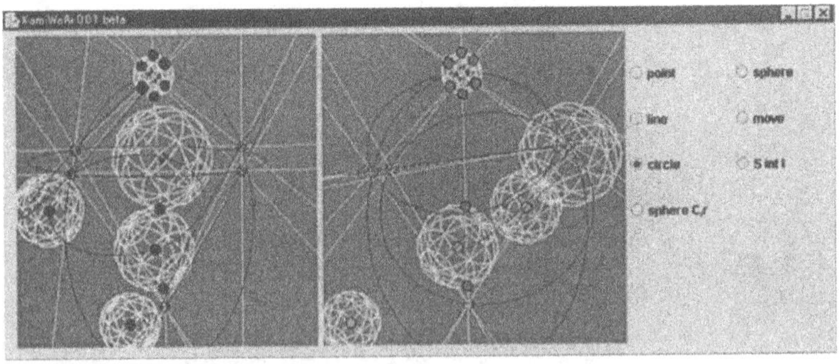

FIGURE 7.3. Screen shot of KAMIWAAI – Interactive 3D sketching with Java

three-dimensional drawings on the screen that can include geometric objects like points, lines, circles, spheres, etc., and which can later be reshaped interactively by moving (dragging) their points over the screen. It is possible to visualize intersections of spheres with lines and other spheres, and to have the intersection sets follow motions of the lines and spheres they depend on.

Upon starting KAMIWAAI in a DOS window, one sees three panels: the front and the side view panels, and a panel with seven radio buttons for creation and manipulation of the objects (see Figure 7.3). The objects are internally implemented as Java classes. As the author states, *"These software objects are geometric entities mathematically defined and manipulated in a conformal geometric algebra,*

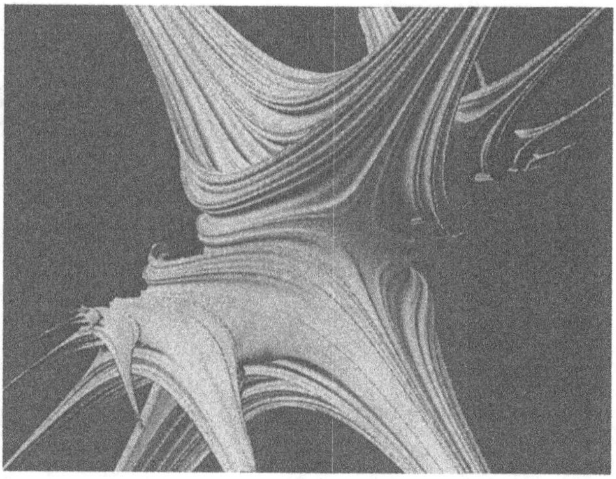

FIGURE 7.4. Plane Fractal: $b = -0.152815 + 0.656528\mathbf{e}_{12} \in C\ell_{2,0}$ created with CLI

combining the five dimensions of origin, three space, and infinity. Simple geometric products in this algebra represent geometric unions, intersections, arbitrary rotations and translations, projections, distance, etc. To ease the coordinate free and matrix free implementation of this fundamental geometric product, a new algebraic three level approach is implemented in KAMIWAAI. According to its author, the underlying `GeometricAlgebra` Java package has recently undergone substantial change, in that there is now a new class PointC for conformal points. All other objects are now constructed in terms of these points. The software is described in [28].

CLI – Clifford fractal generator

Another very interesting and visual software is a Clifford fractal generator called CLI for MS-DOS[5] by Jordi Vives Nebot [40]. It can generate Julia and Mandelbrot 2- and 3-dimensional Clifford fractals. For example, pictures on the cover of [6] were created using this program. They represent sections $x + y\mathbf{e}_{12} + z\mathbf{e}_1$ of two Julia-type 3-dimensional Clifford fractals viewed from an isometric perspective. Each image is a result of 10 iterations of the map $c \to c^2 + b$, where c and b are elements of the Clifford algebra $C\ell_{p,q}$, $p + q = 2$. Convergence of the resulting sequences of Clifford elements was determined by using a spinorial norm (the scalar part of the Clifford number times it Clifford-conjugate). We display these fractals, along with the metrics and base points b that have been used to create them, in figures (7.4) and (7.5).

[5] According to its author, a new Windows version of the software is expected in a few months.

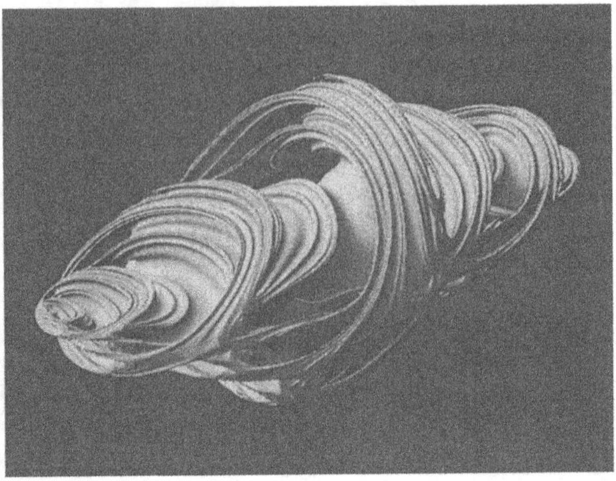

FIGURE 7.5. Quaternionic Fractal: $b = -0.152815 + 0.5\mathbf{e}_1 + 0.656528\mathbf{e}_{12} \in C\ell_{0,2}$ created with CLI

7.1.4 *Symbolic programs*

CLIFFORD – A Maple 8 package for $C\ell(B)$

CLIFFORD and its associated supplementary packages, the most important of which is BIGEBRA, are available at [2]. It is a general purpose Maple package[6] for computations in $C\ell(B)$ where B is an arbitrary bilinear form $B : V \times V \to \mathbb{R}$, $\dim V \leq 9$. Computations with general Clifford matrices, quaternions, octonions, and octonionic matrices are also implemented.[7] The most general element of $C\ell(B)$ is represented as a Grassmann polynomial, although other representations, including Clifford polynomials, are available. All packages are written in the Maple programming language and their code is open. The main module contains 110 various procedures beginning with two user-selectable algorithms that implement the Clifford (geometric) product: CMULNUM based on the Chevalley's recursive formula and CMULRS based on a Rota–Stein cliffordization process [19]. The main goal behind the development of CLIFFORD was a desire to compute not only in $C\ell_{p,q}$, which is the Clifford algebra $C\ell(Q)$ of the quadratic form of signature (p, q), but more generally in $C\ell(B)$ for any bilinear form B. The Clifford product in $C\ell(B)$ is defined in Section 11 of Lecture 1 as

$$\mathbf{x}u = \mathbf{x} \,\lrcorner\, u + \mathbf{x} \wedge u \tag{1.1}$$

for any $\mathbf{x} \in V$ and $u \in C\ell(B)$, and then it is extended by linearity and associativity to all of $\bigwedge V$. The left contraction $\mathbf{x} \,\lrcorner\, u$ depends on B and acts as a derivation on $\bigwedge V$.

It is an amazing consequence of (1.1) that the standard defining relation on the

[6] CLIFFORD is available for Maple V, 6, 7, and 8 from [2].

[7] Computations with octonions are enhanced by a supplementary package OCTONION.

Clifford algebra generators $\{\mathbf{e}_1, \mathbf{e}_2, \ldots, \mathbf{e}_n\}$, $n = \dim V$, namely

$$\mathbf{e}_i\mathbf{e}_j + \mathbf{e}_j\mathbf{e}_i = 2B(\mathbf{e}_i, \mathbf{e}_j) = 2g(\mathbf{e}_i, \mathbf{e}_j), \quad i \neq j, \tag{1.2}$$

where $g = g^T$ is the symmetric part of B, is independent of the antisymmetric part $F = -F^T$ of $B = g + F$. This is because, for any $\mathbf{x}, \mathbf{y} \in V$,

$$\begin{aligned}\mathbf{x}\mathbf{y} &= \mathbf{x} \rfloor \mathbf{y} + \mathbf{x} \wedge \mathbf{y} = B(\mathbf{x}, \mathbf{y}) + \mathbf{x} \wedge \mathbf{y} \\ &= g(\mathbf{x}, \mathbf{y}) + (F(\mathbf{x}, \mathbf{y}) + \mathbf{x} \wedge \mathbf{y})) = g(\mathbf{x}, \mathbf{y}) + \mathbf{x} \dot{\wedge} \mathbf{y}\end{aligned} \tag{1.3}$$

where $\mathbf{x} \dot{\wedge} \mathbf{y} = F(\mathbf{x}, \mathbf{y}) + \mathbf{x} \wedge \mathbf{y}$ is the antisymmetric *dotted* wedge in $C\ell(B)$ [5, 33]. Due to the presence of the extra F-dependent terms in (1.3), similar terms that depend on F appear in the Clifford product uv of any two general Clifford elements $u, v \in C\ell(B)$, or in the reversion of u, making detailed computations in such algebras very complicated.[8] This is especially true when B is purely symbolic, i.e., non-numeric. To compute the Clifford product in this case it is much more efficient to use the algorithm CMULRS. Both algorithms are described in the help pages of the CLIFFORD package (see also [3]). The BIGEBRA package allows the user to compute tensor products of Clifford and Grassmann algebras, and perform calculations in co-algebras and Hopf algebras. This supplementary package comes with its own extensive help pages. It has also been described in [4] and [19], and online computations with both packages can be performed via [20].

GA Package for Maple V from Cambridge

Another Maple package for computations with Clifford (geometric) algebras is the GA package created by the Geometric Algebra Research Group at The University of Cambridge [8]. Computations can be carried out in arbitrary dimensions and signature. The package has four built-in signatures but additional signatures can be programmed by the user. It is based on the geometric algebra axioms from [26]. The package implements computations only in $C\ell_{p,q}$, that is, geometric algebras of a non-degenerate quadratic form. The default metric of the package is the 'mother' algebra from [17]. All basis vectors with positive index have square 1 and those with negative index square to -1. There are three built-in algebras:

(a) SA: The Pauli Algebra $C\ell_3 \cong \mathbb{C}(2)$ with three generators `e[1], e[2]` `e[3]`, all of which square to 1.

(b) STA: The Spacetime Algebra $C\ell_{1,3} \cong \mathbb{H}(2)$ with four generators `e[0]`, `e[1], e[2], e[3]` where `e[0]` squares to 1 and the remaining vectors square to -1.

(c) MSTA: The Multiparticle Spacetime Algebra consisting of many copies of the STA. The generators `e[0], e[4], e[8], ...`, are such that all

[8] By a suitable choice of the basis in V, the symmetric part g of B can be assumed to have been diagonalized.

e[k], where $k = 0 \mod 4$, square to 1, and all other generators square to -1.

The user may add a new metric to the package using the add_metric procedure. However, first the user must do some programming to define the squares of the new generators. Following [26], the package has a built-in inner product, outer product, commutator product, and a scalar product (the scalar part of the geometric product). Projections on homogeneous blade components are implemented. Additional procedures allow one to create multivectors with symbolic coefficients and manage multivector expressions. Geometric calculus includes a multivector derivative mderiv and a multivector differential mdiff. There is only one help page available in the form of a Maple worksheet.

GLYPH – A Maple V package for $C\ell(B)$

GLYPH is a symbolic Maple V package based on the approach presented in the GA package, but it allows for computations with Clifford algebras $C\ell(B)$ of an arbitrary bilinear form B. For some metric spaces, such as the n-dimensional Euclidean and pseudo-Euclidean spaces, the geometric algebras $C\ell_{n,0}$ and $C\ell_{0,n}$, respectively, have been pre-computed and their multiplication tables are stored in the package, while others must be defined by the user. The algebras SA, STA, and MSTA are built-in. Elements of $C\ell(B)$ are represented as Clifford polynomials in generic multivectors. Left and right contraction, the scalar product, the exterior (wedge) and Clifford products are programmed. Algebraic functions that are available include Clifford and exterior inverses, Clifford exponentials and exterior exponentials, Hodge dual, reversion, grade involution, and Clifford conjugation (see Lecture 1 and [35]). The procedure csolve solves systems of equations in Clifford polynomials. The representation of a Clifford polynomial as a matrix in a complex universal Clifford algebra is built in. Computations with quaternions and octonions are possible. The package includes help pages that are accessible through a Maple browser. GA can be downloaded from [44], along with documentation.

CLIFFORD – A Mathematica package

This package, for calculations in the Clifford algebra $C\ell_{p,q}$ with Mathematica, is described in [13] and it is available at [12]. It is used to perform computations in crystallography using geometric algebras in higher dimensions. The package contains 31 functions and permits symbolic operations with no restriction on the dimension n $(= p + q)$. According to its authors, the two key procedures in the package are an algorithm for the geometric product between multivectors (blades) and a grade operator. The former is based on a unique representation of the blades $\mathbf{e}_1^{m_1}\mathbf{e}_2^{m_2}\cdots\mathbf{e}_n^{m_n}$ as n-tuples $(m_1, m_2, \ldots, m_n)$ of zeros and ones that are easily handled by Mathematica.[9] The function `GeometricProduct` (alias `Gp`) is linear and it accepts symbolic, constant and numeric coefficients. The signature of the bilinear form <x, y> can be set by using `$SetSignature=p` (by default `p=20`). Additional functions available include `OuterProduct` (aliased as `Oup`) and `MultivectorInverse` (aliased as `Mvi`).

A Mathematica package for Clifford algebras

A Mathematica package for differential geometric computations with Clifford algebras is described in [41] and it is posted at [42]. Its Clifford product function `CP` is based on the product definition given in [7, p. 186]. One of the most important functions inside the program is `Nabla`. It is the Clifford algebra generalization of Hamilton's ∇ operator. It appears in Maxwell vacuum equations written in $C\ell_{1,3}$ or $C\ell_{3,1}$ as

$$\texttt{Nabla[BivectorField]=0.}$$

The package uses a Grassmann multivector basis obtained by the exterior product of orthogonal basis vectors. The most general element of $C\ell_{p,q}$ is referred to as *cliffor*. The function `CliffordAlgebra` (alias `CA`) takes as an input a list of squares of the orthogonal basis elements and defines the corresponding working environment. For example, `CA[1,-1-1-1]` defines $C\ell_{1,3}$. Several implemented algebra automorphisms are `Rev` for the reversion, `GradeInvol` for the grade involution, and `GRev` for Clifford conjugation. The package permits one to compute with the complexified Clifford algebras $\mathbb{C} \otimes C\ell_{p,q}$. The exterior and interior products are encoded following the approach of [23]. A regular representation of the algebra in terms of $2^n \times 2^n$ matrices where $n = p + q$ is used to implement the Clifford product in a Matlab environment for all $C\ell_{p,q}, n = p + q \leq 6$. However, the Clifford product `CP` is not implemented by matrix multiplication of square matrices, but rather in terms of matrix multiplication of column vectors. The numerical computations in Clifford algebras are performed with Matlab.

In the area of differential geometry, this package carries computations with *cliffor fields* on a manifold. For example, the function

$$\texttt{MetricTensor}[\{h_0, h_1, \ldots, h_n\}, \{q_0, \ldots, q_n\}]$$

defines a metric tensor on the manifold and the orthogonal sets of coordinates that will be used. The h_i are the well-known Lamé's coefficients, and q_i are the names of the coordinates. In particular, the package is intended for differential geometric computations in the four-dimensional Euclidean space and in Minkowski space.

---

[9] Selecting a suitable internal representation of Grassmann multivectors (blades) is of critical importance for the performance of a symbolic package. In CLIFFORD there is no "hidden" internal representation of basis multivectors. For example, the blade $\mathbf{e}_1 \wedge \mathbf{e}_2 \wedge \mathbf{e}_3$ is just encoded as the string $e1we2we3$. The operation of multiplication of two basis multivectors is carried out by an appropriate concatenation and sorting operation on the corresponding strings. Lists provide an alternative, and perhaps a better way to internally represent multivectors.

In addition to the `Nabla` operator and the Laplace operator `Nabla[Nabla]` acting on cliffors, additional tools such as `Ricci` gives Ricci rotation coefficients, `ExtD` gives the exterior differential of a cliffor and `Codif` returns its codifferential.

Two sets of example files related to Maxwell and Dirac equations are supplied by the authors. In the first set, a non-standard solution to the Maxwell equations [45] is treated and it is shown how invariants of an electromagnetic field are computed. In the second set, it is shown how the standard matrix form of Dirac's equation can be put into exact correspondence with four different sets of Hestenes-type geometric equations. This can be done irrespectively of the signature chosen for the Minkowski space-time, $C\ell_{1,3}$ or $C\ell_{3,1}$. The equality of each real component of the Dirac bispinor equation with a geometric component of the Hestenes-type can be tested with the software.

LUCY – A Clifford algebra approach to spinor calculus

LUCY, named after William Kingdon Clifford's wife Lucy [14, 16], is a Maple 8 package available for downloading from [47]. It is a program for calculations involving real or complex spinor algebra and spinor calculus on manifolds in any dimension. It offers the freedom to adopt arbitrary bases in which to do calculations. Translations between the purely (real or complex) Clifford algebraic language of $C\ell(V, \mathbf{g})$, for a symmetric metric tensor $\mathbf{g}$, and the familiar matrix language are built in. As the authors state in [46], "LUCY *enables one to explore the structure of spinor covariant derivatives on flat or curved spaces and correlate the various spinor-inner products with the basic involutions of the underlying Clifford algebra. The canonical spinor covariant derivative is based on the Levi-Civita connection and a facility for the computation of connection coefficients has also been included*". A self-contained account of the facilities available, together with a description of the syntax, illustrative examples for each procedure, and a brief survey of the algorithms that are used in the program are described in [46]. The main (grade) involution, complex conjugation, and the anti-involution of reversion are implemented in $C\ell(V, \mathbf{g})$. LUCY accepts elements that belong to the dual space V^*. For example, to designate $\{X_a\}$ as the basis naturally dual to $\{e^a\}$, i.e., $e^a(X_b) = \delta^a{}_b$ with respect to the Kronecker symbol, one uses the procedure `Cliffordsetup` with X, labeling the ambient dual basis, as an extra fourth argument:

$$\texttt{Cliffordsetup(}\, u..w, \{(a_1, b_1) = g^{a_1 b_1}, (a_2, b_2) = g^{a_2 b_2}, \cdots\}, e, X\,\texttt{)}$$

Instead of using the contravariant metric components $g^{ab} = \mathbf{g}(e^a, e^b)$, one can use the covariant matrix components g_{ab}, where g_{ab} is defined by $g^{ac} g_{cb} = \delta^a{}_b$, so that $g_{ab} = \mathbf{g}^*(X_a, X_b)$ where $\mathbf{g}^*$ the induced metric on V^*. Thus, the program is capable of computing with the Clifford algebra of the cotangent space at a point p of an n-dimensional manifold M. For any tangent vector fields X, Y and Clifford element α, computation of the covariant derivative $\nabla_X Y$ and $\nabla_X \alpha$ with respect to the *Levi-Civita* (pseudo-)Riemannian connection ∇ induced by a user-defined plane wave metric g can be done with LUCY. Computations with spinor left and right $C\ell$-modules generated by primitive idempotents for simple Clifford algebras $C\ell$, as well as computations with dual spinors are implemented. For example, the procedure `Spinorsetup` returns a symbolic primitive idempotent P and two lists of Clifford elements $\{\alpha_i\}$ and $\{\beta_i\}$ called the left and right *kernel elements*, respectively. These elements are such that $\{\beta_i P\}$ spans the left (spinor) module $C\ell P$ while $\{P\beta_i\}$ spans the right (spinor) module $PC\ell$. The matrix representation of a spinor module constructed with LUCY follows the approach of the so called "total matrix algebra" from [10]. Appropriate conversion functions exist which convert a symbolic element ω from $C\ell$ into a matrix in the chosen spinor module and back.

The *spinor adjoint* map with respect to a chosen anti-involution is defined as the procedure `Spinoradjoint` following [48]. It is needed for the introduction of various spinor inner products (see also [32] and Lecture 2 of this volume). Likewise, *spinor conjugation* for any involution is encoded as `Spinorconjugation` in a similar manner. Computation of the *spinor covariant derivative* S_Y with respect to any tangent vector field Y is a type-preserving directional derivative on spinors, and it is implemented in LUCY as `Spinorderiv` following [10, 58]. For more on multivector calculus and applications of multivectors in special relativity see [9].

Calculating the spectral basis with Mathematica

Let $a \in \mathbb{C}_n$ be any multivector in the *complex Clifford algebra* $\mathbb{C}_n$ of an n-dimensional complex vector space $\mathbb{C}^n$. The *minimal polynomial* of a is the unique monic polynomial $\psi(x)$ of least degree $m = \sum_{i=1}^{r} m_i$ such that

$$\psi(a) \equiv \prod_{i=1}^{r} (a - \alpha_i)^{m_i} = 0$$

for the distinct complex numbers $\alpha_1, \alpha_2, \ldots, \alpha_r \in \mathbb{C}$ called the *eigenvalues* of a. The linearly independent powers $\{1, a, \ldots, a^{m-1}\}$ of a over the complex numbers make up the *standard basis* of the commutative subalgebra $\mathcal{A}(a)$ of $\mathbb{C}_n$. Once the minimal polynomial $\psi(x)$ of $a \in \mathbb{C}_n$ is known, the following *spectral* decomposition of a exists:

$$a = \sum_{i=1}^{r} (\alpha_i + q_i) p_i,$$

where p_i and q_j satisfy the rules

$$\{p_1 + \ldots + p_r = 1,\ p_i p_j = \delta_{ij} p_i,\ q_k^{m_k - 1} \neq 0 \ \text{ but } \ q_k^{m_k} = 0,\ p_k q_k = q_k\}.$$

The p_i's are *mutually annihilating idempotents* which partition the unity and the q_j's are *nilpotents* with the respective *indexes* m_j's. The set

$$\{p_i, q_i^k \mid i = 1, \ldots, r,\ k = 1, \ldots, m_i - 1\}$$

is called the *spectral basis* of the algebra $\mathcal{A}(a)$.

Knowing the spectral decomposition of a makes it very easy to extend the definition of any function which is defined at the spectral points α_i up to the $m_i^{th} - 1$ derivative. The following Mathematica Package will calculate the spectral basis of any element a, given its minimal polynomial $\psi(x)$, [54].[10]

```
(* Power Series Method for Evaluating Structure Equation *)
                 (* A Mathematica Package *)
(* Given list m={m1,m2,...,mr} of non-decreasing positive
   integers, psi[m]=(a-x[1])^m1 (a-x[2])^m2...(a-x[r])^mr
   is the minimal polynomial of the list m. *)
psi[m_List]:=Product[(a - x[j])^(m[[j]]),{j,Length[m]}]
psi[m_List,i_]:=psi[m]/((a-x[i])^(m[[i]]))
(* Calculates p[i]=pqseries[m,i,0] and q[i]^k=pqseries[m,i,k]
   using power series for i=1,2, ... , Length[m] *)
pqseries[m_,i_,k_]:=
 Together[Normal[Series[1/psi[m,i],{a,x[i],m[[i]]-(1+k)}]]]*
 (a-x[i])^k * psi[m,i]
(* Calculates p[i] and q[i] for i=1,2, ... , Length[m] *)
pqseries[m_List]:=
  Do[p[i]=pqseries[m,i,0];q[i]=pqseries[m,i,1];
                                 Print[i], {i,Length[m]}]
```

Other packages for Mathematica are described in [53]. One such package contains formulas for obtaining solutions to the classical cubic equation [52]. The unipodal numbers have been studied in [25, 51].

7.1.5 *Online interactive calculators, animations, and 3D computer games*

In this section we review online Clifford algebra "calculators", geometric algebra animations available on the web, and some specialized applications of these algebras in 3D computer games.

CLIFFORD – BIGEBRA interactive

The web page [20] contains an interactive window that gives access to the CLIFFORD and BIGEBRA packages mentioned above. Sample help pages for the BIGEBRA package exist for the vee product, tangle solver, and Grassmann and Clifford co-products. In addition, a two-dimensional example of a quantum Clifford algebra is provided, and a worksheet that shows how the axioms for a Grassmann–Hopf algebra are checked. Perhaps the most interesting feature of that web page is a link to a Maple interactive window which runs CLIFFORD and BIGEBRA. By entering commands to that page one can see how the packages work. The window is a Maple V worksheet where the user can type in commands. Typing

[10] Everything still works if the characteristic polynomial (given a matrix representation of the multivector a) is used instead, but powers of the q_i's might have lower indexes than expected.

?Clifford and pressing the **calculate** button displays the introductory page of CLIFFORD from the Maple browser. To define the Clifford algebra of the Euclidean plane $C\ell_2$ and its multiplication table, one can enter the following set of commands:

```
> restart:with(Cliff5):with(linalg):                  ## line (1)
> B:=diag(1,1);                                       ## line (2)
> cbas:=cbasis(2);                                    ## line (3)
> M:=matrix(4,4,(i,j)->cmul(cbas[i],cbas[j]));        ## line (4)
```

that give

$$B := \begin{bmatrix} 1 & 0 \\ 0 & 1 \end{bmatrix}$$

$$cbas := [Id,\ e1,\ e2,\ e1we2]$$

$$M := \begin{bmatrix} Id & e1 & e2 & e1we2 \\ e1 & Id & e1we2 & e2 \\ e2 & -e1we2 & Id & -e1 \\ e1we2 & -e2 & e1 & -Id \end{bmatrix}$$

A multiplication table in the Clifford algebra $C\ell(B)$ in dimension 2 where B has an antisymmetric part can be similarly computed as follows:

```
> B:=matrix(2,2,[1,b,-b,1]);                          ## line (6)
> M:=matrix(4,4,(i,j)->cmul(cbas[i],cbas[j]));        ## line (7)
```

that gives

$$B := \begin{bmatrix} 1 & b \\ -b & 1 \end{bmatrix}$$

$$M := \begin{bmatrix} Id & e1 & e2 & e1we2 \\ e1 & Id & e1we2 + b\,Id & e2 - b\,e1 \\ e2 & -e1we2 - b\,Id & Id & -b\,e2 - e1 \\ e1we2 & -b\,e1 - e2 & e1 - b\,e2 & -(1+b^2)\,Id - 2\,b\,e1we2 \end{bmatrix}$$

Likewise, one can try any command in CLIFFORD or BIGEBRA.

Online Geometric Calculator

The web page [15] contains a calculator for the Clifford algebra $C\ell_3$ of $\mathbb{R}^3$. It is similar to CLICAL except that it is limited to the Euclidean 3-space. Computations are performed with vectors, bivectors, trivector, quaternions, and, of course, with real and complex numbers. It is interesting to note that the calculator uses Reverse Polish Notation (RPN) like Hewlett Packard calculators. RPN was invented in the 1920s by the Polish mathematician Jan Łukasiewicz [57]. Inner, geometric, and outer products are available as well as projections onto the scalar, 1-vector, 2-vector and 3-vector parts of any Clifford number $u \in C\ell_3$. A nice glossary of terms, references, bibliography and links are provided. The most general element u can be expressed in the standard basis {1,x,y,z,xy,xz,yz,xyz} where $x = \mathbf{e}_1, y = \mathbf{e}_2, z = \mathbf{e}_3, xy = \mathbf{e}_1\mathbf{e}_2 = \mathbf{e}_1 \wedge \mathbf{e}_2, \ldots, xyz = \mathbf{e}_1\mathbf{e}_2\mathbf{e}_3 = \mathbf{e}_1 \wedge \mathbf{e}_2 \wedge \mathbf{e}_3$. Entering 1-2x+4y-xy+xyz (window **a**), by successive usage of stacks, and pressing **enter** button moves it into the top stack window **b**. If we now enter `1-4y-xy` into the bottom window (again by using the stacks) and press the × button, the Clifford product of the **b** and **a** stacks, namely `-16+6x+6xy+xyz+4xz+2y+z`, will appear in the bottom window **a**. Additional functions include $\cos u$, $\sin u$, $\tan u$, $\sqrt{u}$, and the magnitude $|u|$ of $u \in C\ell_3$, replacing u with its multiplicative inverse, dividing u by its magnitude, and isolating the even and the odd parts of u.

Interactive and animated Geometric Algebra with CINDERELLA

CINDERELLA [30] is a very powerful software tool for studying and teaching Euclidean, spherical, and hyperbolic geometries, and for creating Java applets that can be easily pasted onto the web pages. It performs internally computations with complex numbers that form one of the few examples of a commutative Clifford algebra. The software clearly has great visualizing and animating capabilities and could have been included with other visualization software already described above. However, we find its greatest strength in its ability to create the applets that are a great pedagogical tool and a propagator of the usefulness of the geometric algebras. For examples, see the CINDERELLA page [31] as well as [29] where many Java applets, created with the software by Eckhard Hitzer and related to Clifford (geometric) algebras, have been posted. On that last web page, notice in particular the applet on calculating and explaining the inverse of a vector with respect to the geometric multiplication; another one on a projection and rejection of a vector from and onto a bivector (see Figure 7.6) and several other applets demonstrating properties of bivectors, trivectors, Grassmann outer product, etc.. Very interesting

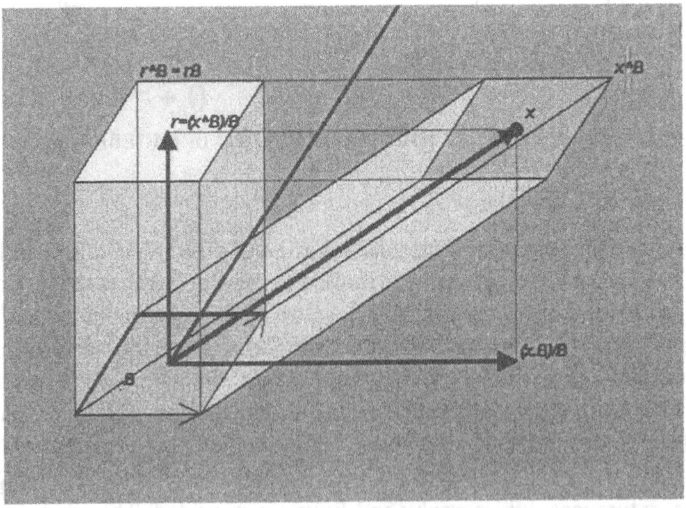

FIGURE 7.6. Projection $(\mathbf{x} \cdot B)B^{-1}$ and rejection $(\mathbf{x} \wedge B)B^{-1}$ components of a vector $\mathbf{x}$ with respect to the bivector plane B (applet created by E. Hitzer with CINDERELLA on [29])

are applets showing rotations in 2D with exponentiated bivectors, and reflections in 2D and 3D — concepts already discussed by Lounesto in Lecture 1. See also an applet on point groups in two dimensions whose mathematics is based on the approach presented in [24]. On the same page, there are also applets showing applications of the geometric algebras to the study of oscillations, circular polarized waves, and simple mechanical structures such as beams with distributed and concentrated loads (created by Luca Redaelli from Milan). Finally, let us also mention the fact that CINDERELLA includes an "automatic" theorem proving feature that does not use, however, symbolic methods to create a formal proof, but, instead, it uses a technique called Randomized Theorem Checking.

Geometric algebras in 3D computer games

In the February 1998 issue of the *Game Developer Magazine* appeared an article *"Rotating Objects Using Quaternions"* by Nick Bobick [11]. This article explains the important role that quaternions play in describing rotations in the 3D space of a computer game or when manipulating a robot arm. The author states that *Implementing a floating camera that tracks and rotates about a real-time 3D character is just one of the many reasons you should understand the importance and implementation of quaternions.* Of course, dual quaternions (also called *biquaternions*) have been used to describe screw motions void of a *gimbal lock.* We refer the reader to Lecture 5 of the present volume (and references therein) as well as to [49] and [50] by the same author. For additional applications of Clifford (geometric) algebras, see [55].

In [9], there is an account of how multivector methods of Clifford (geometric) algebras are used in programming 3D computer games. Following [26] Ian C. G. Bell discusses multivector blades, the geometric product, *step-ordered* and *bitwise-ordered* multivector bases (used by programmers), regular matrix representation of multivectors, the dual of a multivector with respect to a pseudoscalar using the *contractive inner product* (the same as the left contraction defined in Lecture 1) and the XOR operation applied to indices. He also discusses the inner and outer products used by Hestenes and coworkers. He also touches upon operations on multivectors such as the standard involution, multivector inverses, the exponential of a general element, spinors, quaternions, join and meet, Minkowski null vectors, inverse coordinate frames and homogeneous coordinates. Another chapter is devoted to multivectors as geometric objects of lines and planes, simplexes, frames, to multivectors as transformations, Lorentz transforms, translations, higher dimensional embeddings, and reflections, to shears and strains, rotations, 3D rotations, 4D rotations, and culminating in rigid body kinematics. Another chapter applies multivector calculus to physics including special relativity.

7.2 REFERENCES

[1] R. Abłamowicz, Clifford algebra computations with Maple, *Proc. Clifford (Geometric) Algebras*, Banff, Alberta Canada, 1995. Ed. W. E. Baylis, Birkhäuser, Boston, 1996, pp. 463–501.

[2] R. Abłamowicz and B. Fauser, CLIFFORD – *A Maple Package for Clifford Algebra Computations with* 'BIGEBRA', 'CLIPLUS', 'DEFINE', 'GTP', 'OCTONION', ver. 8, December 2002, downloadable from http://math.tntech.edu/rafal/cliff8/.

[3] R. Abłamowicz and B. Fauser, Mathematics of CLIFFORD – A Maple Package for Clifford and Graßmann Algebras, paper presented at ACA'2002: The 8th International Conference on Applications of Computer Algebra, June 25-28, 2002, Volos, Greece (Submitted to *Journal of Symbolic Computation*, Special Issue on Applications of Computer Algebra).

[4] R. Abłamowicz and B. Fauser, Clifford and Grassmann Hopf algebras vie the BIGEBRA package for Maple, paper presented at ACA'2002: The 8th International Conference on Applications of Computer Algebra, June 25-28, 2002, Volos, Greece (Submitted to *Journal of Symbolic Computation*, Special Issue on Applications of Computer Algebra).

[5] R. Abłamowicz and P. Lounesto, On Clifford algebras of a bilinear form with an antisymmetric part, in *Clifford Algebras with Numeric and Symbolic Computations*, eds. R. Abłamowicz, P. Lounesto, J.M. Parra, Birkhäuser, Boston, 1996, pp. 167–188.

[6] R. Abłamowicz, P. Lounesto, and J. M. Parra, (Eds.), *Clifford Algebras with Numeric and Symbolic Computations*, Birkhäuser, 1996.

[7] E. Artin, *Geometric Algebra*. N. Y., Interscience Pub., 1957.

[8] M. Ashdown, GA *Package Ver. 1.1 for Maple V*, Astrophysics Group, Cavendish Laboratory, University of Cambridge, 2003. Downloadable from http://www.mrao.cam.ac.uk/~clifford/software/GA/.

[9] Ian C. G. Bell, *Maths for (Games) Programmers, Multivector Methods and SpaceTime*, 2003. Availabe at URL: http://www.iancgbell.clara.net/maths/geoalg.htm.

[10] I. M. Benn and R. W. Tucker, *An Introduction to Spinors and Geometry with Applications in Physics*, Adam Hilger, Bristol, 1987.

[11] N. Bobick, *Rotating Objects Using Quaternions*, Game Developer Magazine, February 1998. See also URL: http://www.gdmag.com/backissue1998.htm#feb98.

[12] O. Caballero and J. L. Aragon, CLIFFORD – *Mathematica Package for Calculations with Clifford Algebra*, (version 1.0, September, 1995), downloadable from http://www.birkhauser.com/detail.tpl?ISBN=0817639071.

[13] O. Caballero and J. L. Aragon, CLIFFORD – *Mathematica Package for Calculations with Clifford Algebra*, (version 1.0, September, 1995), described in A. Gomez, J. L. Aragon, O. Caballero, and F. Davila, The Applications of Clifford Algebras to Crystallography Using Mathematica in *Clifford Algebras with Numeric and Symbolic Computations*, R. Abłamowicz, P. Lounesto, and J. M. Parra, (Eds.), Birkhäuser, 1996, pp. 251–266.

[14] M. Chisholm, *Such Silver Currents: The Story of William and Lucy Clifford, 1845–1929*, Lutterworth Press, London, 2002.

[15] R. E. Critchlow Jr., *A Geometric Calculator*, 2003. Go to URL: `http://www.elf.org/calculator/`.

[16] M. Demoor and M. Chisholm, Eds., *Bravest of women and finest of friends: Henry James's Letters to Lucy Clifford*, No. 80, English Literary Studies, University of Victoria, Victoria, 1999.

[17] C. Doran, D. Hestenes, F. Sommen and N. Van Acker, Lie groups as spin groups, *J. Math. Phys.* **34** (8) (1993), 3642–3669.

[18] L. Dorst and S. Mann, GABLE - *A Matlab Geometric Algebra Tutorial*, downloadable from `http://carol.wins.uva.nl/~leo/GABLE/` at University of Amsterdam, Amsterdam, The Netherlands, 2003.

[19] B. Fauser, *A Treatise on Quantum Clifford Algebras*, Habilitationsschrift, Fachbereich Physik, Universität Konstanz, Konstanz, Germany, January 2002 (math.QA/0202059).

[20] B. Fauser, CLIFFORD - BIGEBRA *Interactive*, Fachbereich Physik, Universität Konstanz, Konstanz, Germany, 2003. URL: `http://clifford.physik.uni-konstanz.de/~fauser/P_maple.shtml`.

[21] D. Fontijne, T. Bouma and L. Dorst, University of Amsterdam, Amsterdam, The Netherlands, 2003. C++ and C implementation of a geometric algebra. Go to URL: `http://carol.wins.uva.nl/~fontijne/gaigen/about.html`.

[22] J. Helmstetter, Monoïdes de Clifford et déformations d'algèbres de Clifford, *J. of Algebra*, Vol. **111** (1987), 14–48.

[23] D. Hestenes, *New Foundations for Classical Mechanics*. Dordrecht, Kluwer, 1986, 1990.

[24] D. Hestenes, Point groups and space groups in Geometric Algebra in *Applications of Geometric Algebra in Computer Science and Engineering*, L. Dorst, C. Doran, J. Lasenby, Eds., Birkhäuser, Boston, 2002, pp. 3–34.

[25] D. Hestenes, P. Reany, G. Sobczyk, Unipodal algebra and roots of polynomials, *Advances in Applied Clifford Algebras*, Vol. 1, No. 1 (1991), 31–51.

[26] D. Hestenes and G. Sobczyk, *Clifford Algebra to Geometric Calculus*, Reidel, 1984.

[27] E. M. S. Hitzer, *KamiWaAi – Interactive 3D Sketching with Java based on $C\ell_{4,1}$ Conformal Model of Euclidean Space*, 2003. Downloadable from `http://sinai.mech.fukui-u.ac.jp/gcj/software/KamiWaAi/`.

[28] E. M. S. Hitzer, *KamiWaAi – Interactive 3D Sketching with Java based on $C\ell_{4,1)}$ Conformal Model of Euclidean Space*, Advances in Applied Clifford Algebras (submitted: February, 2003).

[29] E. M. S. Hitzer, Interactive and animated Geometric Algebra with CINDERELLA, 2003. Available for viewing at URL: `http://sinai.mech.fukui-u.ac.jp/gcj/software/GAcindy/GAcindy.htm`

[30] U. H. Kortenkamp and J. R. Gebert, *The Interactive Geometry Software* CINDERELLA, Springer Verlag, New York, 1999. See the MAA Online Review by Ed Sandifer posted at `http://www.maa.org/reviews/cinderella.html`.

[31] U. H. Kortenkamp and J. R. Gebert, CINDERELLA, The Interactive Geometry Software, 2003. URL: `http://www.cinderella.de/en/index.html`.

[32] P. Lounesto, Scalar products of spinors and an extension of Brauer–Wall groups, *Found. of Physics*, Vol. **11**, Nos. 9/10 (1981), 721–740.

[33] P. Lounesto, Crumeyrolle's bivectors and spinors, in *Clifford Algebras and Spinor Structures*, R. Abłamowicz and P. Lounesto, (eds.), Kluwer, Dordrecht, 1995, pp. 137–166.

[34] P. Lounesto, Counterexamples in Clifford algebras with CLICAL, in *Clifford Algebras with Numeric and Symbolic Computations*, R. Abłamowicz, P. Lounesto, and J. M. Parra, (Eds.), Birkhäuser, 1996, pp. 3–30.

[35] P. Lounesto, *Clifford Algebras and Spinors*, Cambridge University Press, Cambridge, 1997, 2001.

[36] P. Lounesto, Counterexamples to theorems published and proved in recent literature on Clifford algebras, spinors, spin groups and the exterior algebra, 2003. Go to URL: `http://www.helsinki.fi/~lounesto/counterexamples.htm`, Institute of Mathematics, Helsinki University of Technology, Helsinki, Finland.

[37] P. Lounesto, R. Mikkola, V. Vierros, CLICAL *User Manual - Complex Number, vector Space and Clifford Algebra Calculator for MS-DOS Personal Computers*, Research Reports A248, Institute of Mathematics, Helsinki University of Technology, Helsinki, Finland, August 1987. Go to URL: `http://www.helsinki.fi/~lounesto/CLICAL.htm`.

[38] P. Lounesto, R. Mikkola, V. Vierros, CLICAL ver. 4, August 1992, downloadable from `http://www.helsinki.fi/~lounesto/` at Helsinki University of Technology, Helsinki, Finland, 2003.

[39] S. Mann, L. Dorst, and T. Bouma, *The Making of a Geometric Algebra Package in Matlab*, Research Report CS-99-27, Computer Science Department, University of Waterloo, 1999.

[40] J. V. Nebot, CLI - *Clifford Fractal Generator*, ver.1, (1996), Dept. de Física Fonamental, Universitat de Barcelona, Diagonal 647, E-08028, Barcelona, Spain, e-mail: `jordi@ffn.ub.es`.

[41] J. M. Parra and L. Rosello, MATHEMATICA *Package for Calculations with Clifford Algebra*, described in *Clifford Algebras with Numeric and Symbolic Computations*, R. Abłamowicz, P. Lounesto, and J. M. Parra, (Eds.), Birkhäuser, 1996, pp. 57–68.

[42] J. M. Parra and L. Rosello, MATHEMATICA *Package for Calculations with Clifford Algebra*, 2003. Downloadable from URL: `http://www.birkhauser.com/detail.tpl?ISBN=0817639071`.

[43] Ch. Perwass, Christian-Albrechts-Universität zu Kiel, Kiel, Germany. Visual Clifford algebra calculator CLUCalc, 2003. Go to URL: `http://www.perwass.de/cbup/clu.html` and downloadable from `http://www.perwass.de/cbup/clucalcdownload.html`.

[44] J. Riel, GLYPH - *Maple V release 5 package for performing symbolic computations in a Clifford algebra*, 2003. Downloadable from `http://bargains.k-online.com/~joer/glyph/glyph.htm`.

[45] W. Rodrigues and J. Vaz, *Subluminal and superluminal solutions in vacuum of the Maxwell equations and the massless Dirac equation*. RP 44/95, IMECC, UNICAMP (Brazil).

[46] J. Schray, R. W. Tucker, Ch. H.-T. Wang, LUCY: A Clifford Algebra Approach to Spinor Calculus, described in *Clifford Algebras with Numeric and Symbolic Computations*, R. Abłamowicz, P. Lounesto, and J. M. Parra, (Eds.), Birkhäuser, 1996, pp. 121–144.

[47] J. Schray, R. W. Tucker, Ch. H.-T. Wang, LUCY*: A Clifford Algebra Approach to Spinor Calculus*, 2003. Downloadable from
`http://www.lancs.ac.uk/depts/physics/staff/chtw.htm`.

[48] J. Schray and C. Wang, A Constructive Definition for Bilinear Forms on Spinors, *Advances in Applied Clifford Algebra*, **6** (2) (1996), 151–158.

[49] J. M. Selig, *Geometrical Methods in Robotics*, Monographs in Computer Science, Springer Verlag, Berlin, 1996.

[50] J. M. Selig, Ed., *Geometrical Foundations of Robotics*, World Scientific Pub. Co., 2000.

[51] G. E. Sobczyk, Jordan form in Clifford algebras, in *Clifford Algebras and their Applications in Mathematical Physics*, Proceedings of the Third International Clifford Algebras Workshop, F. Brackx, R. Delanghe, H. Serras, Eds., Kluwer, Dordrecht, 1993.

[52] G. Sobczyk, Hyperbolic number plane, *The College Mathematics Journal*, September, 1995.

[53] G. Sobczyk, *A Unipodal Algebra Package For Mathematica*, described in *Clifford Algebras with Numeric and Symbolic Computations*, R. Abłamowicz, P. Lounesto, and J. M. Parra, (Eds.), Birkhäuser, 1996, pp. 189–200.

[54] G. Sobczyk, The generalized spectral decomposition of a linear operator, *The College Mathematics Journal*, 27–38, January, 1997.

[55] G. Sommer, (Ed.), *Geometric Computing with Clifford Algebras - Theoretical Foundations and Applications in Computer Vision and Robotics*, Springer Verlag, Berlin, 2001.

[56] P. Stein, C++ Template Classes for Geometric Algebras, Nklein Software Co., Minneapolis, MN, 2003. Downloadable from URL:
`http://www.nklein.com/products/geoma/`.

[57] The Museum of HP Calculators, 2003. Go to URL:
`http://www.hpmuseum.org/rpn.htm`.

[58] R. W. Tucker, A Clifford calculus for physical field theories, in *Clifford Algebras and their Applications in Mathematical Physics*, J. S. R. Chisholm, A. K. Common, Eds., NATO ASI Series (C) **183**, pp. 177–199.

Index

D

P

S